6. wissenschaftliche Konferenz
der Gesellschaft Deutscher Naturforscher und Ärzte
Rottach-Egern 1971

in Conjunction with the
Second International Symposium

on

Metabolic Interconversion of Enzymes

Edited by
O. Wieland   E. Helmreich   H. Holzer

With 227 Figures

Springer-Verlag Berlin · Heidelberg · New York 1972

ISBN-13: 978-3-540-05919-6   e-ISBN-13: 978-3-642-80698-8
DOI: 10.1007/978-3-642-80698-8

This work is subject to copyright. All rights are reserved, whether the whole or part of the material is concerned, specifically those of translation, reprinting, re-use of illustrations, broadcasting, reproduction by photocopying machine or similar means, and storage in data banks.
Under § 54 of the German Copyright Law, where copies are made for other than private use, a fee is payable to the publisher, the amount of the fee to be determined by agreement with the publisher.
The use of general descriptive names, trade names, trade marks etc. in this publication, even if the former are not especially identified, is not to be taken as a sign that such names, as understood by the Trade Marks and Merchandise Marks Act, may accordingly be used freely by anyone.
© by Springer-Verlag Berlin · Heidelberg 1972. Library of Congress Catalog Card Number 72-85775.

Preface

The rapid development and recent advances in this field seemed to justify a further meeting dealing primarily with protein interconversions and their biological significance rather soon after the first symposium of this kind, held under the direction of Professor G. BONSIGNORE of Genoa, in Santa Margherita Ligure, Italy, in May 1970. The proceedings of the Second International Symposium on Metabolic Interconversion of Enzymes, held in Rottach-Egern, in the Bavarian Alps, from October 27–30, 1971, under the auspices of the "Gesellschaft Deutscher Naturforscher und Ärzte" bear witness to the high quality of the research reports and the lively discussions which ensued. There are contributions by a number of distinguished scientists and experts in the field who came from several European countries, Canada, Israel, Japan, the Soviet Union and the United States in order to participate in this symposium. We wish to express our sincere gratitude to the speakers for all their efforts.

The main burden of the editorial work was carried by Dr. G. LÖFFLER, Munich, ably assisted by Dr. A. R. FERGUSON, Freiburg, Dr. R. H. HASCHKE, Würzburg, Dr. A. HASILIK, Freiburg, Dr. L. HEILMEYER Jr., Würzburg, Dr. E. SIESS, Munich, and by the secretaries, Miss E. BAUER, Munich and Mrs. G. WEIDEMANN, Würzburg. To the publishers, Springer-Verlag Berlin Heidelberg New York, and especially to Dr. H. MAYER-KAUPP, we are greatly indebted for their cooperation.

Without the help of all those named above and several other colleagues not mentioned by name this volume would not have appeared on time. We hope the reader will find it a useful guide to this important field of biological regulation.

Würzburg, Freiburg, München
November 1972

E. HELMREICH
H. HOLZER
O. WIELAND

Contents

Welcoming Remarks. O. WIELAND 1
Metabolic Interconversion of Enzymes: Introductory Remarks. C. F. CORI 3
Comparative and Evolutionary Aspects of the Control of Phosphorylase.
    E. H. FISCHER, PH. COHEN, M. FOSSET, L. W. MUIR, J. C. SAARI 11
Electron Microscopy of Muscle Phosphorylases b and a. N. A. KISELEV, F. YA. LERNER,
    N. B. LIVANOVA 29
Discussion 50
Studies on Glycogen Phosphorylase in Solution and in the Crystalline State.
    N. B. MADSEN, K. O. HONIKEL, M. N. G. JAMES 55
Probing and Mapping the Pyridoxal 5'-Phosphate Site of Glycogen Phosphorylase.
    S. SHALTIEL, M. CORTIJO, Y. ZAIDENZAIG 73
Association-dissociation Properties of $NaBH_4$-reduced Phosphorylase b. D. J. GRAVES,
    J.-I. TU, R. A. ANDERSON, T. M. MARTENSEN, B. J. WHITE 103
The Mechanism of Action of Cyclic AMP in the Activation of Phosphorylase Kinase.
    E. G. KREBS, J. A. BEAVO, C. O. BROSTROM, J. D. CORBIN, T. HAYAKAWA,
    D. A. WALSH 113
Discussion 118
Phosphorylase a as a Glucose Receptor. W. STALMANS, TH. DE BARSY, M. LALOUX,
    H. DE WULF, H.-G. HERS 121
Properties of Purified Glycogen Synthetase b from Liver. H. L. SEGAL, Y. SANADA,
    S. R. MARTIN 133
Discussion 146
The Mechanism of Activation of the Lac Operon. G. ZUBAY 149
Discussion 157
Regulatory Mechanism of Enzyme Catabolism. N. KATUNUMA, E. KOMINAMI,
    S. KOMINAMI, K. KITO, T. MATSUZAWA 159
Discussion 175
Interconversion of Two Forms of Leucyl-tRNA Synthetase. P. ROUGET, F. CHAPE-
    VILLE 179
Discussion 192
Diphtheria Toxin-catalyzed Adenosine Diphosphoribosylation of Aminoacyl Trans-
    ferase II from Rat Liver. T. HONJO, K. UEDA, T. TANABE, O. HAYAISHI 193
Discussion 208
Regulation of Transcription by T4 Phage Induced Chemical Alteration and Modifi-
    cation of Transcriptase (EC2.7.7.6). D. RABUSSAY, R. MAILHAMMER, W. ZILLIG 213
Discussion 228
On the Mechanism of Action and Metabolic Control of the Multifunctional Enzyme
    Complex that Catalyzes Adenylylation and Deadenylylation of *Escherichia Coli*
    Glutamine Synthetase. E. R. STADTMAN, M. BROWN, A. SEGAL, W. A. ANDERSON,
    B. HENNIG, A. GINSBURG, J. H. MANGUM 231

Inactivation of Glutamine Synthetase in Intact *E. coli* Cells. H. HOLZER, H. SCHUTT, P. C. HEINRICH 245

Studies on the Mechanism of the ATP: Glutamine Synthetase Adenylyltransferase Reaction. D. WOLF, R. WOHLHUETER, E. EBNER 253

The Role of Enzyme Inactivation in the Regulation of Glutamine Synthesis in Yeast: *in vivo* Studies Using $^{15}NH_3$. A. P. SIMS, A. R. FERGUSON 261

Discussion 277

Molecular Aspects of the Regulation of the Mammalian Pyruvate Dehydrogenase Complex. L. J. REED, T. C. LINN, F. HUCHO, G. NAMIHIRA, C. R. BARRERA, TH. E. ROCHE, J. W. PELLEY, D. D. RANDALL 281

Regulation of the Mammalian Pyruvate Dehydrogenase Complex: Physiological Aspects and Characterization of PDH-Phosphatase from Pig Heart. O. WIELAND, E. SIESS, H. J. v. FUNCKE, C. PATZELT, A. SCHIRMANN, G. LÖFFLER, L. WEISS 293

Discussion 310

Enzyme-dependent Activation of Pyruvate Formate-lyase of *Escherichia coli*. J. KNAPPE, H.-P. BLASCHKOWSKI, R. EDENHARDER 319

Discussion 327

Lactose Synthetase: Structure and Function. R. L. HILL, R. BARKER, K. W. OLSEN, J. H. SHAPER, I. P. TRAYER 331

UDP-glucose: D-fructose-6-phosphate 2-glucosyltransferase: Metabolic Control of the Enzyme Activity. M. A. R. DE FEKETE 347

Discussion 352

Mechanism of Interaction between Arginase and Ornithine Transcarbamoyl Transferase of *Saccharomyces cerevisiae*. J. M. WIAME, F. MESSENGUY, M. PENNINCKX 357

Discussion 370

Enzymatic Modification of Basic Chromosomal Proteins in Developing Trout Testis. G. H. DIXON, G. S. BAILEY, E. P. M. CANDIDO, A. J. LOUIE, M. M. SANDERS, M. T. SUNG 375

Discussion 377

Active Site-directed Side Chain Modification. H. FASOLD, F. W. HULLA, A. KENMOKU 381

Discussion 390

Metabolic Interconversion of Enzymes: Relation to the Hysteretic Response. C. FRIEDEN 391

Discussion 401

Why Are Enzymes Interconvertible? R. H. HASCHKE, L. HEILMEYER, Jr., E. HELMREICH 405

General Discussion 416

List of Participants 439

Index of Contributors

Anderson, R.A., 103[+]
Anderson, W.A., 231[+]
Antonov, 353

Bailey, G.S., 375[+]
Barker, R., 331[+]
Barrera, C.R., 281[+]
Blaschkowski, H.P., 319[+]
Blavo, J.A., 113[+]
Brostrom, C.O., 113[+]
Brown, M., 231[+]
Buc, H., 277, 401, 423, 434
Bücher, Th., 370

Candido, E.P.M., 375[+]
Chapeville, F., 179[+], 192, 209, 210
Cohen, P., 11[+]
Corbin, J.D., 113[+]
Cori, C.F., 3[+], 51, 118, 146, 313, 352, 403, 428
Cortijo, M., 73[+]

De Barsy, T., 121[+]
Decker, K., 327, 354, 378, 379
De Fekete, M.A.R., 354, 355, 372, 347[+]
De Wulf, H., 121[+]
Dixon, G.H., 375[+], 157, 209, 310, 370, 377, 378, 379, 426, 429

Ebner, E., 253[+]
Edenharder, R., 319[+]

Fasold, H., 381[+], 118, 390, 402
Ferguson, A.R., 261[+]
Fischer, E.H., 11[+], 50, 51, 52, 118, 144, 177, 192, 211, 278, 354, 377, 390, 416, 426, 429, 430
Foset, M., 11[+]
Frieden, C., 391[+], 353, 401, 402, 403, 404, 424, 430, 431
Funcke, v.H.J., 293[+]

Gancedo, C., 228, 352, 371, 372, 373, 434
Ginsburg, A., 231[+]
Graves, D.J., 103[+], 51, 312, 422

Hartmann, G., 157, 228, 278
Haschke, R.H., 405[+], 209, 210
Hasilik, A., 192, 313, 434
Hayaishi, O., 193[+], 208, 209, 210, 211
Hayakawa, T., 113[+]
Heilmeyer, L., jr., 405[+], 52, 378, 417, 418, 419, 420, 423
Heinrich, P.C., 245[+]
Helmreich, E., 405[+], 119, 146, 208, 277, 327, 390, 401, 403, 421, 422, 428
Hennig, B., 231[+]
Henning, U., 353, 373, 426
Hers, H.G., 121[+]
Hess, B., 51, 314, 315, 316, 317
Hill, R.L., 331[+], 352, 353, 354, 379
Holzer, H., 245[+], 50, 119, 176, 177, 192, 278, 279, 355, 431, 435, 436, 437

Honikel, K.O., 55[+]
Honjo, T., 193[+]
Hucho, F., 281[+]
Hulla, F.W., 381[+]

James, M.N.G., 55[+]

Katunuma, N., 159[+], 175, 176, 371, 430, 437
Kenmoku, A., 381[+]
Kiselev, N.A., 29[+]
Kito, K., 159[+]
Knappe, J., 319[+], 327, 328, 329
Kominami, E., 159[+]
Kominami, S., 159[+]
Krebs, E.G., 113[+], 52, 118, 119, 175, 210, 311, 313, 314, 379, 402, 428, 429, 435, 438

Laloux, M., 121[+]
Lerner, F.Ya., 29[+]
Linn, T.C., 281[+]
Livanova, N.B., 29[+]
Löffler, G., 293[+]
Louie, A.J., 375[+]
Lynen, F., 279, 311, 353, 355, 425, 432, 436

Madsen, N.B., 55[+], 421, 435
Mailhammer, R., 213[+]
Mangum, J.H., 231[+]
Martensen, T.M., 103[+]
Martin, S.R., 133[+]
Matsuzawa, T., 159[+]
Messenguy, F., 357[+]
Muir, L.W., 11[+]

Namihira, G., 281[+]

Olsen, K.W., 331[+]

Palm, D., 404
Palmer, N., 52, 53, 54
Patzelt, C., 293[+]
Pelley, J.W., 281[+]
Penninckx, M., 357[+]
Pette, D., 175, 313, 390, 403

Rabussay, D., 213[+], 228, 229, 438
Randall, D.D., 281[+]
Reed, L.D., 281[+], 310, 313, 314
Reinauer, H., 310, 311
Roche, T.E., 281[+]
Rouget, P., 179[+]

Saari, J.C., 11[+]
Sanada, Y., 133[+]
Sanders, M.M., 375[+]
Schirmann, A., 293[+]
Schutt, H., 245[+]
Segal, A., 231[+]
Segal, H.L., 133[+], 147, 175, 208, 277, 311, 328, 372, 402, 433
Seubert, W., 312
Shaltiel, S., 73[+], 119, 352, 354, 377, 379
Shaper, J.H., 331[+]
Siess, E., 293[+]
Sims, A.P., 261[+]
Sols, A., 50, 146, 210, 278, 310, 312, 329, 354, 355, 371, 437
Stadtman, E.R., 231[+], 52, 228, 277, 278, 279, 327, 328, 370, 401, 402, 416, 424, 425, 426, 429, 431, 432, 433, 434, 435, 436, 437
Stalmans, W., 121[+], 146
Sung, M.T., 375[+]

Tanabe, T., 193[+]
Trayer, I.P., 331[+]
Tu, Jan-I., 103[+]

Ueda, K., 193[+]

Wallenfels, K., 377
Walsh, D.A., 113[+]
Weiss, L., 293[+]
White, B.J., 103[+]
Wiame, J.M., 357[+], 370, 371, 373
Wieland, O., 1[+], 293[+], 118, 211, 310, 311, 312, 313, 377, 403, 426
Wieland, Th., 377
Wohlhueter, R., 253[+], 433
Wolf, D., 253[+]

Zaidenzaig, Y., 73[+]
Ziegler, H., 354
Zillig, W., 213[+]
Zubay, G., 149[+], 157, 228, 229

[+]Symposium paper

Welcoming Remarks
O. Wieland

It's my pleasure to welcome you in behalf of the organizers of the Second International Symposium on Interconvertible Enzymes. I want to thank you all for taking the time to come here and to take part in this meeting.

I would like to take this occasion and to express my sincere appreciation to Professor Bonsignore and our Italian colleagues who initiated the International Symposia on Interconvertible Enzymes at beautiful St. Margarita, Ligure, last year. In St. Margarita we commited ourselves to organize the second meeting.

Without the generous help of the Gesellschaft Deutscher Naturforscher und Ärzte, however, we could not have raised the necessary funds. Therefore, I wish to thank in the name of all three of us the Society and their Secretary General, Professor Auhagen, who have made that meeting possible. With their financial aid we could invite distinguished colleagues from abroad and Europe, among them Professor Carl F. Cori, who might be called the "father" of interconvertible enzyme research.

As you may see from the program we made an effort to broaden the topic of interconvertible enzymes and to include protein-protein interaction because we believe that to be the basis for the understanding of these regulatory phenomena.

Finally I wish to state, that the smooth cooperation between Freiburg, Würzburg and Munich was a decisive factor in planning the final program. Everyone surely tried his best. If this meeting should be a success we sincerely hope that a third Symposium on Interconvertible Enzymes may soon follow.

Metabolic Interconversion of Enzymes: Introductory Remarks

Carl F. Cori

Enzyme Research Laboratory, Massachusetts General Hospital and
Harvard Medical School, Department of Biochemistry, Boston, Massachusetts/USA

As an introduction to this symposium I would like to make a few remarks about metabolic regulations. In general, the regulatory control of metabolism in higher organisms falls into two large categories:

(1) Regulation of enzyme turnover

(2) Regulation of enzyme activity.

The actual level of enzymes in the tissues depends on the rate of enzyme synthesis relative to the rate of enzyme degradation. Each of these parameters can vary independently of the other. Enzyme induction and feedback inhibition are examples of conditions under which the rate of enzyme synthesis increases or decreases. Similarly, the rate of enzyme degradation may be increased or decreased under different conditions. These changes in enzyme turnover, although relatively slow, are nevertheless of great metabolic importance, for example, in the regulation of hepatic gluconeogenesis.

Whereas changes in enzyme turnover are measured in hours, regulation of enzyme activity itself is a matter of minutes or even seconds. Two principal mechanisms involved in the control of enzymatic rates will be discussed, although there are others known and undoubtedly several others unknown.

(1) Changes in the concentration of reactants

(2) Interconversion of enzymes between an inactive and active form.

In principle, one could think of rapid changes of enzyme activity in terms of increase or decrease in substrates, cofactors, activators and inhibitors over their effective range of concentrations. Such effects are known to occur _in vivo_--and this is to be emphasized, since many effects that can be produced on isolated enzymes _in vitro_ are probably of no importance in the intact organism. Examples that apply here are the activation of a protein kinase by an increase in cyclic AMP and the activation of the myosin-ATPase of muscle by an influx of calcium ions during passage of the nerve impulse. In both cases mechanisms exist for the rapid decrease in the concentration of these effectors--cyclic AMP is split by a phosphodiesterase and calcium ions are removed by a pump in the sarcoplasmic reticulum. The examples just given emphasize the fact that regulation of enzyme activity through changes in the concentration of effectors requires that the concentration of the effector itself be regulated. This fact is often neglected in discussions of the effect of concentration of reactants on enzymatic rates.

In discussing the second mechanism for the control of enzymatic rates, the metabolic interconversion of two forms of the same enzyme, I would like to ask a number of questions in the hope that some of them will be answered by other speakers at this symposium. The first question concerns the glycogen phosphorylase and synthetase of the liver. There can be little doubt that the glycogen level of the liver is regulated by the relative activity of these 2 enzymes and that during rapid glycogen synthesis the phosphorylase must be inactive and during rapid degradation the synthetase activity must be low. In fact, such a reciprocal regulation of the activity of the two enzymes seems insured by the fact that the phospho-form is the active form of phosphorylase and the inactive form of synthetase and vice versa. Thus, epinephrine or glucagon, through activation of the respective kinases lead to the phosphorylation of both enzymes, with opposite effects on their activity. A glucose load, on the other hand, stimulates dephosphorylation of both enzymes. The dephospho-form of liver phosphorylase, unlike that of muscle, is only slightly activated by AMP, and no other agent that could serve as activator under physiological conditions has been detected. It should therefore be relatively easy to find conditions under which the phosphorylase activity in a freshly prepared liver homogenate is low, but this is not the case. Most investigators find that more than one half of the phosphorylase is in the active form. It seems unlikely that this is the situation in vivo when liver glycogen is being synthesized at a rapid rate after glucose administration, for example, at the rate of about 1% per hour in a previously fasted rat. In order to get basal values for phosphorylase in resting muscle, it was necessary to work out special methods of rapid freezing and inactivation of phosphorylase kinase and phosphatase, but this does not seem to work in liver. I thus come finally to my first question: Is the high phosphorylase activity found in liver homogenates of fed animals an artifact, a "Schönheitsfehler" as one would say in German and of no consequence or is something important being overlooked here. A subsidiary question is: Are there two mechanisms for the activation of phosphorylase kinase in liver such as exist in muscle, one by cyclic AMP and the other by calcium ions.

The second problem I want to discuss has to do with the activation of phosphorylase *b* during muscular contraction. Resting muscle contains 95% or more phosphorylase *b*. On stimulation of frog sartorius with single shocks up to rates of 1 shock every 2 seconds the lactate production rises progressively with the rate of stimulation to values 50 times the basal rate, but no phosphorylase *a* accumulates even after 30 minutes of stimulation. It is possible that some phosphorylase *a* is formed during each contraction and that it disappears again during the pauses between the contraction, but this does not tell us which form of phosphorylase is responsible for the increase in glycogen breakdown. At faster rates of stimulation, from 1 to 10 per second, phosphorylase *a* builds up to a plateau which is proportional to the rate of stimulation and which represents a steady state between the action of phosphorylase kinase and phosphatase. Finally, when the muscle is tetanized almost all of the phosphorylase b is converted to a with a half time of less than one second. The activation of the kinase in stimulated muscle can be ascribed to the influx of calcium ions which as Krebs (1) has shown are essential for kinase activity.

The question whether the conversion of b to a is obligatory for the breakdown of glycogen in muscle could be definitely answered by working with a mutant strain of mice, discovered by Lyon, that lacks phosphorylase b kinase. Some 8 years ago Lyon brought some of these mice to our laboratory in St. Louis, and he and Danforth (2) carried out the experiments shown in Fig. 1 which I have redrawn from the original data in order to illustrate the time relationship between the rise in phosphorylase a and the breakdown of glycogen. We see the immediate rise of phosphorylase a when

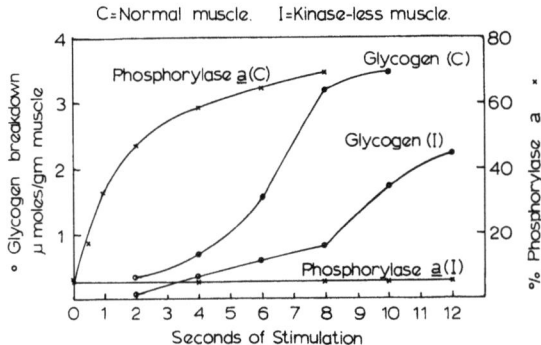

Fig. 1. The data of Danforth and Lyon (2) for the increase in hexosephosphate and lactate during stimulation have been used to calculate the rate of glycogen breakdown

the normal muscle is tetanized and the complete absence of such a rise in the kinase-less I strain. We also see that although there is a lag period, glycogen breakdown occurs sooner and reaches a greater speed and a higher plateau in the control muscles as compared to the muscles from the I strain, but there is no doubt that the kinase-less muscle can break down glycogen during stimulation. The fact that they can do so should be emphasized, because it implies a mechanism for the activation of phosphorylase b itself which may also be operative in normal muscle.

Here I would like to urge that more experiments of this type be done with these kinase-less mice. They are a unique biological material, useful for solving a number of problems. One would like to know, for example, how effective epinephrine is in these muscles in stimulating glycogen breakdown and how much lactate would be produced with single shock stimulation at different rates. This would make it possible to say more definitely how important the b ⇌ a transformation is for the energy production in a working muscle. The kinase-less mice appear normal under the sheltered conditions of the laboratory. In the wild state, the kinase mechanism could confer selective advantage when a rapid and intense energy output of muscle is required.

My second question is: By what mechanism does phosphorylase b get activated during contraction and inactivated again after contraction. I have tried to answer this question myself, but as you will see this is not an easy problem.

In Fig. 2 we have used available kinetic data for muscle phosphorylase b to simulate in vivo conditions. The arrow marks the concentration of inorganic phosphate

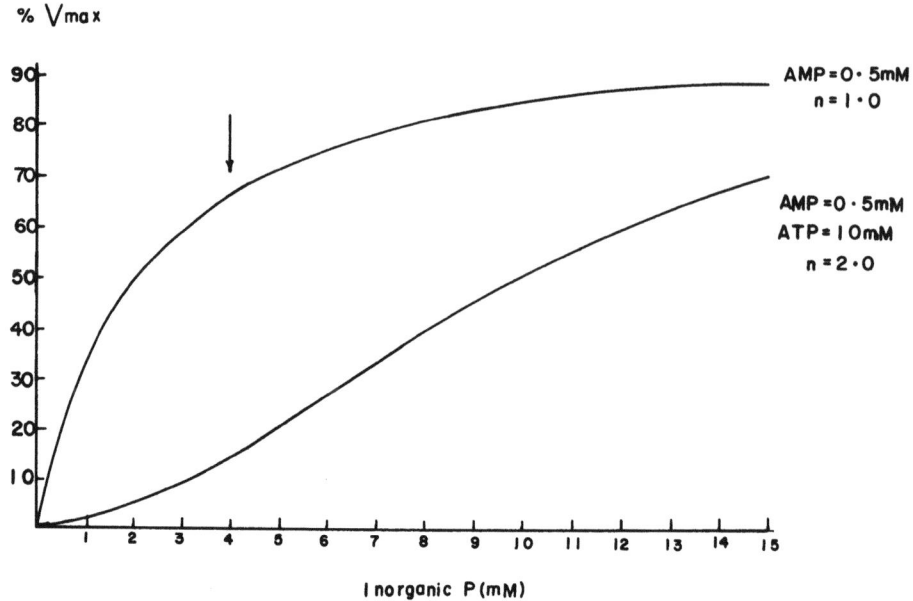

Fig. 2. Simulated curves for phosphorylase b activity in muscle at physiological concentrations of reactants. The arrow marks the concentration of inorganic phosphate in a resting anaerobic frog muscle. The Hill equation was used to calculate the curve where n equals 2

in resting anaerobic frog muscle. AMP is about 0.5 mM, which is 10 times the Km value, ATP+ ADP is 10 mM and glucose-6-phosphate 0.2 mM The top curve is based on the assumption that in stimulated muscle phosphorylase b is in contact with glycogen, inorganic P and AMP, but not with the inhibitory nucleotides ADP and ATP. Under these conditions one gets normal saturation kinetics when inorganic P is increased. It seems unlikely that this is the situation in muscle, since one must allow for a large increase in phosphorylase activity during stimulation. If we assume that the enzyme is in contact with 10 mM ATP we get the lower curve which has been calculated for a Hill coefficient of 2, the maximum cooperativity for an enzyme with 2 subunits. Here we have at least a fairly inactive enzyme at the concentration of inorganic P in muscle, but it is also clear that in order to counteract the ATP-inhibition, one would need large increases in the concentration of inorganic P and AMP or corresponding decreases in the concentration of ATP + ADP. However, during single shock stimulation for rates up to 24 per minute for 30 minutes there was no significant change in inorganic P, AMP and the sum of ADP + ATP, while lactate production rose 40 fold over the basal level (3).

It is possible that at the beginning of a series of contractions there are larger changes in concentration, but during a steady-state period of work they have largely disappeared. However, arguments based on total tissue concentration are unsatisfactory, since they neglect the unequal distribution of nucleotides in the different compartments of muscle. Caspersson, in particular, has shown that the isotropic region of muscle contains much more nucleotides than the anisotropic region and that there is a reversible shift of nucleotides into the anisotropic region during contraction (4). In some manner, by an on-off mechanism, phosphorylase b must come in contact with 5'-AMP during contraction and be separated again during relaxation. Since phosphorylase is the pacemaker of the glycolytic chain of reactions in muscle, my question of how it becomes activated is really concerned with the problem of the mechano-chemical coupling of muscle. My feeling is that progress will come from work with a model system which retains some of the properties of structured muscle. In this respect, Fischer (5) working with a glycogen-muscle phosphorylase complex has already shown that it has different properties from the usual soluble enzyme system of phosphorylase.

My third and last question has to do with the glycogen synthetase. This enzyme has a number of properties which are similar to those of phosphorylase. The D form of synthetase, like phosphorylase b, is dependent on an activator and is inhibited by ATP. The I form of synthetase, like phosphorylase a, is insensitive to inhibitors and is independently active. The effect of the activators, glucose-6-P and 5'-AMP respectively, is to increase the affinity for substrate and this plays a role even with the independently active forms, if the substrate concentration is very low. another similarity between the I and a forms of the respective enzymes is that their affinity for the activator is much greater than that of the D and b forms.

Normal liver contains most of the synthetase in the D form. It is postulated that the activity of this enzyme is modulated mainly by the D $\rightleftharpoons$ I interconversion. This argument was originally put forward by Mersmann and Segal (6) and is based on the following concentration relationships (Table 1). Segal calls the D form synthetase b and the I form synthetase a, in analogy to muscle phosphorylase. During incubation at 20° most of the b form is converted to a and this makes possible measurements of the activity of both forms in the same homogenate. At non-saturating UDP-glucose and glucose-6-P concentrations, such as exist in liver, the b form is

Table 1. Activity of liver glycogen synthetase at physiological concentrations of reactants

The b and a forms were measured in the same homogenate before and after conversion of b to a by preliminary incubation at 20° for 50 minutes. (Data from Mersmann and Segal (6).

Type of animal	UDPG mM	G-6-P mM	Pi mM	Rate (cpm) b	Rate (cpm) a	Ratio a/b
Fasted	0.25	0.05	3.0	357	4950	14
Fed	0.50	0.25	3.0	480	8190	17

quite inactive. Although both the UDP-glucose and the glucose-6-P concentrations rise in the fed animal, the a form is still 17 times more active than the b form because of its much greater sensitivity to the stimulatory effect of glucose-6-P and inorganic P at physiological concentrations of these reactants. These results led Segal to propose that changes in the concentration of reactants play a minor role in the regulation of the enzyme. If we accept this, with the reservations made previously about the validity of total tissue concentrations for enzyme activity, one arrives at the conclusion that the b ⇌ a interconversion, controlled by the opposing action of phosphatase and kinase, is the principal regulatory mechanism.

It seems clear, from what has been said, that the main targets of regulatory control are the separate kinases and phosphatases that act on phosphorylase and synthetase. In Table 2 are summarized some recent findings. A distinction has

Table 2. Control of glycogen metabolism in liver (L) and muscle (M)

Agent	Phosphorylase		Synthetase		References
	b → a kinase	a → b phosphatase	a → b kinase	b → a phosphatase	
Cyclic AMP (L,M)	activates		activates		14, 15
Calcium (M)	activates	inhibits			1, 5
Glycogen (M)	accelerates			inhibits	9, 11
Glucose (L)		accelerates		removes phosphorylase a inhibition	12, 13

been made between activation which results in an increase in the amount of active enzyme due to b → a transformation and acceleration or inhibition which results in a change in rate at a given enzyme level. The latter need not be direct effects on the enzyme. For example, glucose as an allosteric modifier of poosphorylase a (7) could make it a better substrate for phosphorylase phosphatase. We hope to hear about this in a paper to be given by Stalmans and his group. A similar mechanism could explain the accelerating effect of an increasing glycogen level on phosphorylase b kinase which is seen both in vivo (8) and in vitro (9). A good correlation also exists between the glycogen levels in vivo (10) and in vitro (11) at which inhibition of synthetase phosphatase occurs. The glycogen level is thus seen to regulate itself in muscle and perhaps also in liver by a feedback control over synthesis as well as degradation. One is reminded once more of the fact that the glycogen of liver and muscle has the property of adsorbing most of the enzymes that are concerned with its metabolism. Working with such a complex from muscle, Fischer et al. (5) found recently that calcium ions at concentrations which activate the phosphorylase kinase simultaneously inhibit the phosphatase. This inhibitory effect is not observed with the purified enzyme.

A final point illustrated in Table 2 is the inhibition of synthetase phosphatase by phosphorylase a, a recent finding reported by the Hers group (12, 13). Here one can point to the similarity of the amino acid sequence of the chain fragment containing the phosphoserine residue in phosphorylase a and synthetase a as a possible mechanism of inhibition. A lag period after glucose administration which is seen both in vivo and in vitro before synthetase a increases and which is correlated to the rate of disappearance of phosphorylase a points to the physiological significance of this regulatory mechanism. This is additional reason to reexamine the question of the level of phosphorylase a in liver homogenates, a problem which was mentioned earlier in this discussion.

The demonstration by Villar-Palasi and Larner (16) that insulin causes an increase in synthetase a activity in muscle and the extension of this finding to liver has led to considerable discussion about the underlying mechanism. Since it is not clear at the present time how insulin produces this effect, it has been omitted from Table 2. Glucocorticoids have likewise been omitted. The effect they produce on enzyme activity takes several hours to develop and seems to involve enzyme induction.

I am sorry that time will not permit me to touch on other metabolically interconvertible enzymes. In a general way I feel that much progress has been made in the last few years and that the subject of metabolic regulation has acquired considerable sophistication, not the least because of the realization that what one wishes to know is how the enzymes function within the complex organization of the cell.

References

1. BROSTROM, C.O., HUNKELER, F.L., and KREBS, E.G., J. Biol. Chem., 246, 1961 (1971).
2. DANFORTH, W.H. and LYON, JR., J.B., J. Biol. Chem., 239, 4047 (1964).
3. HELMREICH, E. and CORI, C.F., Advances in Enzyme Regulation, Vol. 3, Pergamon, New York.
4. CASPERSSON, T. and THORELL, B., Acta Physiol. Scandinav., 4, 97 (1942).
5. HASCHKE, R.H., HEILMEYER, JR., M.G., MEYER, F. and FISCHER, E.H., J. Biol. Chem., 245, 6657 (1970).
6. MERSMANN, H.J. and SEGAL, H.L., Proc. Nat. Acad. Sci., 58, 1688 (1967).
7. HELMREICH, E., MICHAELIDES, M.C. and CORI, C.F., Biochemistry, 6, 3695 (1967).
8. DANFORTH, W.H. in Control of energy metabolism, B. Chance, R.W. Eastabrook and J.R. Williams (Editors), Academic Press, New York, 1965, p. 287.
9. KREBS, E.G., LOVE, D.S., BRATVOLD, G.E., TRAYSER, K.A., MEYER, W.L. and FISCHER, E.H., Biochemistry, 3, 1022 (1964).
10. DANFORTH, W.H., J. Biol. Chem., 240, 588, (1965).
11. VILLAR-PALASI, C. and LARNER, J., Fed. Proc., 25, 583 (1966).
12. De WULF, H., STALMANS, W. and HERS, H.G., Eur. J. Biochem., 15, 1 (1970).
13. STALMANS, W., De WULF, H., LEDERER, B. and HERS, H.G., Eur. J. Biochem., 15, 9 (1970).
14. De LANGE, R.J., KEMP, R.G., RILEY, W.D., COOPER, R.A. and KREBS, E.G., J. Biol. Chem., 243, 2200 (1968).
15. SODERLING, T.R. and HICKENBOTTOM, J.P., Fed. Proc., 29, 601 (1970).
16. VILLAR-PALASI, C. and LARNER, J., Biochem. Biophys. Acta, 39, 171 (1960).
17. BISHOP, J.S. and LARNER, J., J. Biol. Chem. 242, 1354 (1967).

Comparative and Evolutionary Aspects of the Control of Phosphorylase*

Edmond H. Fischer, Philip Cohen, Michel Fosset, Larry W. Muir, and John C. Saari

Department of Biochemistry, University of Washington,
Seattle, Washington/USA

The complex nature of the enzymatic control of glycogen breakdown makes this system ideally suited for studies concerned with the evolution of a regulatory mechanism. First, the anaerobic pathway of carbohydrate metabolism must have evolved very early since it is generally assumed that life on Earth emerged under reducing conditions. Indeed, glycolysis is essentially ubiquitous to all forms of life and is found in every species from unicellular organisms to the complex tissues of higher plants and animals. Second, phosphorylase (E.C. 2.4.1.1), which is directly involved in glycogen breakdown, is regulated by both covalent and noncovalent modification, if not also by protein-protein interaction. Covalent control of muscle phosphorylase involves several activating enzymes acting successively on one another, various nucleotides and divalent metal ions, and is closely integrated with other physiological processes since glycolysis is initiated both by hormone release and the nervous stimulation which triggers contraction. As a consequence, the enzyme must contain a variety of sites to account for these multiple interactions. It was therefore surmised that a study of phosphorylase in earlier species might indicate how control of its activity originally arose and evolved with time.

All phosphorylases so far investigated fall within a rather homogeneous class of enzymes sharing several properties in common (for review, see 1). All have subunit molecular weights of $ca.$ 100,000 and stoichiometric amounts of pyridoxal 5'-P; their optimum pH of activity is between 5.8 and 6.8 and most have turnover numbers of the order of 8000 μmoles substrate converted/min/μmole enzyme monomer. They all have rather similar amino acid compositions as judged by the divergence factors of Harris and Teller (2), as seen in Table I. All animal enzymes, in contrast to those of the protista and plants, appear to be activated by phosphorylation of the protein; some are also activated by AMP. Phosphorylases from rabbit, human, rat and frog muscle, and that from human platelet, tetramerize readily whereas the enzymes from the rabbit liver, and rabbit heart isozyme I do not.

---

* Supported by grants from the National Institutes of Health (AM 07902), the National Science Foundation (GB 20482), and the Muscular Dystrophy Association of America.

TABLE I

AMINO ACID DIVERGENCE FACTORS OF VARIOUS PHOSPHORYLASES[1]

	RABBIT (3)	RAT (3)	HUMAN (4)	FROG (5)	DOGFISH (6)	LOBSTER (7)	INSECT[2] (8)	YEAST (9)	POTATO (10)	RABBIT LIVER (11)
RABBIT	0	10	17	19	26	37	75	49	57	35
RAT	10	0	14	22	24	33	74	47	60	35
HUMAN	17	14	0	27	27	34	76	52	62	36
FROG	19	22	27	0	21	35	69	45	55	30
DOGFISH	26	24	27	21	0	26	66	34	47	24
LOBSTER	37	33	34	35	26	0	66	38	45	34
INSECT	75	74	76	69	66	66	0	75	74	63
YEAST	49	47	52	45	34	38	75	0	47	38
POTATO	57	60	62	55	47	45	74	47	0	47
RABBIT LIVER	35	35	36	30	24	34	63	38	47	0

The homology among phosphorylases is estimated according to the procedure of Harris and Teller (2) in which a divergence factor (D) is calculated from the amino acid composition of related proteins according to the function:

$$D = [\Sigma(X_{i,A} - X_{i,B})^2]^{1/2} \times 1000$$

where $X_{i,A}$ is the mole fraction of amino acid i in protein A, and $X_{i,B}$ is the mole fraction of amino acid i in protein B.

---

[1] Unless otherwise stated, animal phosphorylases are from the muscle.

[2] The higher divergence factors found for insect muscle phosphorylase when compared to the other enzymes is almost entirely due to extraordinarily high cysteine content reported for this protein.

Two species were selected for this study: the Pacific dogfish (*Squalus sucklii*) since it is representative of one of the most primitive vertebrates, having diverged from the main line leading to mammals *ca*. 450 million years ago, and yeast, since it is a unicellular eukaryotic organism which lends itself to genetic manipulation.

Comparative and Evolutionary Aspects of Dogfish Phosphorylase. Dogfish phosphorylase was obtained in homogeneous form as judged by polyacrylamide gel electrophoresis and ultracentrifuge analysis. Table II compares its physical and catalytic properties to those of rabbit skeletal muscle phosphorylase (12). The tissue concentration of both enzymes is comparable. As isolated, dogfish phosphorylase is in a form totally dependent on AMP for activity and therefore designated as b by analogy with the mammalian system. Dogfish and rabbit phosphorylases are closely similar in specific activities, 280/260 nm absorbance ratios, absorbance indices, amino acid composition, number of rapidly reacting SH groups, sedimentation constants, subunit molecular weight and molecular weights of the native enzymes. Both contain one molecule of pyridoxal 5'-P per subunit and the interactions between this cofactor and the protein are very similar. Dogfish phosphorylase b can be converted to the a form by incubation with rabbit muscle phosphorylase kinase, Mg^{2+} and ATP, and reconverted to the b form with either rabbit muscle or liver phosphorylase phosphatase; moreover, the rates of these interconversions are identical to those found with the "natural" substrates (rabbit b and a) (Figure 1). One molecule of phosphate is incorporated per subunit in the b to a conversion, and resides on a unique seryl residue of the dogfish enzyme (see below).

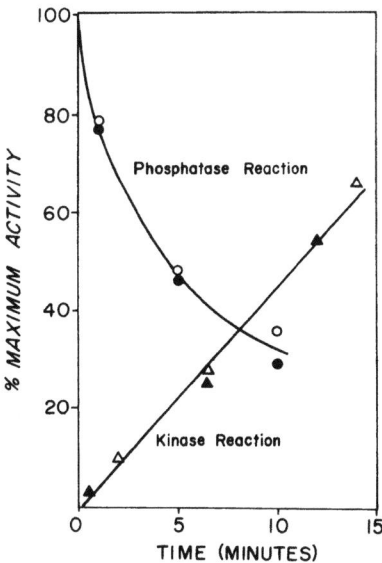

FIGURE 1. Conversion of dogfish (Δ) and rabbit (▲) muscle phosphorylase b to phosphorylase a by rabbit muscle phosphorylase kinase at pH 8.6 and conversion of dogfish (O) and rabbit (●) muscle phosphorylase a to b by rabbit muscle phosphorylase phosphatase at pH 7.0. Reaction mixtures contained 3.0 mg/ml of phosphorylase; all assays were carried out at 30° in the absence of AMP

TABLE II

COMPARATIVE PROPERTIES OF DOGFISH AND RABBIT MUSCLE PHOSPHORYLASES

Property	Effector	Dogfish b	Dogfish a	Rabbit b	Rabbit a
260/280 nm absorbance ratio		0.56		0.53	
$A^{1\%}_{280}$		12.9		13.1	
Specific activity* (Units/mg)	−AMP +AMP	<1 62	48 62	<1 88	54 80
Apparent specific volume		0.746		0.746	
Subunit MW		99,000		100,000	
MW native enzyme	−AMP +AMP	194,000 196,000	249,000	198,000 Assoc.	370,000
$S_{20,w}$ (S)	−AMP +AMP	8.9 8.9	10.0	8.8 12.4	13.7
PLP (moles/10^5 g)		1.0		1.0	
Phosphate incorporated in b → a (mole/10^5 g)		1.0		1.0	
Crystallization (pH 7.0 ± Mg^{2+}, 0°)	±AMP	−	−	+	+

*Activity measurements were carried out in 0.1 M maleate buffer (pH 6.5). A unit of activity is expressed as μmoles of P_i released/min at 30°. Other experiments were carried out in 50 mM glycerophosphate − 1.0 mM EDTA (pH 7.0) with mercaptoethanol varying from 1 to 50 mM. For sedimentation velocity and sedimentation equilibrium of rabbit phosphorylase a, 0.2 M NaCl was also included.

Differences between the dogfish and rabbit enzymes include their association-dissociation properties, solubility at low temperature, and K_m for glycogen as a primer. Thus dogfish phosphorylase a is only 20-25% associated from dimer to tetramer when rabbit phosphorylase a is completely tetramerized, and dogfish phosphorylase b remains as a dimer in the presence of AMP under conditions where rabbit phosphorylase b is 70-80% associated. There appears to be a distinct correlation between the ability of the proteins to tetramerize and crystallize at low temperature; since the Pacific dogfish lives at $ca.$ 5°, its inability to crystallize must be of physiological importance.

The many similarities between the dogfish and rabbit phosphorylases suggested that several areas of the molecules concerned with specific interactions might be conserved. Therefore in terms of the evolution of the control mechanism and the huge size of the enzyme subunit, it was of interest to compare their amino acid sequences. Indeed any alterations in the control of the enzyme must be intimately related to the rate of evolution of the phosphorylase molecule itself. To this end, carboxymethylated, $NaBH_4$-reduced and [^{32}P]-labeled dogfish phosphorylase a was cleaved with CNBr, and fractionated by various chromatographic procedures to separate the radioactive, phosphopyridoxyl, and small (less than 15 residues) peptides. Table III shows the partial or complete sequences of 9 small CNBr peptides of dogfish and rabbit phosphorylase, two of which CB-3' and CB-5 together comprise the sequence around the pyridoxal 5'-P binding site (16) and are identical in both enzymes (6,13). Table IV shows the sequence of an 11-residue peptide containing the site phosphorylated in the b to a conversion isolated after chymotryptic digestion of the large [^{32}P]-labeled CNBr peptide; all known phosphopeptides obtained from other phosphorylases are also listed. The dogfish peptide differs from the homologous sequence from the rabbit muscle enzyme only by the conservative substitution of an arginine for a lysine at position 2.

Although it was not wholly unexpected that the sequences around the phosphopyridoxyl lysine and phosphoserine would be extremely similar in view of their importance in catalysis and regulation, similarities in the other small CNBr peptides were more surprising. These are unusual in possessing two methionines close together (on average, one methionine per 38 residues is expected) but there is no other reason at present to suspect that they should be invariant. Nevertheless, of the seven peptides (excluding CB-3' and CB-5 in Table III), five corresponding to 44 residues were identical in composition and partial or complete sequence to CNBr peptides isolated from rabbit phosphorylase, while a sixth (CB-6, 8 residues) was probably homologous in the two proteins. The homology between these six peptides of more than 90% is therefore little different from that found for the two "functional site" peptides described above. If this degree of identity is maintained in the remainder of the sequence, it would correspond to little more than one amino acid change per 100 residues per 100 million years of divergence of these two species, or a slower rate of evolution than found thus far for other proteins, with the exception of histones. This high degree of conservation may be related to the multiplicity of functional sites found in this molecule, since the enzyme interacts with three substrates, effec-

TABLE III

PARTIAL OR TOTAL SEQUENCES OF SMALL CYANOGEN BROMIDE PEPTIDES OBTAINED FROM DOGFISH AND RABBIT MUSCLE PHOSPHORYLASES

	Dogfish Phosphorylase	Rabbit Phosphorylase	Number of Residues (Dogfish)	Residues In Common
CB 1	Leu-Hse	Leu-Hse	2	2
CB 2	Absent	Ala-Lys-Hse		
CB 3'	Pxy ↓ Lys-Phe-Hse	Pxy ↓ Lys-Phe-Hse	3	3
CB 4	Asx-Gly-Ala(*Asx,Glx,Val*)Hse	Asx-Gly-Ala(*Asx,Glx,Val*)Hse	7	7
CB 5	Gly-Arg-Thr-Leu-Gln-Asn-Thr-Hse	Gly-Arg-Thr-Leu(*Gln,Asn,Thr*)Hse	8	8
CB 6	Leu-Val-Asx(*Asx,Glx,Val*)Hse	Val(*Val,Asx,Thr,Glx,Ala,Leu*)Hse	8	4
CB 7	Leu-Asx-Gly-Ala(*Leu,Thr₂,Gly,Ile*)Hse	Leu-Asx-Gly-Ala-Leu(*Thr₂,Gly,Ile*)Hse	10	10
CB 8	Ile(*Gly,Gly,Lys,Ala,Ala,Pro,His,Gly,Tyr*)Hse	Ile-Gly-Gly-Lys-Ala-Ala-Pro(*His,Gly,Tyr*)Hse	11	11
CB 10	Ala-Glx-Glx-Ala(*Asx,Glx₂,Gly₂,Ile,Phe₃*)Hse	Ala-Glx-Glx-Ala(*Asx,Glx₂,Gly₂,Ile,Phe₃*)Hse	14	14
CB A	N-Acetyl-Ser-Lys-Pro-Lys-Ser-Asp-Hse	Absent		
			63	59

tors such as AMP and glucose-6-P, the cofactor pyridoxal 5'-P and two enzymes involved in its phosphorylation and dephosphorylation. In addition, two types of subunit contact have been identified allowing for the formation of phosphorylase dimers and tetramers. Whether or not these sites overlap, a substantial portion of the surface of phosphorylase must be devoted to specific recognitions.

The seventh small CNBr peptide was found only in the dogfish and not in the rabbit enzyme; it has a blocked amino terminus, and its sequence was found to be N-Acetyl-Ser-Lys-Pro-Lys-Ser-Asp-Met (6). It was sequenced by a tryptic subdigest which cleaved opposite the second lysyl residue, and by the use of a 2 min exposure of

TABLE IV

SEQUENCES AROUND THE PHOSPHOSERINE AND PYRIDOXAL LYSINE RESIDUES IN PHOSPHORYLASES

Source	1	2	3	4	5	6	7	8	9	10	11	12	13	14	Reference
Rabbit Muscle	Ser	Asp	Gln	Glu	Lys	Arg	Lys	Gln	Ile	$\overset{P}{Ser}$	Val	Arg	Gly	Leu	(14)
Dogfish Muscle				Glu	$\underline{Arg}$	Arg	Lys	Gln	Ile	$\overset{P}{Ser}$	Val	Arg	Gly	Leu	(6)
Rat Muscle	Ser	Asp	Gln	$\underline{Asp}$	Lys	Arg	Lys	Gln	Ile	$\overset{P}{Ser}$	Val	Arg	Gly	Leu	(3)
Human Muscle							Lys	Gln	Ile	$\overset{P}{Ser}$	Val	Arg			(15)
Rabbit Liver							$\underline{Arg}$	Gln	Ile	$\overset{P}{Ser}$	$\underline{Ile}$	Arg			(11)
Rabbit Muscle*	Met	$\overset{Pxy}{Lys}$	Phe	Met	Gly	Arg	Thr	Leu	(Gln,	Asn,	Thr)	Met			(16)
Dogfish Muscle	Met	$\overset{Pxy}{Lys}$	Phe	Met	Gly	Arg	Thr	Leu	Gln	Asn	Thr	Met			

Differences from the rabbit muscle sequence are underlined.

---

* The sequence of a further 30-residues on the NH_2-terminal side of the rabbit pyridoxal peptide has been determined (16).

the resulting NH$_2$-terminal tetrapeptide to 6 N HCl at 100° which specifically unblocked serine. The seryl substituent was identified as an acetyl group following hydrazinolysis. The suggestion that it might be derived from the amino terminus of the protein was confirmed by the isolation of the tetrapeptide N-Acetyl-Ser-Lys-Pro-Lys from a tryptic digest of the whole protein. Recently, in conjunction with a study of the total sequence of rabbit and dogfish muscle phosphorylases carried out in collaboration with Drs. Neurath and Walsh of this Department, a tryptic peptide was isolated from the [^{32}P]-CNBr fragment of rabbit phosphorylase which had an acetylated NH$_2$-terminal and a composition similar to that of the dogfish NH$_2$-terminal peptide (Ser$_2$,Arg,Pro,Leu,Asp, Glx$_2$,Lys) (Titani et al., unpublished data). This, together with other preliminary data on the sequence of the rabbit CNBr phosphopeptide suggests that the acetylated NH$_2$-terminal peptides and the phosphopeptides listed in Table III might follow one another, so that the phosphoseryl residue would occupy position 14 in the molecule (865 residues):

<u>Dogfish</u>   5   10   P   15

N-Acetyl-Ser-Lys-Pro-Lys-Ser-Asp-Met | Glu-Arg-Arg-Lys-Gln-Ile-Ser-Val-Arg-Gly-Leu...

<u>Rabbit</u>   P

(N-Acetyl,Ser,<u>Arg</u>,Pro,<u>Leu</u>)Ser-Asp-<u>Gln</u>-Glu-<u>Lys</u>-Arg-Lys-Gln-Ile-Ser-Val-Arg-Gly-Leu...

Differences between the dogfish and rabbit enzymes are underlined; two of the changes are explained by single base changes, while two require two base changes. The vertical bar between residues 7 and 8 in the dogfish sequence indicates that the overlap at this point has not been confirmed.

The close proximity of the phosphoseryl residue to the amino terminus is of interest in terms of the changes in structure and activity that occur following phosphorylation of the protein; knowledge of its position will of course be helpful in the total X-ray analysis of the protein presently being carried out in Oxford by Louise Johnson and David Phillips. It was anticipated that the phosphoseryl residue would lie in an exposed region, since the kinase and phosphatase must both have access to it. Previous work (17) had shown that tryptic attack converts phosphorylase <u>a</u> to a form of the enzyme inactive in the absence of AMP (phosphorylase <u>b</u>'); it was later established (14,18) that during this reaction, a phosphorylated hexapeptide (residues 11-16) is released while the rest of the molecule remains essentially untouched.

Table V compares the amino terminal sequences of dogfish skeletal muscle phosphorylases with those of other glycolytic enzymes and muscle proteins. As can be seen, only two of the glycolytic enzymes possess a free amino group; in every case where the nature of the blocking group has been determined, it turned out to be an N-acetyl derivative. The functional significance of this property remains to be clarified. One possibility is that acetylation is essential to polypeptide chain initiation in higher organisms. However, in contrast to bacteria, the amino-terminal residue varies and threonine, alanine, glycine and aspartic acid have been found in addition to serine.

TABLE V

NH$_2$-TERMINAL SEQUENCES OF GLYCOLYTIC ENZYMES AND OTHER MUSCLE PROTEINS

Enzyme	Source	N-Terminus	Reference
Phosphorylase	rabbit muscle	(N-Acetyl,Ser,Asp,Pro,Leu,Ser,Asp,Glu)	(6)
	dogfish muscle	N-Acetyl-Ser-Lys-Pro-Lys-Ser-Asp-Met	(19)
Phosphoglucomutase	rabbit muscle	Blocked	(20)
Phosphofructokinase	rabbit muscle	Blocked	(21)
Aldolase	rabbit muscle	Proline	(22)
Triose Phosphate Isomerase	rabbit muscle	Alanine	(23)
Glyceraldehyde 3-Phosphate Dehydrogenase	lobster muscle	N-Acetyl-Ser-Lys-Ile-Gly-Ile-Asp-Gly	(24)
	pig muscle	Val-Lys-Val-Gly-Val-Asp-Gly	(25)
Phosphoglycerate Kinase	human erythrocyte	N-Acetyl-Serine	(26)
Enolase	rabbit muscle	N-Acetyl-Alanine	(27)
	yeast	Ala-Gly-Lys-Val-Gly-Asp-Thr	(28)
Pyruvate Kinase	rabbit muscle	Acetylated	(29)
Lactate dehydrogenase	chicken muscle H$_4$	N-Acetyl-Ala-Ala-Thr	(30)
	dogfish muscle M$_4$	N-Acyl-Thr-Ala-Leu	
	rat liver M$_4$	N-Acetyl-Ala-Ala	
Glycerol 3-Phosphate Dehydrogenase	rabbit muscle	Acetylated	(31)
Alcohol Dehydrogenase	horse liver	N-Acetyl-Ser-Thr-Ala-Gly-Lys-Val-Ile	(32)
Myosin	rabbit muscle	N-Acetyl-Ser-Ser-Asp-Ala-Asp	(33)
Actin	rabbit muscle	N-Acetyl-Asp-Glu-Thr-Glu-Asp-Thr-Ala	(34)
Cytochrome C	rabbit heart muscle	N-Acetyl-Gly-Asp-Val-Glu-Lys-Gly-Lys	(35)

Alternatively, acetylation could protect these enzymes against degradation by amino peptidases released by rupture of lysosomes within the cell during its lifetime.

The long-range goal of this research is to study the regulation of glycogen metabolism in the dogfish in comparison with the mechanisms known to be operative in the rabbit. At the present time, attempts are being made to isolate the enzymes responsible for the control of phosphorylase activity in dogfish muscle and characterize their behavior in response to hormonal and nervous stimulations.

Comparative Aspects of Yeast Phosphorylase. Yeast was selected as the second source of material for this study because it can accumulate large quantities of glycogen when grown anaerobically on high carbohydrate media (36,37); when switched to a minimal medium under anaerobic conditions, it quickly utilizes this polysaccharide suggesting control at the phosphorylase level. Furthermore, while several studies of yeast glycogen phosphorylase have appeared (38-40), this enzyme had not yet been obtained in pure form and its mechanism of regulation was unknown. In fact, no regulation of activity by phosphorylation of the protein had been firmly established in either bacteria (41,42) or plants (43,44).

Initial attempts at purifying yeast phosphorylase were frustrated because of the presence of proteases which accompanied this enzyme during purification. Isolation of phosphorylase was finally achieved by inclusion of diisopropylphosphorofluoridate or phenylmethanesulfonyl fluoride or both at each step of the purification. The extract obtained by grinding baker's yeast with glass beads in an Eppenbach colloid mill (Model MV-6-3, Gifford Wood, Inc., Hudson, N.Y.) was purified by a combination of streptomycin and ammonium sulfate precipitations, various DEAE and Sephadex chromatographies and gel filtration on Sephadex G-200 (6). In the final DEAE-Sephadex column, phosphorylase emerged in two active peaks (Figure 2).

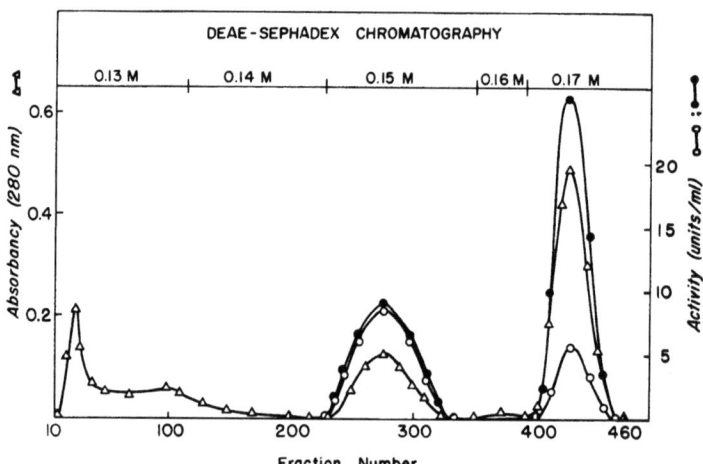

FIGURE 2. DEAE-Sephadex A-50 was equilibrated against 0.13 M sodium succinate, 1 mM EDTA at pH 5.8. Stepwise elution was carried out at a flow rate of 35 ml/hr and 11 ml fractions were collected. Absorbancy at 280 nm (Δ-Δ) and phosphorylase activity both before (O-O) and after (●-●) kinase activation of individual fractions are illustrated

The enzyme obtained from each was homogeneous by gel isoelectric focusing and electrophoresis in the presence of sodium dodecyl sulfate and by sedimentation velocity in the ultracentrifuge. Both forms had identical amino acid compositions, subunit molecular weights of 103,000 and contained one molecule of pyridoxal 5'-P per enzyme subunit. The two forms, however, differed in their kinetic properties in that the material in the first peak had a specific activity of 135 ± 10 units/mg (as compared to 90 for crystalline rabbit muscle phosphorylase), whereas the material in the second peak had a specific activity of 25 ± 5 units/mg. Neither form was affected by 2', 3' or 5'-AMP. By analogy with the rabbit muscle system, these two forms were designated _a_ and _b_, respectively. This assumption was supported by earlier experiments indicating that partially purified fractions of yeast phosphorylase could be activated by incubation with Mg^{2+} and ATP. Later, when more purified enzyme became available, uptake of ^{32}P from γ-labeled ATP could also be demonstrated. As expected, pure phosphorylase _b_ did not show any activation or incorporation of phosphate with Mg^{2+} and ATP alone; the required phosphoprotein kinase was eventually purified from yeast extract as will be discussed later.

When the individual fractions of the DEAE-Sephadex columns were incubated with Mg-ATP and the specific kinase, the results shown in Figure 2 were obtained: whereas no increase in activity occurred with the enzyme present in the first peak (presumably phosphorylase _a_), the specific activity of the material in the second peak rose to that of the first. This increase in activity was accompanied by incorporation of *ca.* one equivalent of phosphate per enzyme subunit; the bound phosphate was acid stable and alkali-labile, as would be expected for a monoester such as phosphoserine. The exact nature of the residue involved remains to be determined.

The properties of the two forms of phosphorylase are listed in Table VI. Phosphorylase _a_ exists as a slightly associated dimer whereas phosphorylase _b_ appears to be tetrameric. In this sense, the yeast enzyme behaves more logically than rabbit muscle phosphorylase since the active conformation of the latter has a strong tendency to tetramerize even though the phosphorylated dimers are the most active (45). There is no immunological cross reaction between antibodies to yeast phosphorylase and the phosphorylases from rabbit and dogfish muscle, and rabbit liver.

Kinetic studies carried out on both forms of the enzyme showed that they have rather similar Michaelis constants when measured in the direction of glycogen synthesis in the presence of both substrates (Figure 3). However, they differ significantly in their affinities for a single substrate in the absence of the other (K_i glycogen and K_i G-1-P), with the _a_ form displaying the higher affinities; this suggests for instance that at low substrate concentration, most of the phosphorylase _a_ might be bound to glycogen while phosphorylase _b_ would be largely free. Glucose-6-P is an effective inhibitor of both phosphorylase _a_ and _b_ (see Table VI), the latter being much more sensitive than the former. During periods of active glycogen synthesis, glucose-6-P can rise as high as 8 mM (46). Under these conditions, glycogen synthetase would be fully active and phosphorylase _b_ essentially inactive. On the

TABLE VI

PROPERTIES OF YEAST GLYCOGEN PHOSPHORYLASE

Property	a	b
$S_{20,w}$	9.6	14.2
M_w	250,000	370,000
Molecular form	Partially associated dimer	Tetramer
Subunit molecular weight	103,000 ± 3,000	103,000 ± 3,000
$A_{280}^{1\%}$	14.9	14.9
PLP Content (moles/10^5 g)	0.95	0
Phosphate incorporated (moles/10^5 g)	0.83	0
Isoelectric point	5.1 ± 0.1	5.1 ± 0.1
pH optimum	5.8	5.8
Specific activity (units/mg)	135 ± 10	25 ± 5
K_I (G-6-P)	11 mM	1 mM

Activity measurements were carried out in 0.1 M sodium succinate buffer, pH 5.8, 30°, in the presence of 0.15 M glucose-1-P and 2% glycogen.

other hand, conversion of phosphorylase b to a would enable the yeast cell to overcome this phosphorylase inhibition if glycogen breakdown were required even at high glucose-6-P levels.

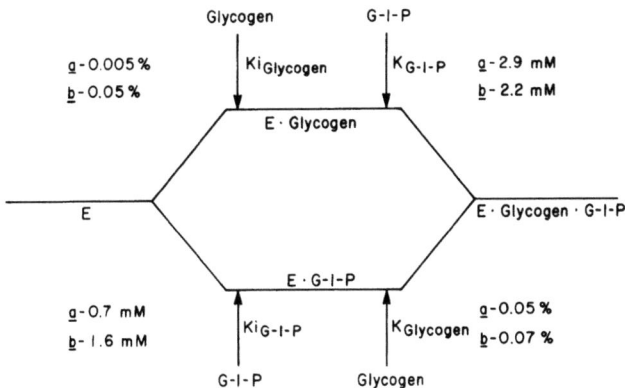

FIGURE 3. Schematic representation of a possible reaction mechanism for yeast phosphorylase. Experimental constants measured in the direction of glycogen synthesis are given

It is likely that the activity displayed by phosphorylase b is intrinsically due to this form of the enzyme rather than to contamination by phosphorylase a: as indicated above, phosphorylase b displays different binding constants and inhibition by glucose-6-P. It is nonetheless possible that this low level of activity is due to the presence of partially phosphorylated hybrids having quaternary structures, kinetic properties, and elution patterns from ion-exchange columns different than those of the homologous, fully phosphorylated species.

Supposing phosphorylase b is indeed $ca.$ 20% active and that the same situation prevails within the cell, it might be of interest to speculate on the significance of this activity. It can be reasonably assumed that at the origin, phosphorylase must have existed only in an active form. Control of this molecule became possible when this activity could be effectively suppressed for instance by subunit interaction: positive effectors or phosphorylation of the protein could then restore the activity of the enzyme, $e.g.$ by bringing about certain favorable changes in conformation. If this were the case, one could visualize yeast phosphorylase b as being a rather primitive enzyme for which control is still imperfect or "leaky"; actually, artifical distortions of the molecule or loosening up of its structure, for instance by addition of 0.4 M Na_2SO_4 further increases this activity.

It is also of interest to note that on the basis of our present knowledge of the yeast system, control of phosphorylase appears to rely primarily on phosphorylation of the protein, not on noncovalent interaction with an effector such as AMP. Therefore, covalent control has appeared very early; it was not introduced as an added sophistication in regulatory processes confined to higher organisms. Phosphorylase from higher plants (*e.g.* potato) seems to be regulated neither by allosteric nor covalent modifications (43,44). *E. coli* maltodextrine phosphorylase is not phosphorylated though slight activations by AMP have been reported (41,42). More complex eukaryotic organisms all appear to be subject to both types of control. How did these regulatory systems originate and how were they further developed? It has been reported (40) that yeast phosphorylase binds AMP, even though fixation of this nucleotide does not affect its activity. It is conceivable that following an initial period of purely unspecific binding, the AMP-binding site was gradually modified in such a way that it ultimately could serve to modulate enzymatic activity.

Yeast also differs from the animal phosphorylases with respect to its susceptibility to the regulatory enzymes. For instance, whereas rabbit muscle phosphorylase kinase and phosphatase will catalyze the interconversion of all animal phosphorylases so far investigated (from rat, dogfish, frog, lobster and insects), they will not touch the yeast enzyme. This finding necessitated the search for the specific enzymes responsible for the regulation of yeast phosphorylase. A protein phosphokinase was fractionated from baker's yeast following a 5000-fold purification according to the scheme shown in Table VII.

The final material is not homogeneous since gel electrophoresis in sodium dodecyl sulfate indicates the presence of approximately five bands ranging in molecular weight from 30,000 to 70,000; taken together with the elution pattern following Sephadex G-200 chromatography, these data suggest that the kinase exists as a monomeric protein of molecular weight *ca.* 55,000.

At no stage of the purification of the yeast kinase could any requirement for cyclic 3',5'-AMP be demonstrated: with or without cyclic-AMP, the enzyme phosphorylates yeast phosphorylase b (but not a), casein and phosvitin. On the other hand, it is totally inactive on rabbit or dogfish muscle phosphorylase or phosphorylase kinase, yeast phosphofructokinase, bovine serum albumin, protamines or histones. It did not alter the glucose-6-P requirement of yeast glycogen synthetase (E. Cabib, personal communication).

The optimum pH of activity of the kinase was 8.0, and ATP, GTP and UTP all appeared to serve as phosphate donors with a relative effectiveness of 1:1:0.3, respectively. While Mg^{2+} gave maximal rates, Co^{2+} and several other divalent cations were also effective, but not Ca^{2+}. Preliminary kinetic studies using casein as substrate yielded Michaelis constants of *ca.* 0.25 mM for ATP and 3 mg/ml for casein.

Of major concern during most of this study was our inability to detect a protein phosphatase that could catalyze the reconversion of phosphorylase a to b; obviously, this reaction must occur to allow for an effective control of glycogen breakdown. Lately, however, evidence was obtained that the a → b conversion proceeds not through the action of a phosphorylase phosphatase, as found in the animal systems, but by direct reversal of the kinase reaction itself. First indications for such a reaction were seen when, on prolonged incubation with the kinase, the level of bound

TABLE VII

PURIFICATION OF YEAST PHOSPHORYLASE KINASE

Step	Specific Activity (Units/mg)	Purification	Recovery (%)
Crude extract	0.5	1	100
Streptomycin sulfate supernate	0.6	1.3	96
CM-Sephadex eluate	21.5	48	91
50% Ammonium sulfate precipitate	85	190	82
CM-Sephadex column	382	850	58
G-200 Sephadex column	854	1900	50
Casein-Sepharose column	2460	5500	13

Activity measured in 0.1 M Tris-HCl, pH 7.5, containing 5 mg/ml casein, 1 mM γ-[^{32}P]-ATP, 5 mM $MgCl_2$, and 0.5 mM EDTA. Kinase was added and aliquots were withdrawn at given times to determine incorporation of ^{32}P into trichloroacetic acid precipitable material. One unit of activity will transfer 10^{-9} moles of ^{32}P from ATP to casein in one minute at 30°. The crude extract from eight pounds of yeast contained $ca.$ 100,000 units of kinase.

phosphate started to decrease. More recently, it was shown that when ^{32}P-labeled yeast phosphorylase $\underline{a}$, casein or phosvitin were incubated with Mg-ADP, and the protein kinase, the bound radioactivity was lost. Omission of any one of the components of the system totally prevented this loss; GDP and, to a lesser extend, UDP could replace ADP in this conversion of phosphorylase $\underline{a} \rightarrow \underline{b}$.

It is likely that the reversible phosphorylation described above is identical to that reported a few years ago by Rabinowitz and Lipmann (47); probably, the same protein kinase is involved. There remains the question as to the exact nature of the protein-phosphate bond (presumably a phosphate monoester) which readily allows the synthesis of ATP from ADP. It will have to be determined whether a particular conformation of the protein, amino acid sequence (*e.g.* the polyserine sequences found in phosphoproteins) (48,49) or structural changes in the phosphate linkage (such as the O → N migrations postulated to occur under mild alkaline conditions) (50,51) can elevate the energy level of a seryl phosphate bond to that of ATP.

Attempts to answer such questions have been hampered by the structural complexity of the phosphoproteins heretofore used as substrates, since they contain a number of phosphate residues of widely varying reactivities. The finding that yeast phosphorylase, which possesses a single phosphorylated site, may also serve as a phosphate donor should provide considerable information on the nature of the ATP-forming reaction.

In view of the fact that the yeast protein phosphokinase appears to exist only in the active form (no cyclic-AMP requirement, for instance), the conditions responsible for the control of glycogen breakdown in this organism are still not fully understood. It will have to be determined if the ratio of phosphorylase $\underline{a}$ to $\underline{b}$ is mainly determined by an ATP/ADP ratio or if other factors are also involved.

REFERENCES

1. FISCHER, E.H., POCKER, A. and SAARI, J.C., Essays in Biochemistry, Vol. 6, pp. 23-68, P.N. Campbell and F. Dickens, Eds., Academic Press, London (1970).
2. HARRIS, C. E. and TELLER, D. C., submitted to J. Theoret. Biol. (1971).
3. SEVILLA, C. L. and FISCHER, E. H., Biochemistry 8, 2161 (1969).
4. YUNIS, A. A., FISCHER, E.H. and KREBS, E. G., J. Biol. Chem. 235, 3163 (1960).
5. METZGER, B. E., GLASER, L. and HELMREICH, E., Biochemistry 7, 2021 (1968).
6. COHEN, P., SAARI, J. C. and FISCHER, E. H., Biochemistry, in press.
7. ASSAF, S. A. and GRAVES, D. J., J. Biol. Chem. 244, 5544 (1969).
8. CHILDRESS, C. C. and SACKTOR, B., J. Biol. Chem. 245, 2927 (1970).
9. FOSSET, M., MUIR, L. W., NIELSEN, L. D. and FISCHER, E. H., Biochemistry, in press.
10. KAMOGAWA, A., FUKUI, T. and NIKUNI, Z., Jap. J. Biochem. 63, 361 (1968).
11. WOLF, D. P., FISCHER, E. H. and KREBS, E. G., Biochemistry 9, 1923 (1970).
12. COHEN, P., DUEWER, T., and FISCHER, E. H., Biochemistry 10, 2683 (1971).
13. SAARI, J. C. and FISCHER, E. H., Biochemistry, in press.

14. NOLAN, C., NOVOA, W. B., KREBS, E. G. and FISCHER, E. H., Biochemistry 3, 542 (1964).
15. HUGHES, R. C., YUNIS, A. A., KREBS, E. G. and FISCHER, E. H., J. Biol. Chem. 237, 40 (1962).
16. FORREY, A. W., SEVILLA, C. L., SAARI, J. C., and FISCHER, E. H., Biochemistry 10, 3132 (1971).
17. KELLER, P. J. and CORI, G. T., J. Biol. Chem. 214, 127 (1955).
18. FISCHER, E. H., GRAVES, D. J., CRITTENDEN, E. R. and KREBS, E. G., J. Biol. Chem. 234, 1698 (1959).
19. JOSHI, J. G. and HANDLER, P., J. Biol. Chem. 239, 2741 (1964).
20. PAETKAU, V. H., YOUNATHAN, E. S. and LARDY, H. A., J. Mol. Biol. 33, 721 (1968).
21. UDENFRIEND, S. and VELICK, S. F., J. Biol. Chem. 190, 733 (1960).
22. KRIETSCH, W. K. G., PENTCHEV, P. G., KLINGENBÜRG, H., HOFSTÄTTER, T. and BÜCHER, T., Eur. J. Biochem. 14, 289 (1970).
23. HARRIS, J. I. and PERHAM, R. N., Nature 219, 1025 (1968).
24. YOSHIDA, A., J. Biol. Chem., in press.
25. WINSTEAD, J. A. and WOLF, F., Biochemistry 3, 791 (1964).
26. BREWER, J. M., FAIRWELL, T., TRAVIS, J. and LOVINS, R. E., Biochemistry 9, 1011 (1970).
27. COTTAM, G. L., HOLLENBERG, P. F., and COON, M. J., J. Biol. Chem. 244, 1481 (1969).
28. BRUMMEL, M. C., SANBORN, B. M. and STEGINK, L. D., Arch. Biochem. Biophys. 143, 330 (1971).
29. ALLISON, W. S., ADMIRAL, J. and KAPLAN, N. O., J. Biol. Chem. 244, 4743 (1969).
30. SANBORN, B. M., BRUMMEL, M. C., STEGINK, L. D. and VESTLING, C. S., Biochem. Biophys. Acta 221, 125 (1970).
31. VAN EYS, J., JUDD, J., FORD, J. and WOMACK, W. B., Biochemistry 3, 1755 (1964).
32. JORNVALL, H., Eur. J. Biochem. 14, 521 (1970).
33. OFFER, G. W., Biochem. Biophys. Acta 111, 191 (1965).
34. COLLINS, J. H., MORKIN, E. and ELZINGA, M., Fed. Proc. 30, 1148, Abs. 558 (1971).
35. NEEDLEMAN, S. B. and MARGOLIASH, E., J. Biol. Chem. 241, 853 (1966).
36. CHESTER, V. E., Biochem. J. 86, 153 (1963).
37. CHESTER, V. E., Biochem. J. 92, 318 (1964).
38. KIESSLING, W., Naturwissenschaften 27, 129 (1939).
39. WHELAN, W. J., Methods Enzymol. 1, 199 (1955).
40. SAGARDIA, F., GOTAY, I. and RODRIQUEZ, M., Biochem. Biophys. Res. Commun. 42, 829 (1971).
41. CHEN, G. S. and SEGAL, I. H., Arch. Biochem. Biophys. 127, 175 (1968).
42. SCHWARTZ, M. and HOFNUNG, M., Eur. J. Biochem. 2, 132 (1967).
43. KAMAGAWA, A., FUKUI, T. and NIKUNI, Z., J. Biochem. 63, 361 (1968).
44. FUKUI, T. and KAMOGAWA, A., J. Jap. Soc. Starch Sci. 17, 117 (1969).
45. WANG, J. H. and GRAVES, D. J., Biochemistry 3, 1437 (1964).
46. ROTHMAN, L. B. and CABIB, E., Biochemistry 8, 3332 (1969).
47. RABINOWITZ, M. and LIPMANN, F., J. Biol. Chem. 235, 1043 (1960).
48. WILLIAM, J. and SANGER, F., Biochim. Biophys. Acta 33, 294 (1959).

49. BELITZ, H. D., Lebensm. Untersuch. Forsch. 127, 341 (1965).
50. TABORSKY, G., Biochemistry 2, 266 (1963).
51. BARGONI, N., RIHANDO, M. T. and TOURN, M. L., Ital. J.Biochem. 15, 43 (1966).

Electron Microscopy of Muscle Phosphorylases *b* and *a*

N. A. Kiselev and F. Ya. Lerner

Institute of Crystallography of the Academy of Sciences of the USSR,
Moscow/USSR

N. B. Livanova

Bach Institute of Biochemistry of the Academy of Sciences of the USSR,
Moscow/USSR

I. Introduction

Muscle glycogen phosphorylase exists in two enzymatically interconvertible forms: phosphorylase *a* and phosphorylase *b*. It has been shown that phosphorylase *a* in solution is a tetramer of molecular weight 400000 (s=13,2S), and phosphorylase *b* is a dimer of molecular weight 200000 (1). In the presence of p-chlormercuribenzoate (2) phosphorylases *a* and *b* dissociate into four and two monomers, respectively, both having the same molecular weight. On conversion of phosphorylase *b* into *a*, in the presence of ATP, Mg^{2+} and phosphorylase *b* kinase, one phosphate group appears to be bonded to the serine residue per monomer of the enzyme (3). If phosphorylase *a* is fully active in the native state without 5'-AMP, phosphorylase *b*, the non-phosphorylated form of the enzyme, is active only in the presence of 5'-AMP or IMP with protamin (4,5). The ultracentrifuge studies have shown that the dimeric form of phosphorylase *b* is converted into the tetrameric form (s=13,2S) in the presence of a mixture of 0,001M-AMP+0,03M cysteine+0,01M Mg^{2+}(6) or 0,001M 5'-IMP+ 0,0001M protamin or protamin only (7). Under the same conditions, at a rather high protein concentration and a low temperature, phosphorylase *b* forms crystals quite readily (6,8); phosphorylase *a* forms crystals at a low temperature and high protein concentration in 0,03M cysteine or 0,03M β-mercaptoethanol as well as in the presence of protamin.

II. Materials and methods

Phosphorylase *b* from rabbit muscle was purified by the method of Fischer and Krebs (6), somewhat modified (9), and used after four recrystallizations at 0° from solution which contained 0,001M 5-AMP, 0,01M $(CH_3COO)_2Mg$ and 0,03M cysteine. Phosphorylase *a* was prepared from phosphorylase b which had been recrystallized four times, using phosphorylase b kinase isolated from rabbit muscle by the method of Krebs and others (10). Phosphorylase a was recrystallized three times from 0,03M cysteine. Before the crystallization with protamin the phosphorylase b preparations were passed through a Norit-cellulose column and dialyzed against 0,04M sodium glycerophosphate buffer, pH 7,0. The final concentration of protamin at the crystallization of phosphorylase b (3 mg/ml) was 0,0001M, and at the crystallization of phosphorylase a - 0,00002M.

The specimens for electron microscopy were prepared by negative staining with uranyl acetate. For the treatment of images of the periodical structures on microphotographs use was made of the method of optical diffraction, and the filtering of the image (13). The optical diffractometer used was of a horizontal type (11) with the helium-neon laser as a light source.

III. Phosphorylase b

a. Particles from solution

While investigating the tetrameric form of phosphorylase b in solution (at protein concentration 0,4 mg/ml and in the presence of 0,0001M protamin and 0,001 5'-AMP or IMP, 20°C) one could mainly observe separate particles, and only in rare cases short tubes and narrow elongated in one direction plane layers of particles (Fig.1). Two distinctly expressed types show up among the particle images. The images of the first type can be characterized by a half-ring on one side, and by a cross (or letter "V") on the other. The length of such particles is 115-120 Å, whereas the dimensions in the transverse direction range from 120 to 90 Å. The images of the second type bear resemblance to figure "8" and are characterized by dimensions of 115-120 Å (in length) and about 70-80 Å (in width). In a still more dilute enzyme solution (0,1 mg/ml) no crystalline formations were observed, the enzyme particles showing the same appearance as that given in Fig.1. On sedimentation under analogous conditions, but at a higher protein concentration (1,5 mg/ml) we observed a mixture of dimers and tetramers, but could not detect the presence of aggregates of higher molecular weight.

b. Crystalline formations. Plane layers

Electron microscopy of negatively stained preparations to which was added protamin, as well as of preparations with protamin and AMP (or IMP) added, revealed three types of crystalline formations: plane layers of particles, tubes, three-dimensional crystals (12).

The plane layers which were now and then observed under conditions unfavourable for crystallization (see above) are shown in Fig.1. They usually have 2-4 rows in width. The rows are displaced lengthwise relative to each other, thereby forming a peculiar step-like structure. The layers of such a type are sometimes found to be connected with short sections of the tubes whose walls represent a monolayer of particles. Sometimes it is seen that the particles are characterized by a half-ring on one side and by a cross (or letter "V") on the other.

The plane layers of particles observable under conditions favourable for crystallization are shown on the micrograph of Fig.2(a). An optical diffraction pattern from a section of the layer is shown in Fig.3(a). The right-hand part of the pattern is indexed on the reciprocal lattice at the points of which the reflections associated with the repeat structure of the layer of particles are located.

To eliminate the background effect and other unfavourable phenomena, the optical filtering of images was used (13). To do this a mask with round holes at the site of reflections (Fig.3(b)) was placed in the diffraction plane of the optical diffractometer. To enhance the contrast of the filtered image the aperture at the site of the central cross was covered with a mesh of about 50% transmission. The results of the filtering are shown in Fig.3(c). Here, the structure of particles stand out clearly as compared with the original image (Fig.2(a)). These particles can also be characterized by a half-ring on one side and a cross on the other. It can be seen that the particles in the neighbouring rows in the layer are oriented in the opposite directions. The unit cell contains two packing units, and has dimensions of about 120 Å x (120 x 2) Å. They are found to vary even within the layers shown in one of the micrographs.

Fig.2(b) illustrates the rare type of aggregation of particles, i.e. the case where they resemble the "8"-type particles from solution.

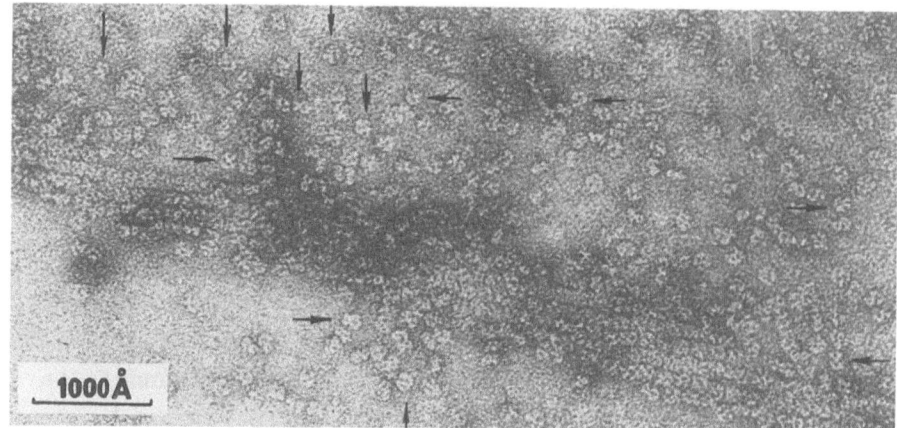

Fig. 1. Phosphorylase b particles from solution and plane layers, observed under conditions unfavourable for crystallization

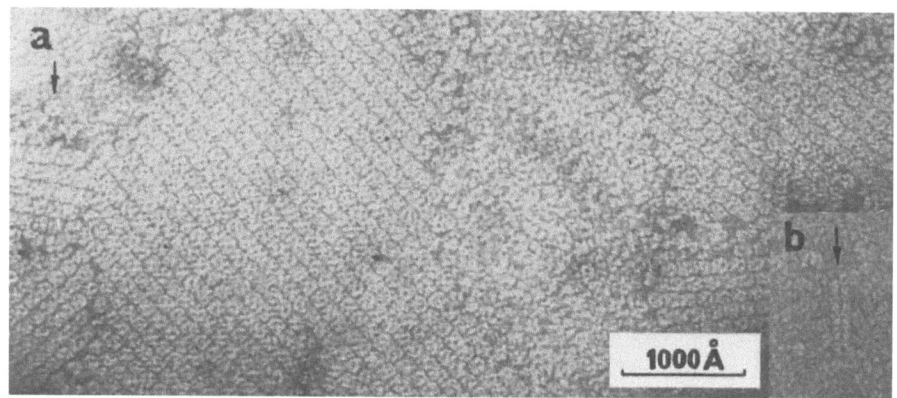

Fig. 2. (a) Plane layers of phosphorylase b.
(b) Rare type of particle aggregation

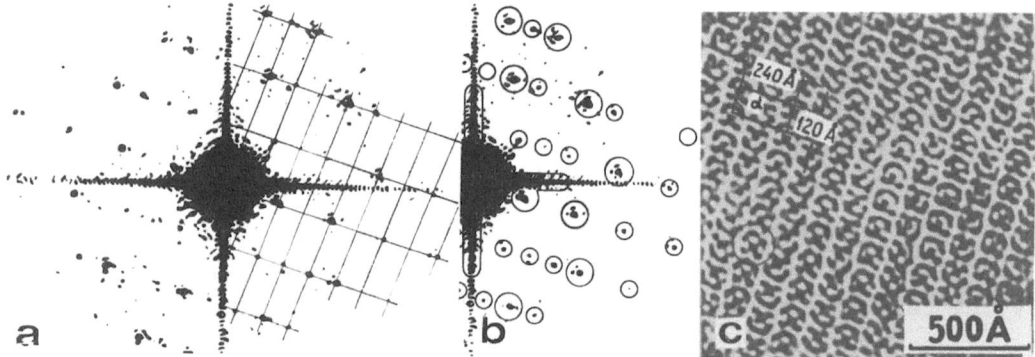

Fig. 3. (a) Optical diffraction pattern from the fragment of a plane layer (the right part is indexed on the reciprocal lattice).
(b) The same (with the mask used for filtering).
(c) Filtered image. Protein is black

c. Tubes

The main form of crystallization of phosphorylase <u>b</u> in preparations with protamin added are tubes with one-layer walls (Fig.4(a,b)) and tubes with two-layer walls (Fig.4(c)). The majority of the tubes with one-layer walls are fairly flattened on the film; this can be explained in terms of comparatively thin (about 70 Å) tube walls as regards their diameter. When the tubes are viewed nearly perpendicular to the axis, one can sometimes observe the rows of particles helically packed together in a wall of the tube with a pitch of about 120 Å.

In tubes with two-layer walls (Fig.4(c)) the wide sections alternate with the narrow ones along the tube, thus giving impression of periodic deformations. One can usually count between the wide sections 35 rows of particles helically packed together. The deformations can be explained on the assumption that there exist certain periodic combinations of the structure of the external and internal walls of the tubes which are more rigid than the others. The external diameter of the tubes at the narrow part is about 475 Å, the internal diameter is 335 Å, the thickness of the walls of the outer tube is about 70 Å.

If the formation of one-layer tubes is defined only by the properties of phosphorylase molecules in the presence of protamin, then the already formed cylindrical surface of the inner layer must play a considerable role in forming the outer tube. Of this is indicative, first of all, the fact that we so far failed to observe the tubes with one-layer walls which could be identified by the parameters analogous to those of the outer tube in two-layer tubes. Thus, these outer tubes do not exist by themselves. On the other hand, the second layer is sometimes seen to cover the inner tube only in part (Fig.4 (d,e,f)). It seems safe to suggest that just here is fixed the process of the outer tube formation. The picture reverse to this, i.e. when the inner tube would partly fill the outer tube, was not revealed. The formation of the second cylindrical layer of particles does not seem to be the end of the process. Fig.4(d,f), for example, shows how the third layer is being formed on the surface of the two-layer tube. Occasionally, one can observe the tubes exhibiting the number of layers, amounting to 4,6 and 8 (Fig.4(g)).

Fig.5(a) shows the diffraction pattern from the one-layer tube of Fig.4(a), indexed on the reciprocal lattices. All the reflections are located in the system of layer lines shown in Fig.5(d). It is known that the number of the layer lines to which there contributes the zero-order Bessel function lying closest to the zero line, is

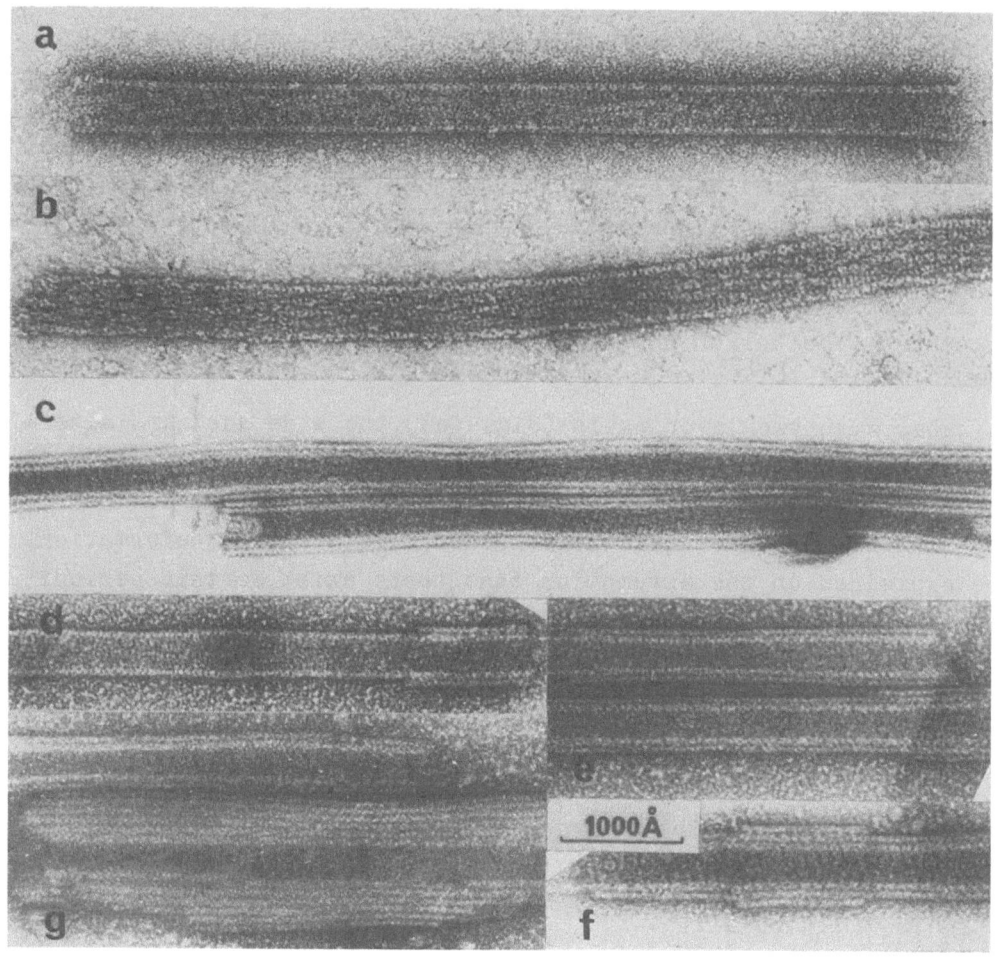

Fig.4. Phosphorylase b tubes.

    (a,b) One-layer walls.
    (c) Two-layer walls.
    (d) Tube partly covered with the second layer.
    (e,f) Tubes with one-, two and three walls.
    (g) Multi-layer tube

equal to the number of packing units falling on the repeat distance (14,15,16). In the diffraction pattern given in Fig.5(a) the meridional reflections of this type are missing; however, starting from symmetry considerations, one would expect the meridional reflection to be located on layer line 27. Actually, some other diffraction patterns on layer line 27 do show a weak but distinctly expressed meridional reflection (Fig.5(c)). The reflection closest to the meridian (i.e. that being characterized by the lowest-order Bessel func-

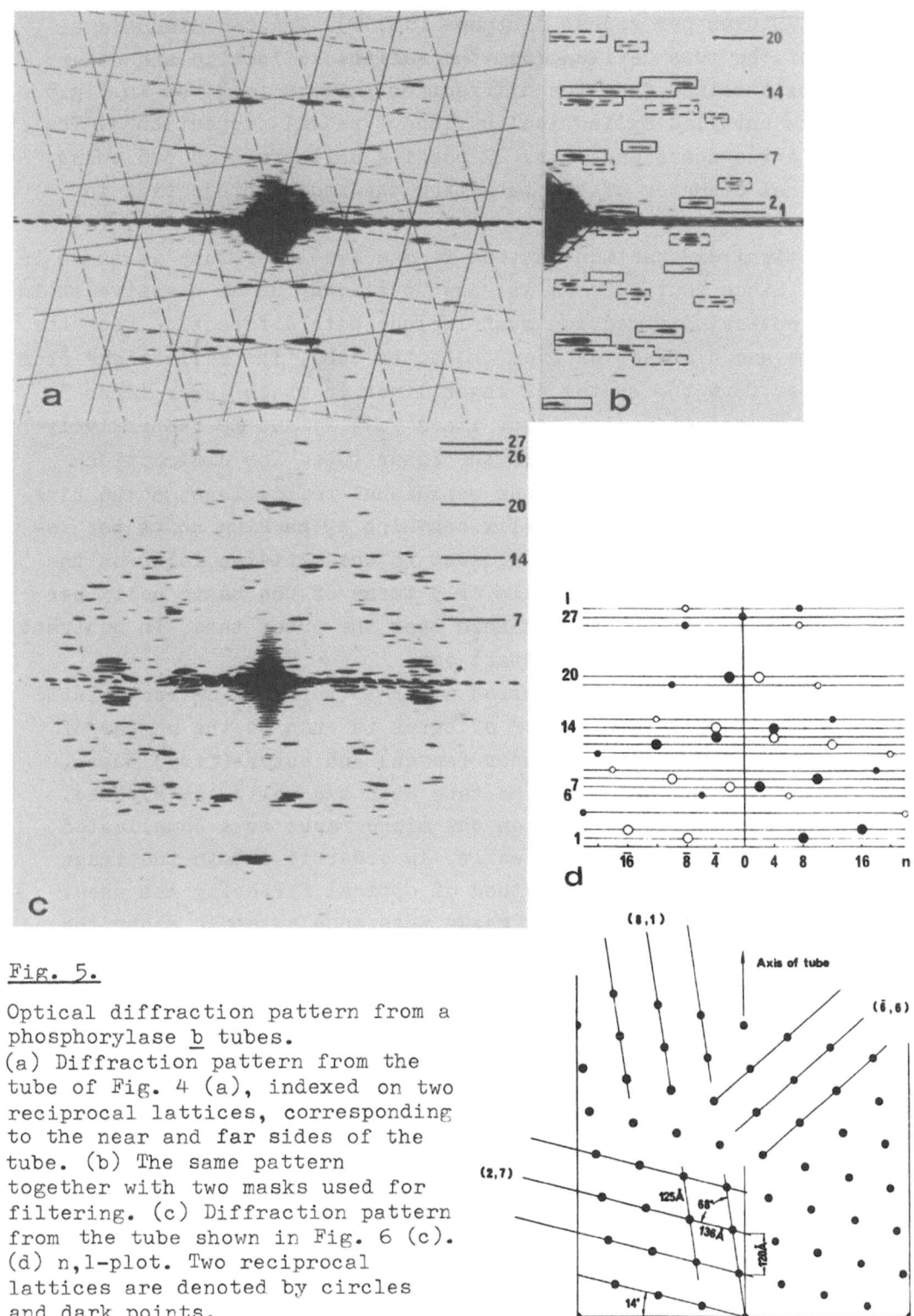

Fig. 5.

Optical diffraction pattern from a phosphorylase b tubes.
(a) Diffraction pattern from the tube of Fig. 4 (a), indexed on two reciprocal lattices, corresponding to the near and far sides of the tube. (b) The same pattern together with two masks used for filtering. (c) Diffraction pattern from the tube shown in Fig. 6 (c). (d) n,l-plot. Two reciprocal lattices are denoted by circles and dark points.
(e) Radial projection of a tube

tion) is located on the seventh layer line, indicating the basic helix with 7 turns per repeat distance (16,17). The two-dimensional lattice of the tube derived from the reciprocal lattice and other parameters obtained from the diffraction pattern is given in Fig.5 (e) as the unrolled cylindrical surface ("radial projection"). The helix is a two-start one, i.e. 27 packing units fall on 3,5 turns. The external diameter of the tube from the radial projection is 320 Å.

The optical diffraction pattern from a two-layer tube is shown in Fig.7(b). Since four tube walls show up already under negative staining, the optical diffraction must include both diffraction from the inner tube and diffraction from the outer tube. The reflections from this latter form the system of layer lines in which layer lines 5, 10,15 and 20 coincide with layer lines 7,14,21 and 28 respectively on the diffraction pattern from the inner tube. The diffraction pattern is characterized by weak meridional reflections on the nineteenth layer line, i.e. the helix contains 19 packing units per repeat distance. The reflection closest to the meridian falls on the fifth layer line, and this tells of 5 turns of the basic helix per repeat distance. It was established that the outer tube, in contrast to the inner one, is a three-start one.

As noted above, the deformation after 35 turns is characteristic of two-layer tubes. This number of turns is even to the number of turns in the repeats of the inner (seven) and outer (five) tubes, and represents the period of the tube with two walls, as a whole.

The image of the tube seen on the micrographs is a complicated superposition pattern of tube walls. In order to obtain the image of one side of the tube the method of optical filtering was used. The apertures in the filtering masks were made extended along the direction of the layer lines (Fig.5(b)). The results of filtering of the tubes are shown in Fig.6. The following two factors seem to be of most importance. First, the particles close in shape to those in the layer (Fig.3(c)) can be met with among other particles. They are characterized as having the half-ring on one side and the cross on the other. In tubes, the particles of this type are usually seen to lie closer to the lateral section. Secondly, the images of the particles at the opposite edges of the tube are turned through 180° with respect to each other, i.e. the "crosses" (or the "half-rings") are oriented in the opposite directions.

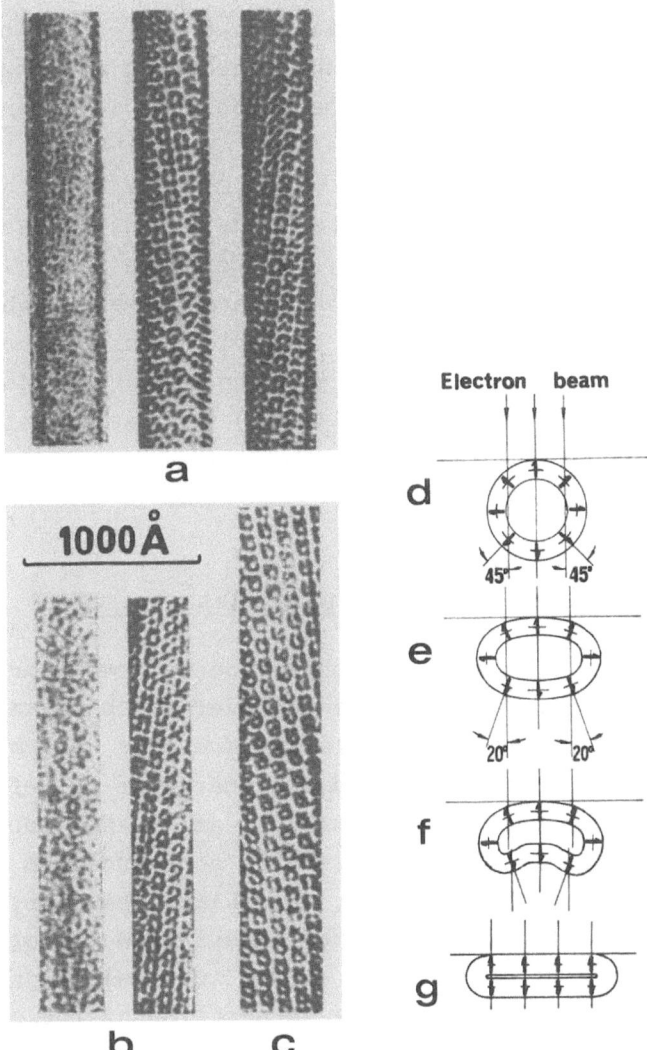

Fig. 6. (a-c) Electron micrographs and filtered images of phosphorylase <u>b</u> tubes. Protein is black.
(d-g) Cross-section of the tubes on a supporting film. Particles forming the wall are conditionally denoted by "I". It is seen that the angle of rotation of particles at the central part of the tube with respect to the observer depends on the degree of flatness.
  (d) The ideal tube.
  (e) The tube, slightly flattened.
  (f) The tube in which one of the walls is bent inwards.
  (g) The tube, completely flattened

d. Three-dimensional crystals

There are two kinds of crystal images. One of them shows crystals made up of the particles packed together in a hexagonal manner (Fig.8(a)). For several crystals the dimensions were varied within the range: the distance between the rows is $a/2 = 94-105$ Å; the periodicity along the rows $b = 113-157$ Å; $a/b = 1,2-1,37$. Fig.8(b) shows the optical diffraction pattern from the image of the crystal (a).

The second-type image of the crystal is shown in Fig.8(c). The particles and the mode of their packing are different when compared with the first-type image, but the periodicity along the rows is here approximately the same ($b = 130$ Å). The distance between the rows $c = 75-85$ Å. Optical diffraction (Fig.8(d)) indicates that the structure repeats after a row (i.e. $C' = 150-170$ Å).

g. Modelling the particles observed

The image of the particles in tubes depends on the orientation of these particles with respect to the observer which is associated with the curvature of the wall. This curvature is determined by the degree of flatness of the tubes and the character of deformation. In Fig.6(d) is shown the cross section of an "ideal" tube. The cross section of the slightly flattened tube is reproduced in Fig.6(e). The tube is symmetrically deformed; in reality, however, one side of the tube may be deformed to a greater extent than the other. Furthermore, the character of deformation of the tube wall adjacent to the supporting film may differ from the deformation of the outer wall (Fig.6(f)). The latter, for example, can be bent "inwards". In the case of the fairly flattened tubes their walls represent the plane layers (Fig.6(g)).

In spite of the fact that the poor quality of the filtered image of the flattened tube does not permit the characterization of the structure of particles (Fig.6(c)), it can be seen, however, that these particles are oriented in the same way. In the case of the less deformed tubes the particles at the opposite edges of the tube are identical, but oriented in the opposite directions. On the other hand, it is known that on passing from one edge to the other the particles are rotated with respect to the observer. It seems logical to suppose that the structure of the particles is such that when they are rotated through a certain angle with respect to the observer their image may change in a mirror way.

Fig.7.(a) Phosphorylase b tube with two-layer walls. (b) Optical diffraction pattern taken from (a).

The layer lines in the left hand part correspond to the diffraction pattern from the outer wall, and the layer lines in the right hand part - to the diffraction pattern from the inner wall

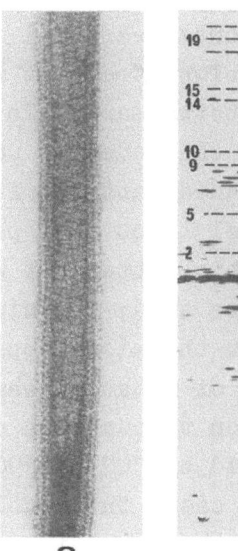
a

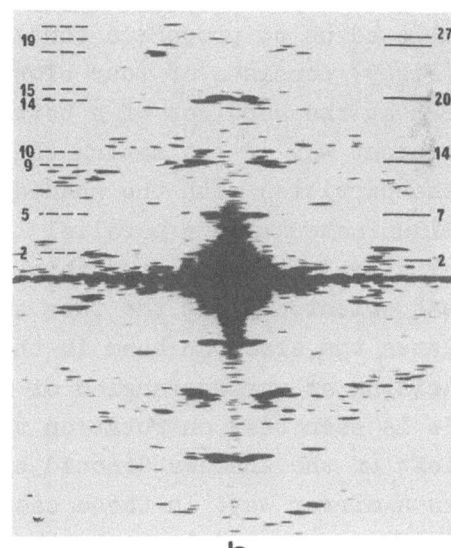
b

a

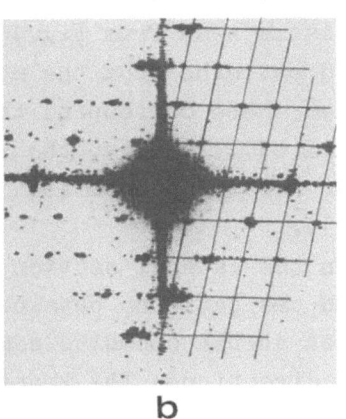
b

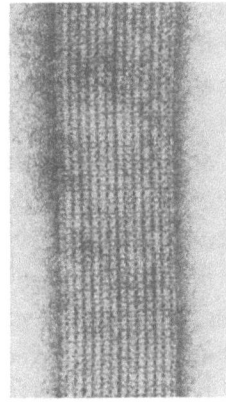
c

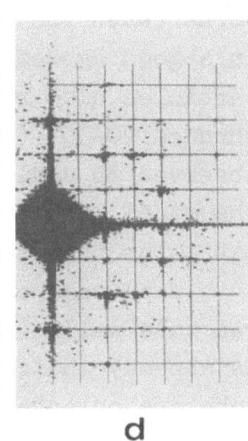
d

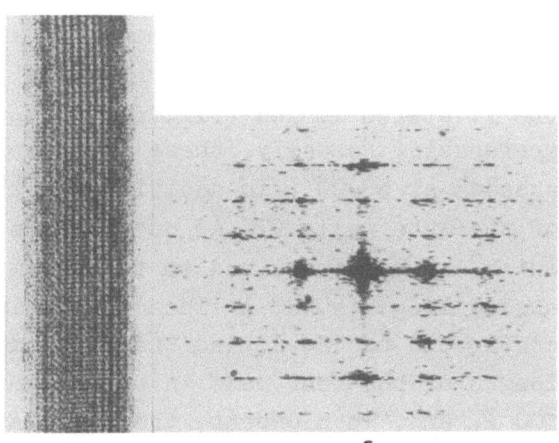
e f

Fig.8. Three-dimensional crystals of phosphorylase b (a-d) and a (e,f).
(a) An image with the hexagonal packing.
(b) Optical diffraction pattern from (a).
(c) An image with the orthogonal packing.
(d) Optical diffraction pattern from (c).
(e) Phosphorylase a crystal.
(f) Optical diffraction pattern from (e)

Analysis of the image of particles in the layer and tubes allowed us to propose a tentative model for the particles. The model (Fig.9) consists of four elongated bent rods arranged with symmetry 222 at the vertices of a tetrahedron. In order to have a more convenient way of estimating the model we used the method of comparing the particles with the shaded patterns from the model which, being illuminated with a parallel light beam perpendicularly to the 2-fold axis, was rotated about this latter (it was assumed that this axis was oriented along the tube axis; in this case the light beam simulates the electron beam in the microscope). The shaded patterns arising at various angles of rotation were photographed (Fig.9(e)). It is seen that on rotation through the equal angles to the right or left of the shadows denoted by "0" or "90", the shaded images repeat in a mirror way. In these cases the model is oriented with respect to the source of light by the operation of the 2-fold axis (Fig.9a,c)

Comparison of the particles in the layer with the shaded patterns (when the model is turned through 40–60°) shows that they are in a good agreement (Fig.10(d)). As is seen from Fig.10(c), the particles also resemble the shaded patterns from the model turned through 20–30° (Fig.10(b)). To them there correspond the particles observed in small aggregates (Fig.2(b)). Finally, the shadows from the model in the "0" position bear resemblance to the "8"-like particles from solution (Fig.10(a)).

The image of the particles in the lateral section of the filtered tube walls (Fig.6) agrees with the patterns obtained at 45–60°. As already indicated, the particles at the opposite edges of the tube are oriented in the opposite directions. The transition from the particle configuration corresponding to the rotation of the model through 45–60° to the particle with mirror configuration can be made either via the "0" position or the "90" position. In both cases the shadow from the model exhibits the characteristic outlines. It may be seen by inspection of the particle images at the center of the filetered images (Fig.10(e)) that the second variant is more preferable. Actually, these particles resemble in many cases the shadows at 80–90°. In addition, in the second variant the thickness of the tube walls (70 Å) corresponds well with the sizes of the model which can be determined on the assumption that the length of the shaded image is 120 Å. Then the maximum width of the model would correspond to 120 Å, and the minimum width – to 75 Å (at 0°). The length of the bent rods in projection onto the plane (at 90°) is 100 Å, and, when unbent, the rods have the length of about 110 Å, $\beta \sim 105°$.

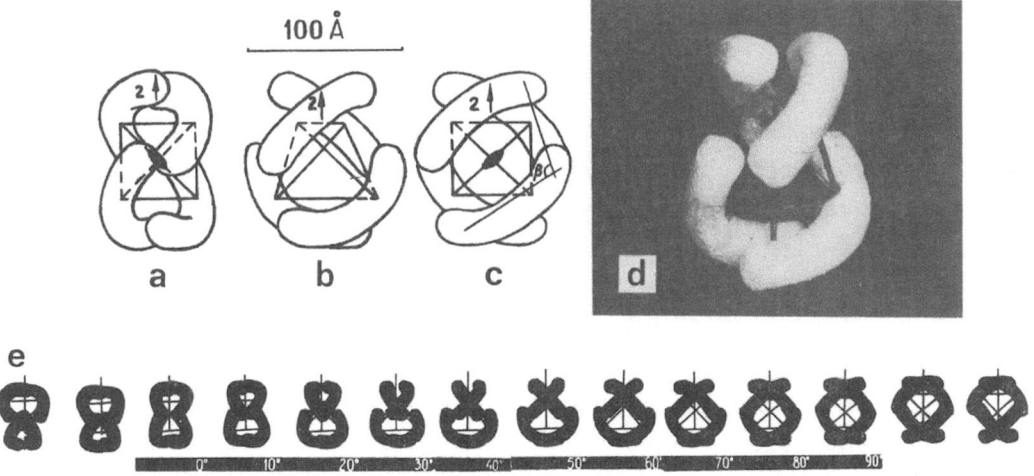

Fig.9. The model of the particles of phosphorylase b and a (a-d) and shaded patterns (e).

The model was rotated with respect to axis 2 and illuminated in the direction perpendicular to this axis. Two other 2-fold axes are marked on (a) and (c). (a) - the model is in the "0", (d) - in the 90° position

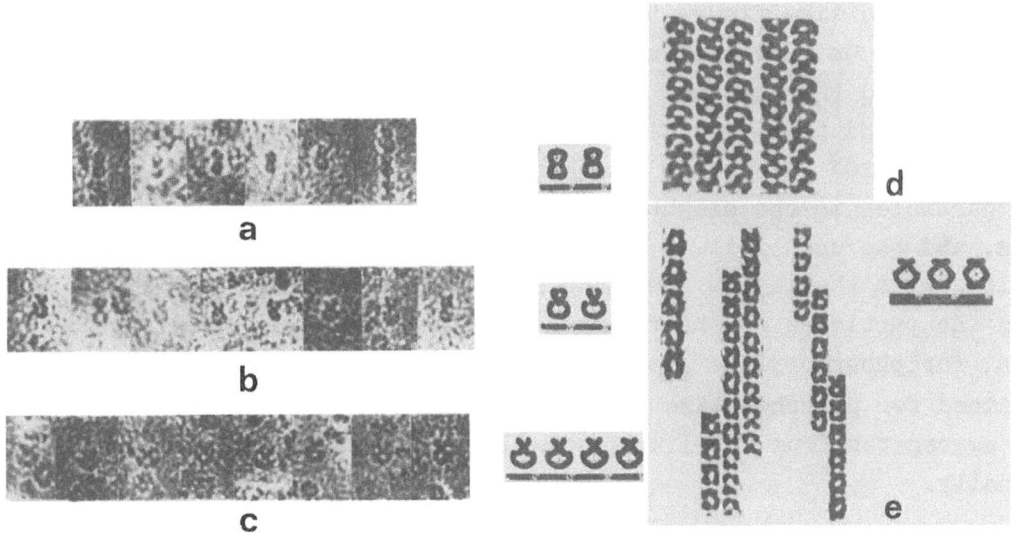

Fig.10. Images of phosphorylase b particles.
(a-c) Particles from solution.
(d) Particles in the layer.
(e) Particles at the center of the image of a tube wall

IV. Phosphorylase a

a. Particles from solution

On micrographs of phosphorylase a from solution without protamin (Fig.11 (a)) one can observe many characteristic particles of two types analogous to those observed in phosphorylase b preparation: 1) with a half-ring on one side and a cross on the other, the length of the particles being 115-120 Å; 2) particles reminiscent to figure "8" with the length of 120 Å and the width up to 70 Å.

Many individual particles of the first and second types can be observed in phosphorylase a preparations with protamin added, at protein concentrations lower than 0,4 mg/ml; one can also detect short tubes and small plane layers (Fig.11(b)).

b. Plane layers

Under conditions unfavourable for crystallization one could observe in preparations with protamin added small sections of layers (Fig.11(b)) reminiscent to layers of phosphorylase b. At protein concentrations higher than 1 mg/ml the layers of entirely different appearance were formed (Fig.12(a)). The particles forming these layers looked like figure "8", i.e. they corresponded to the second-type images. On micrographs the periodicity along the rows was 110-120 Å and the distance between them 60-70 Å. From the optical diffraction pattern (Fig.12(c)) the unit cell is characterized by dimensions 120Å (along the rows) and 140Å (across the rows). Thus, the particles in the neighbouring rows somewhat differ in configuration, and the unit cell is seen to contain two particles. Under conditions when the layers just mentioned were observed, the layers from the particles of the first type were encountered very rarely. Thus, for phosphorylase a the picture appears to be reverse to that obtained for phosphorylase b when these layers predominate whereas the aggregates from "8"-like particles can be recognized only occasionally.

c. Tubes

In phosphorylase a preparations with protamin added one could mainly observe the tubes with one-layer walls (Fig.13(a)).

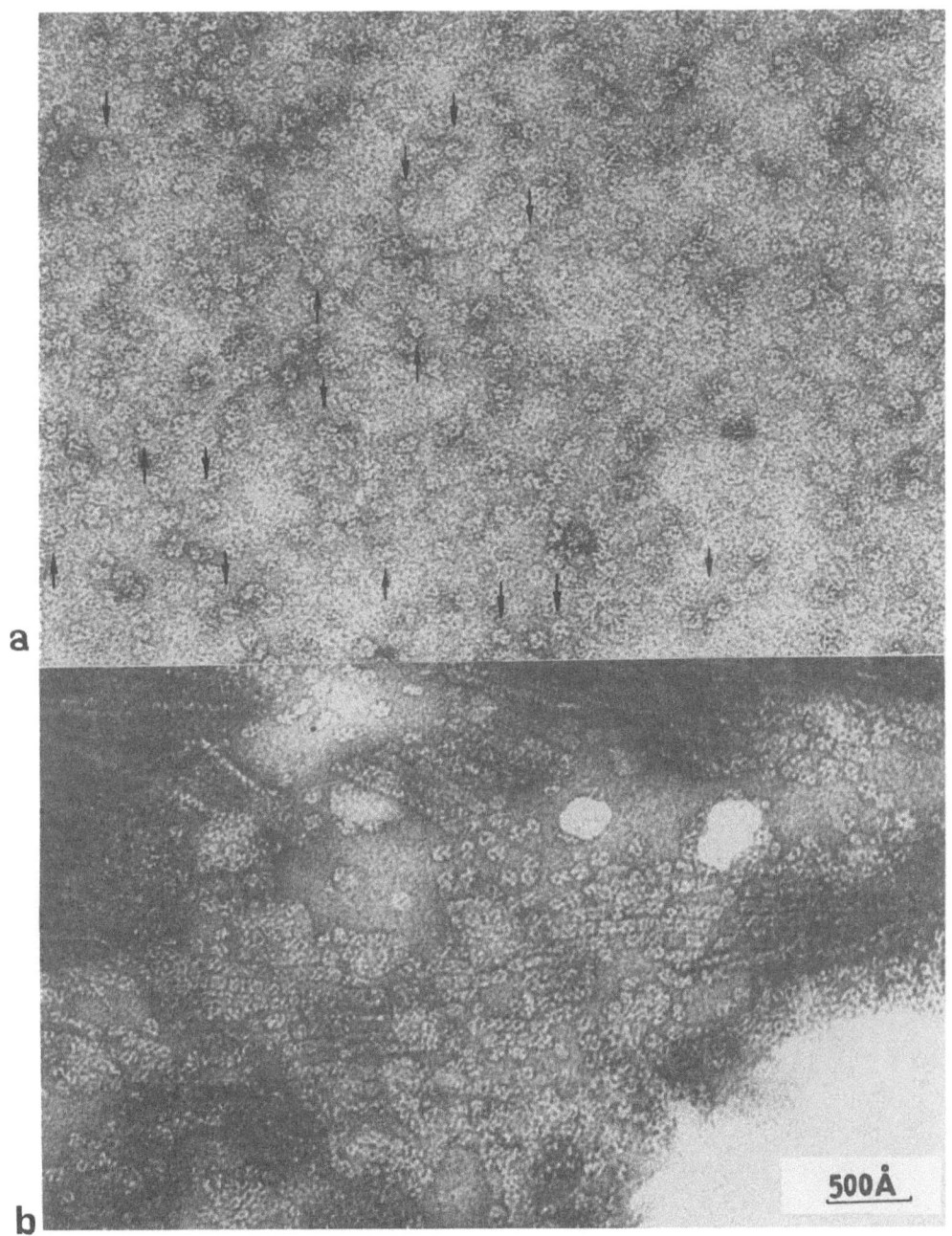

Fig.11. Phosphorylase a.
(a) Phosphorylase a particles from solution (without protamin). Some particles with a half-ring and a cross as well as "8"-type particles are marked with arrows.
(b) Phosphorylase a particles from solution (with protamin) and layers observed under conditions unfavourable for cryst

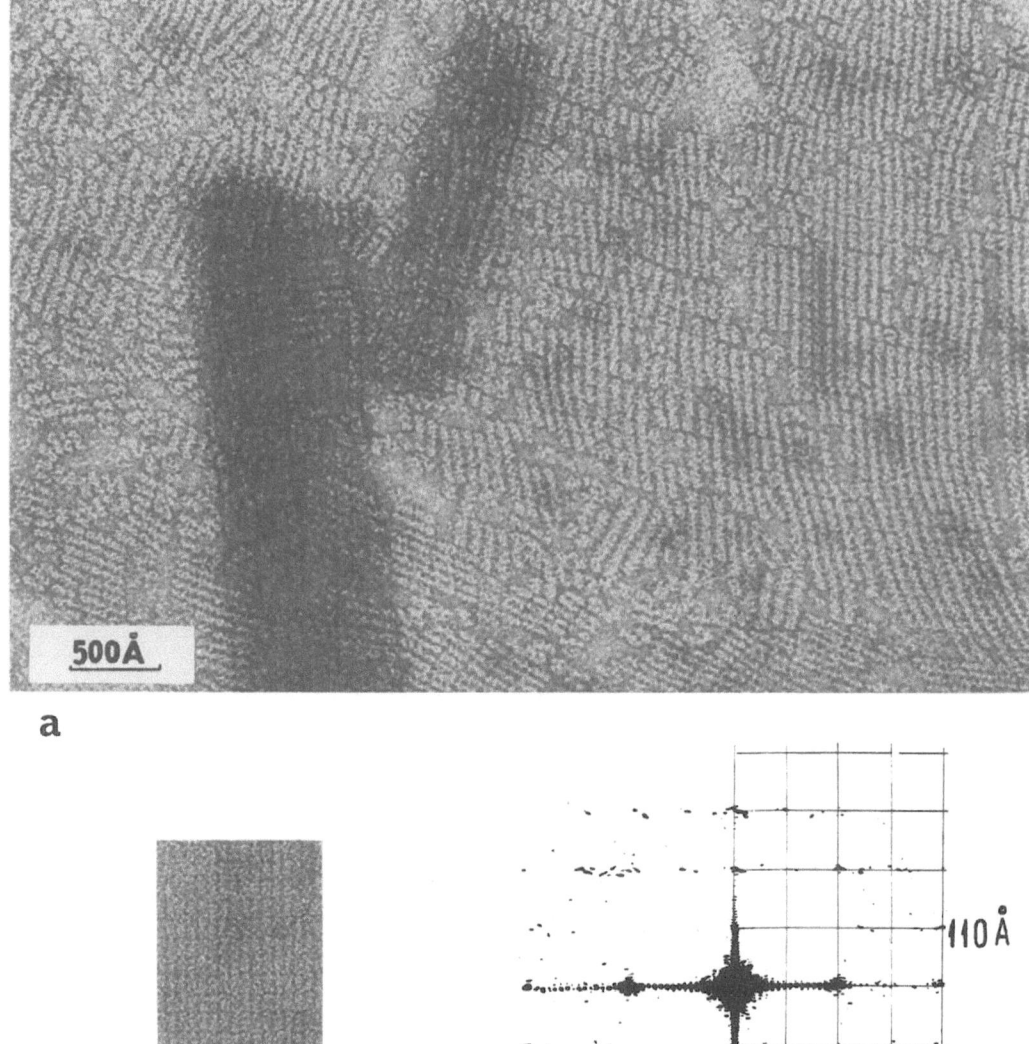

Fig. 12. (a) Plane layers of particles of phosphorylase a.
(b) Section of the plane layer.
(c) Optical diffraction pattern taken from a section (b)

The optical diffraction pattern from such a tube is shown in Fig.13(b). It differ markedly from the diffraction patterns from phosphorylase b tubes, and is characterized by strong meridional reflections on layer line 26. On the repeat period there fall 9 turns of the basic helix which is a three-start one. The diameter of the tubes under consideration is about 320 Å, the thickness of the walls being 70 Å. It is likely that there are variants for the tubes with one-layer walls, differing in diameter and parameters of the helical packing of particles in the walls. For instance, in Fig.13(d) is shown a fairly flattened tube with width of about 600 Å, whereas the maximum width of the tubes of 320 Å in diameter can reach, on flattening, not more than 500 Å.

The tubes with two-layer walls (Fig.13(e)) occur in phosphorylase a preparations very rarely. The diameter of the inner tube is approximately equal to that of the tubes with one-layers walls.

The filtered image of one of the walls of the tube shown in Fig.13(a) is given in fig.13(c). The images of the particles at the edges of the wall are characterized by a half-ring and a cross, i.e. they resemble the particles of the first type. The particles at the opposite edges are oriented in the different directions. The distribution of the particle images resembles that observed on the filtered image of one of the phosphorylase b tubes (Fig.6(a)), and is based on the assumption that the given tube wall is slightly bent inwards (as shown in Fig.6(f)). The phosphorylase a tube under discussion is also slightly flattened. Therefore, in this particular case, the analogous interpretation is possible, also.

d. Three-dimensional crystals

The three-dimensional crystals of phosphorylase a could be observed in preparations without protamin. The fragment of one of such crystals is shown in Fig.8(e). It can be said from outward looks that it resembles one of the types of crystals (Fig.8(c)) observed in preparation of phosphorylase b with protamin. The periodicity along the rows is about 115 Å, and in the transverse direction - about 65 Å. The optical diffraction pattern (Fig.8(f)) indicates that the cell dimensions in the transverse direction are, in reality, twice as large (130 Å). Thus the analogy with the three-dimensional crystals of phosphorylase b is evident.

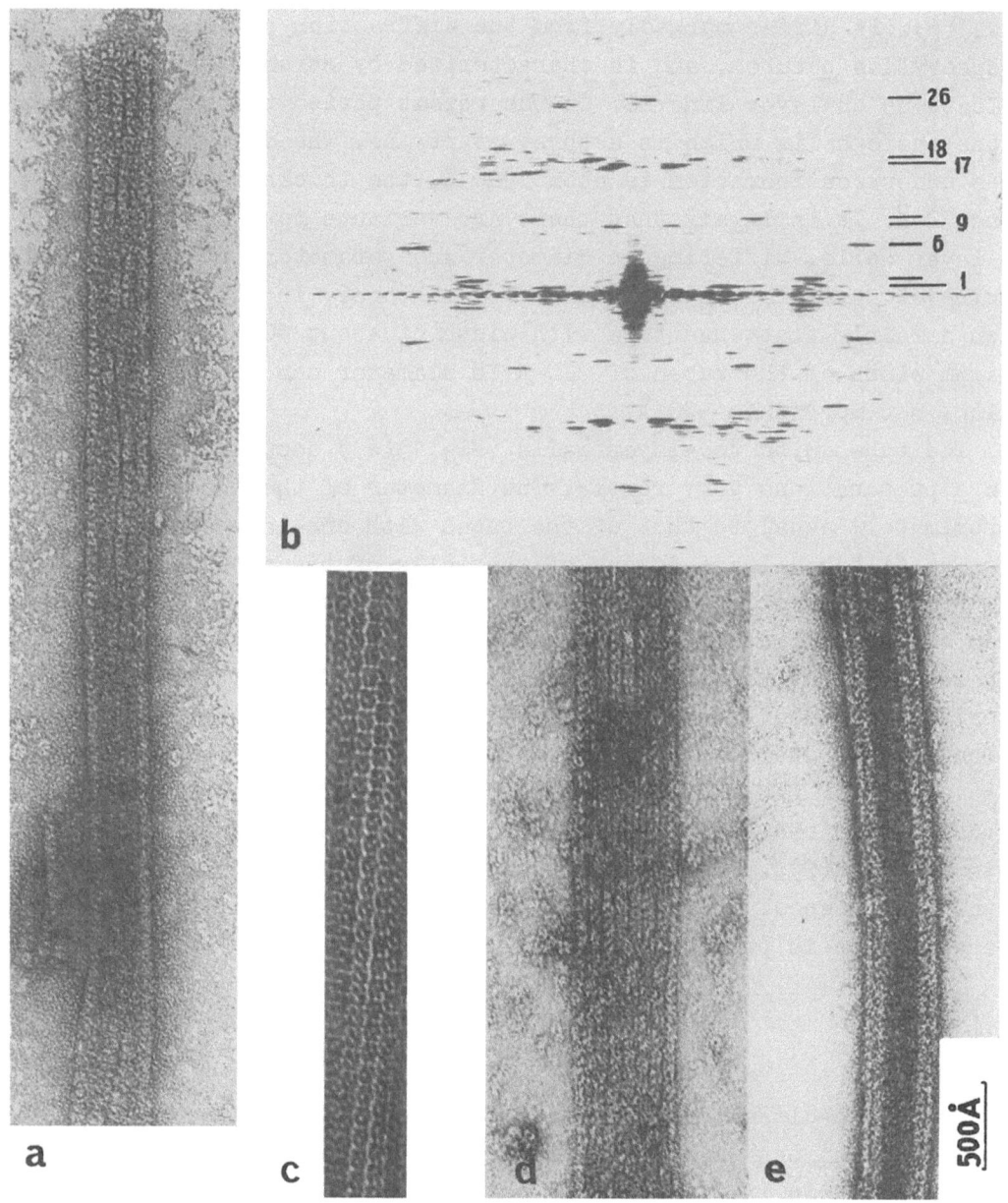

Fig. 13.   Phosphorylase a tubes.
(a) Tubes with one-layer walls.
(b) Optical diffraction pattern from the tube shown in (a).
(c) Filtered images of one of the tube walls (a).
(d) The wide flattened tube.
(e) A tube with two-layer walls

e. Modelling phosphorylase a particles

We have now available the images of particles: from solution (in both preparations with protamin and without it), in plane layers and tubes. All these images can be interpreted with the aid of the model proposed for phosphorylase b particles. Actually, the first type of particles from solution corresponds to the shaded patterns from the model turned through 20-60°. To these patterns there correspond the particles images on the filtered images of the tubes. The second type of particles from solution and particles in the layer agrees with the shadows from the model in position "0" or at a small angle of rotation.

V. Discussion

Considering the possible type of symmetry for particles of finite dimensions, arranged in the ordered manner relative to the central point, Klug (17) points out that in this case one of the three types of point-group symmetry, that which contains a set of rotation axes passing through the point, is the only possible type of symmetry: 1) cyclic symmetry; 2) dihedral symmetry (the 2-fold axes combined at right angles with any of the n-fold axes). The tetrahedron with the point-group symmetry 222 can serve as an example, when n=2; 3) cubic symmetry.

The enviroments of the subunits in the phosporylase tetramer are supposed to be equivalent. This suggests either the cyclic 4-fold symmetry with one class of bonds, or 222 symmetry. The dissociation scheme of the molecules indicates two types of bonding the subunits. In this case the cyclic molecule can possess only the 2-fold axis (the environments of the subunits are nonequivalent).

On studying the phosphorylase b molecules from solution Chignell, Gratzer & Valentine (18) come to the conclusion that the phosphorylase dimers are elongated particles with dimensions 110x65x55 Å, whereas the tetramers, in projection, are characterized by the rhombic shape (β =107°) and dimensions 110x110 Å. In fact, the model is cyclic with the 2-fold axis.

In our investigations the particles are characterized by symmetry 222, i.e. it would seem that another symmetry variant possible for the phosphorylase molecule is being realized. What can represent the particles under observation? Here two variants are possible: either each such particle is equivalent to one phosphorylase molecule in the tetrameric form or each particle represents an aggregate from two tetramers, i.e. it contains two molecules. Consider, first, the second supposition. We have here to admit that the tetramer represents two elongated particles (dimers) packed side-by-side.

It seems most unlikely that such specific aggregates composed of two tetramers can exist. In any case, the sedimentation analysis fails to reveal such aggregates in preparations of phosphorylases b and a with protamin added, and in phosphorylase a without protamin. Therefore, the supposition that the particles observed represent phosphorylase molecules in the form of tetramers seems well grounded.

The particles indicated were observed in both preparations of phosphorylase b and preparations of phosphorylase a. Their structure, at the given level of resolution, can be interpreted satisfactorily with the aid of the same model; in other words, both the phosphorylase b molecule (in tetrameric form) and the phosphorylase a molecule can be represented as four elongated subunits arranged with 222 symmetry. The difference in structure of these forms of phosphorylase manifests itself in the character of mutual aggregation of molecules and in the parametrs of the crystal lattice formed by them.

Acknowledgements

We express deep gratitude to Prof. B.K.Vainshtein for the important assistance in the work, and to Prof. M.N.Lyubimova-Engelgardt for continued interest.

REFERENCES

1. Cohen Ph., Duewer Th., Fischer E.H. Biochemistry, 10, 2683 (1971).
2. Madsen N.B. & Cori C.F. J.Biol.Chem., 223, 1055 (1956).
3. Nolan C., Novoa W.B. & Fischer E.H. Biochemistry, 3, 542 (1964).
4. Cori G.T., Colowick S.P. & Cori C.F. J.Biol.Chem., 123, 381 (1938)
5. Krebs E.G. Biochim. Biophys. Acta, 15, 508 (1954).
6. Fischer E.H. & Krebs E.G. J. Biol. Chem., 231, 65 (1958).
7. Silonova G.V., Lissovskaya N.P. & Pikhelgas V.Ya. Doklady Akademii Nauk SSSR, 169, 483 (1966)
8. Madsen N.B. & Cori C.F. Biochim. et Biophys.Acta, 15, 516 (1954).
9. Lissovskaya N.P., Livanova N.B. & Silonova G.V. Biochimiya, 29, 1012 (1964).
10. Krebs E.G., Kent A.B. & Fischer E.H. J.Biol.Chem., 231, 73 (1958)

11. Kosourov G.I., Livshits I.E. & Kiselev N.A. **Kristallografiya**, *16*, 813 (1971).
12. Kiselev N.A., Lerner F.Ya. & Livanova N.B. **Molekulyarnaya Biologiya**, *5*, 642 (1971) (Mol.Biol. in Russian).
13. Klug A. & De Rosier D.J. **Nature**, *212*, 29 (1966).
14. Watson I.D. **Biochim. et Biophys. Acta**, *13*, 10 (1954).
15. Vainshtein B.K. Diffraction of X-Rays by Chain Molecules, Amsterdam: Elsevier. 1966.
16. Mikhailov A.M. **Kristallografiya**, *15*, 818 (1970).
17. Franklin R.E. & Klug A. **Biochim. et Biophys.Acta**, 19, 403 (1956).
18. Chignell D.A., Gratzer W.B. & Valentine R.C. **Biochemistry**, *7*, 1082 (1968).
19. Klug A. In "Symmetry and Function of Biol. System at the Macromolecular Level". Ed.: A.Engström a. B.Strandberg Almqvist & Wiksell, Stockholm. 1970.

Discussion:

Holzer

Dr. Fischer, have you looked for the role of AMP instead of ADP in the reversion of your reaction, and have you looked for GDP, as you mentioned that GDP is a phosphorylatable agent? My second point concerns the pH-dependency of the reversal of your reaction. We have found reversal of the adenyl-transferase reaction in glutamine synthetase by strongly pH dependent pyrophosphorylation. The equilibrium constant at pH 6 is about 1, at pH 7 it is 10, at pH 5 it can not be measured, because the enzyme is not active, but it really could be 0,1. Thus in yeast changes in the pH connected with glycolysis could be involved in the regulation of the system.

Fischer

AMP does not serve as a phosphate acceptor in the reverse reaction. But both ADP and GDP serve, and to a much lesser extent UDP. We have not investigated the reverse reaction. As indicated by the reaction: substrate + ATP = substrate-phosphate + ADP + H^+ indeed both the forward and reverse reaction are strongly pH dependent. If my memory is correct, Rabinowitz and Lipman have found for the reverse reaction an optimum around pH 6,0 whereas the forward reaction has an optimum around 8. We do not know whether this reaction in yeast is determined by the hydrogen ion concentration or the ATP/ADP ratio or by any other factors.

Holzer

At what pH have you measured your reaction?

Fischer

Around pH 7,5 - 8. We have not made a study of the pH optimum.

Sols

Could you tell us if the ratio of ATP to ADP would favour the reverse reaction?

Fischer

No, we cannot. The reverse of the reaction was found very recently and we have not yet looked for the group which is the phosphate donor or for the pH dependency. Concerning the significance of the *b* to *a* conversion, I think we can come back to one of the questions asked by Dr. Cori. Why would we have a *b* to *a* conversion in yeast? You remember that phosphorylase *b* in yeast is inhibited by glucose-6-phosphate with a K_i of 1 millimolar, whereas for phosphorylase *a* the K_i is 11 mM.

Perhaps conversion of yeast phosphorylase b to a might enable phosphorylase to operate under conditions of high glucose-6-phosphate if this is needed, and Cabib has shown that under conditions of high glycogen the concentration of glucose-6-phosphate will rise as high as 8 mM. This would knock out effectively the activity of phosphorylase b and not that of phosphorylase a.

Cori

Do you consider that the potato is too slow to have a b to a conversion system?

Fischer

Yes, maybe this is the only explanation, why a well organized eucaryotic organism like the potato does not have any covalent or not covalent type of metabolic control.

Hess

In the search for the mechanism of the overall regulation of glycogen metabolism in yeast and other systems it should be distinguished between a trigger function, which initiates a steady state, and the mechanism, which establishes an overall chemical potential in terms of mass action. A trigger does not initiate a steady state if there is no mass action in terms of a substrate load available. Conversely, the build up of a substrate load for a pathway remains silent if a trigger doesn't release this load. In looking for a suitable control metabolite for stabilizing a triggered state it might be reasonable to look for a glycolytic intermediate. In a study of the substrate levels in gluconeogenesis and glycolysis we recently found that in yeast the fructose-1,6-diphosphate-level changes between o,o2 mM under gluconeogenetic conditions and 2,o mM under glycolytic conditions. Such a large shift of the steady state levels of fructose-1,6-diphosphate in dependence on the direction of the flux through this system effects the state of pyruvate kinase and might well effect the state of other important steps of the system such as phosphofructokinase, fructose-1,6-diphosphatase as well as the phosphorylase system itself (see C.J. Barwell and B. Hess, FEBS-letters 1972, in press).

Fischer

One might conclude from these findings, that the control does not occur at the ATP level.

Graves

From the information on the primary structure of phosphorylase can you make any speculation, why some of the phosphorylases are dimers and som- tetramers. Do you think that the N-terminals are involved?

Fischer

Obviously something of this kind is involved, but we have to wait for the crystalographic data.

Krebs

To comment to Dr. Cori's talk, phosphorylase kinase is effected by glycogen, when it is phosphorylating casein so that a least part of the effect of glycogen on the acceleration of the phosphorylase reaction is a kinase-glycogen interaction rather than an interaction of glycogen with phosphorylase. Another comment, with regard to Dr. Fischer's talk, is that mammalian tissue does contain a protein kinase that appears to have some of the specificities and properties that you described; that is that protein kinase is not effected by cyclic AMP, it is active for casein and it is inactive with respect to histone and protamine. It would be very interesting to try this on the yeast enzyme.

Heilmeyer

I would like to comment on the role of AMP in phosphorylase. Two years ago we started to work, in collaboration with Dr. Fischer, on the complex of glycogen and protein and we investigated the role of AMP in such a complex. Unfortunately we could not work with AMP itself, because its converted to IMP due to contamination by desaminase. So we chose AMP-anoxide which is stable in this system and we found, that it activated to a much smaller extent in the glycogen-protein complex as compared to crystalline phosphorylase.

Stadtman

Is it true that the protein kinase, which is involved in phosphorylase and glycogen synthetase does not work on phosphorylase directly?

Krebs

It is true that the protein kinase that acts on glycogen synthetase does not act on phosphorylase.

Palmer

Dr. Fischer discussed the control of glycogen phosphorylase from an evolutionary and comparative viewpoint. As example of divergent evolution, bacterial and plant α-1,4-glucan phosphorylases are relatively unsophisticated enzymes, lacking, as far as is known, multiple metabolically interconvertible forms. Presumably, this apparent simplicity results from the fact that in vivo the enzymes do not assume a rate-control function. Functional definition of the role of α-1,4-glucan phosphorylases demands knowledge of the catabolic sequences in which the enzymes are implicated. Our understanding of α-glucan catabolism in bacteria and plants is, at best, fragmentary.

As the result of recent studies at the University of Miami (T. N. Palmer, G. Wöber & W. J. Whelan, unpublished results), we are now in a position to propose two alternative pathways of α-1,4-glucan catabolism in Escherichia coli. (Figure 1). The central feature of the model is the dual functionality of the enzymes of the maltose operon. Hitherto, these enzymes, namely maltose permease, amylomaltase and maltodextrin phosphorylase, have been thought to be exclusively involved in the pathway of utilization of extracellular maltose. Our studies show unequivocally that the enzymes are specifically orientated towards the metabolism of maltotriose and higher maltosaccharides. Maltotriose and higher maltosaccharides, as opposed to maltose, are the preferred inducers and substrates of the system.

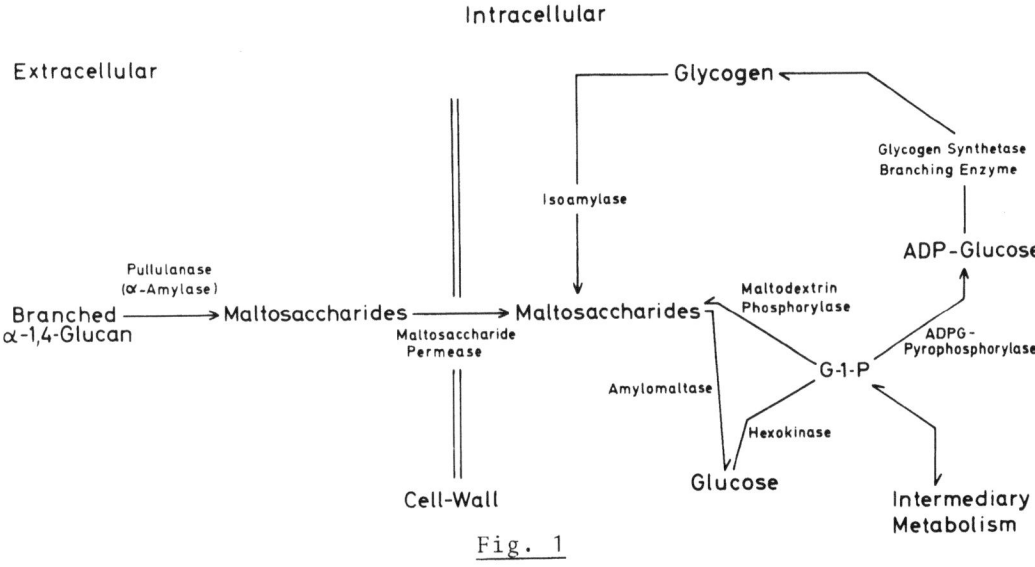

Fig. 1

Maltosaccharides arise by two alternative routes. The extracellular catabolic route involves a pullulanase activity. The pullulanase (located in a lipoprotein-lipopolysaccharide particulate complex in the cell-wall), functioning in concert with an α-amylase activity, hydrolyses the α-1,6-linkages of exogenous branched α-1,4-glucans to produce maltosaccharides. The maltosaccharides enter the cell by the action of maltosaccharide permease. The nomenclatural change from maltose to maltosaccharide permease is justified on the grounds of the redefined specificity of the enzyme. The enzyme catalyses the uptake of maltosaccharides of DP two to at least twenty-two. The permease is rate-controlling. The pullulanase activity is subject to marked product inhibition. Thereby, the rate of α-1,6-linkage hydrolysis is governed by maltosaccharide permease activity.

The intracellular catabolic route proceeds via the action of an isoamylase. The isoamylase and pullulanase activities, both α-1,6-linkage-specific glucanohydrolases, differ in specificity. They are functionally distinct entities. Indirect evidence leads us to regard the isoamylase as the rate-controlling enzyme in the intracellular catabolic route.

Maltosaccharides, produced by either the intra- or extra cellular catabolic routes, are converted to glucose and glucose-1-phosphate by the combined actions of amylomaltase and maltodextrin phosphorylase. The two catabolic pathways are subject to specific and partially independent induction by maltosaccharides.

In summary, we propose the scheme as a general catabolic system in bacteria. Both Streptococcus mutans and Aerobacter aerogenes metabolize exogeneous branched α-1,4-glucans via pathways identical in every respect to the extracellular pathway proposed. I also wish to draw attention to the apparent homology between the enzymes of starch catabolism in plants and those of the proposed extracellular pathway in E. coli. Plants contain pullulanase, transglycosylase (D-enzyme) and phosphorylase activities. It is tentatively proposed that the two systems are functionally homologous, except that the bacterial pathway includes a permease activity.

The proposed catabolic pathway constitutes another example of a catabolic pathway in branched α-1,4-glucan metabolism in which glycogen phosphorylase is not the rate-controlling enzyme. Indeed, glycogen phosphorylase has no direct role. Other catabolic routes not involving glycogen phosphorylase are those involving neutral α-glucosidases, lysosomal α-glucosidase and α-amylases.

Studies on Glycogen Phosphorylase in Solution and in the Crystalline State

N. B. Madsen, K. O. Honikel, and M. N. G. James

Department of Biochemistry, University of Alberta, Edmonton, Alberta/Canada

The original studies of Kent, Krebs and Fischer (1) showed that the coenzyme of phosphorylase, pyridoxal-P, is bound through the aldehyde group to an ε-amino group of lysine. In order to explain the absorbance of the bound pyridoxal-P at about 330 nm, these authors suggested that another group (called X) added to the original Schiff base formed between the ε-amino and the aldehyde. This group X is still unknown but it should be nucleophilic and a hydroxyl, amino or sulfhydryl group would be appropriate.

More recently, Shaltiel and Fischer (2) and Hedrick et al. (3) have shown that when the pyridoxal-P in phosphorylase b is excited with light at 330 nm, it shows a fluorescence emission with a maximum at 535 nm (Fig. 1), a "Stokes shift" of 11,500 cm^{-1}. By contrast, pyridoxal-P and other free vitamin B_6 compounds exhibit a Stokes shift of 5,000 - 7,000 cm^{-1} (4). Thus pyridoxal-P when excited at 330 nm emits at 430 nm (Fig. 1) while pyridoxamine-P, suggested by Kent (5) to be a more realistic model of the bound pyridoxal-P, shows an emission maximum at 400 nm (Fig. 1). The increased Stokes shift in the enzyme is not explained by a hydrophobic environment because pyridoxal-P shows a fluorescence emission between 415 and 430 nm in different organic solvents.

Quite recently, Johnson, Tu, Bartlett and Graves (6) reported an explanation for the anomalous fluorescence behaviour of the coenzyme in phosphorylase. They propose that the pyridoxal-P is bound to the apoenzyme in the form of a Schiff base to the ε-amino group of a lysine. But in contrast to other pyridoxal-P containing enzymes such as decarboxylases (7) which absorb between 410 and 430 nm, the absorbing wavelength of the Schiff base, the coenzyme in phosphorylase absorbs at about 330 nm. They suggest that the Schiff base in phosphorylase is buried in a hydrophobic region of the enzyme. From consideration of other investigations with pyridoxal analogs and derivatives in organic solvents such as methanol and dioxane (8, 9, 10), the Iowa group concludes that the Schiff base in phosphorylase exists in the ground state

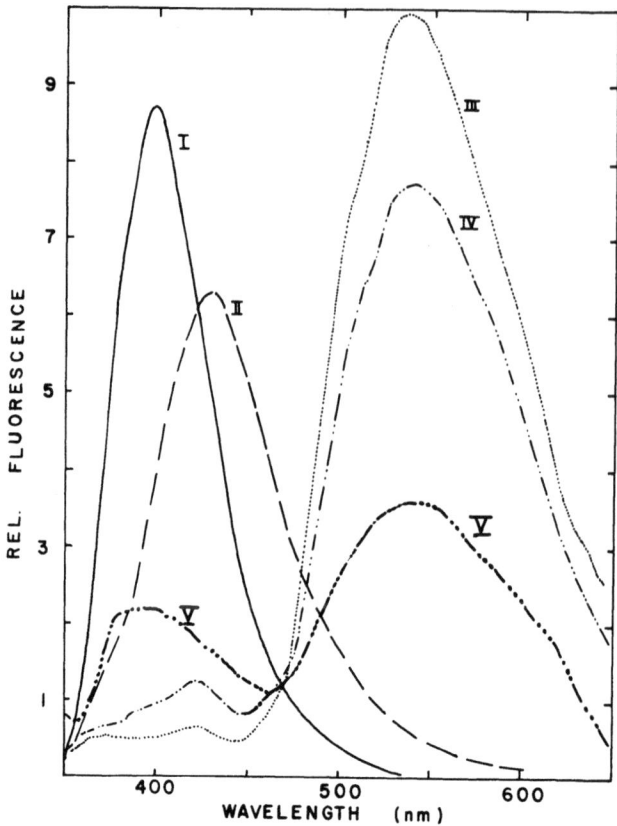

Fig. 1. Fluorescence emission spectra of pyridoxal-P, pyridoxamine-P, phosphorylase b, and the L-lysine ethylester adduct of pyridoxal-P. Excitation wavelength 330 nm. (I) 3.5×10^{-6} M pyridoxamine-P in 4×10^{-2} M potassium phosphate buffer, pH 7.0, at 30°C; (II) 1.2×10^{-5} M pyridoxal-P in 4×10^{-2} M potassium phosphate buffer, pH 7.0, at 20°C; (III) phosphorylase b (2.6 mg/ml = 2.8×10^{-5} M pyridoxal-P) in 5×10^{-3} M glycerophosphate, 5×10^{-3} M mercaptoethanol, and 1.5×10^{-3} M EDTA, pH 6.8, at 20°C; (IV) III + 10^{-3} M AMP and 8.9×10^{-2} M P_i; (V) 4.5×10^{-5} M pyridoxal-P + 4.5×10^{-4} M L-lysine ethylester in 97% dioxane (30°C)

in a form where the phenolic group in position 3 on the pyridine ring system is undissociated (Scheme 1A, #1). On radiation at 330 nm the proton of this phenolic group might be transferred to the imine nitrogen of the Schiff base (Scheme 1A, #2). This radiation product would then have the same configuration as a Schiff base which absorbs at 415 nm. But this form, created on excitation at 330 nm, does not emit at about 430 nm, it emits at the wavelength of the Schiff base at 535 nm. We do not accept this hypothesis because we feel, as suggested by the investigations of Metzler (11), that a proton on the phenolic group would in fact be shared between the oxygen and the imine nitrogen, forming a hydrogen bond.

Scheme 1, Part A. #1 and #2 are structural formulas for the enzyme bound pyridoxal-P, as proposed by Johnson et al. (6). R is the apo-enzyme, to which pyridoxal-P is bound. #3 to #5 are the structures we propose for the adduct of pyridoxal-P and different amino compounds in chloroform. R' is the remainder of the amino compound, exclusive of its amino group. Em means the maximum of fluorescence emission when excited at the stated absorbance maximum (Abs.). Part B: Reaction between an aldehyde and amine forming a Schiff base

Bentley's more recent X-ray diffraction studies on a Schiff base between pyridoxal and aniline confirm Metzler's view (12). In the planar structure of the heterocyclic ring and the Schiff base, the distance between the phenolic oxygen and the imine nitrogen was 2.56 Å, indicating a strong hydrogen bond. While all other hydrogens in the structure could be located, that between the oxygen and the nitrogen could not be seen, suggesting that it has no fixed position and oscillates between the two basic groups. This study contradicts the proposal of Johnson et al. (6) as well as the structure proposed by Martell and his collaborators (8, 9, 10).

Shaltiel and Cortijo (13) arrive, in general, to the same conclusions as the Iowa group. Additionally, they report that an adduct between n-hexylamine and pyridoxal-P in chloroform shows the unusual fluorescence behaviour typical of phosphorylase. With their reported conditions, however, we could not repeat their experiment and we found only a very small fluorescence emission at 530-550 nm on excitation at 335 nm (Fig. 2B). In the paper of the Israeli group the emission peak of the pyridoxal-P - hexylamine adduct (10^{-5} M) seems as high as that of phosphorylase b (1.5×10^{-5} M pyridoxal-P at the reported enzyme con-

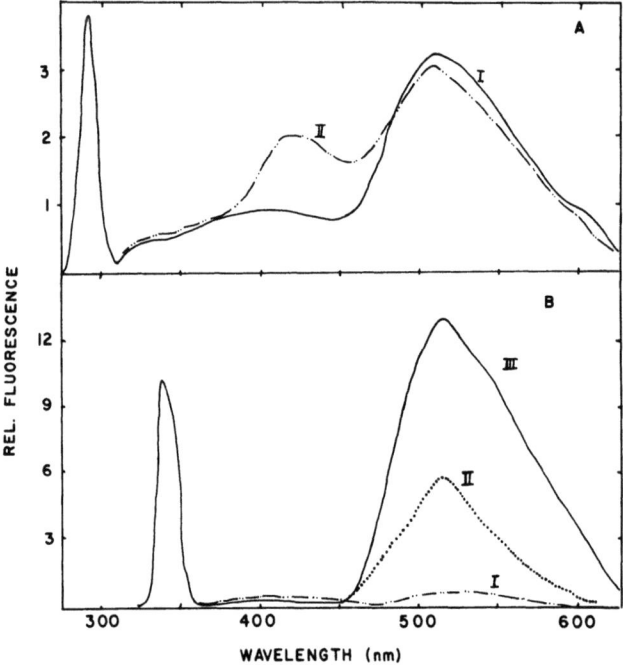

Fig. 2. Fluorescence emission spectra of the pyridoxal-P - hexylamine adduct in chloroform.

Fig. 2A: Excitation wavelength 290 nm: (I) 2.5×10^{-5} M pyridoxal-P, 10^{-4} M hexylamine and 0.4 M acetic acid; (II) 2.5×10^{-5} M pyridoxal-P, 10^{-4} M hexylamine, 0.4 M acetic acid, and 0.55 M butanol, 30 min after mixing.

Fig. 2B: Excitation wavelength 335 nm: (I) 2.5×10^{-5} M pyridoxal-P, 10^{-4} M hexylamine in chloroform, 12 hours after reaction in darkness (quantum yield 0.002 at the 530 peak); (II) conditions of curve I plus 0.2 M acetic acid, spectrum taken immediately after mixing (quantum yield 0.023); (III) conditions of curve I plus 0.4 M acetic acid, spectrum taken immediately after mixing (quantum yield, 0.04)

centration). In every case, they took the maximum of the fluorescence emission as 100% fluorescence intensity. The quantum yield, however, of the adduct under the reported conditions is 0.002, which is only 1/10 of the quantum yield shown by phosphorylase $\underline{b}$ of 0.02. The low quantum yield of the product of pyridoxal-P and hexylamine increases immediately on adding glacial acetic acid and at a concentration of 0.2 M it is increased more than 10-fold to 0.023, near the quantum yield of the coenzyme in phosphorylase $\underline{b}$ (Fig. 2B).

The fluorescence emission at 515 nm (excitation at 335 nm) is stable for several hours. The quantum yield of the adduct increases in a linear fashion with the concentration of acetic acid, indicating the proportionate creation of a new compound.

The absorbance spectrum of the reaction product of pyridoxal-P and hexylamine in chloroform shows, under the conditions of Shaltiel and Cortijo (13), a peak at 335 nm and a shoulder at 425 nm. On adding acetic acid to 0.4 M the 335 nm peak decreases and the shoulder at 425 nm increases to a peak (Fig. 3). Additionally, a peak at 290 nm starts to appear and is seen here in the form of a slight shoulder. The explanation of the increased absorbance at 425 nm is that on adding glacial acetic acid the Schiff base between pyridoxal-P and hexylamine is formed.

This result is in accord with the chemical mechanism of forming a Schiff base, which occurs in a two-step reaction. In the first step a

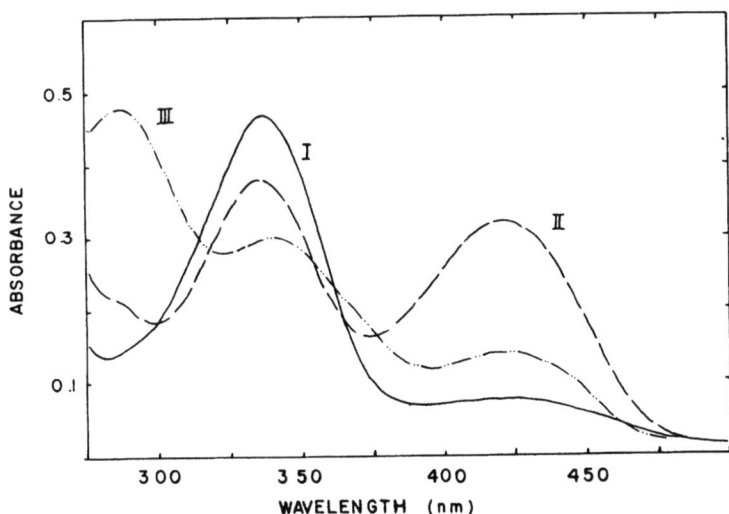

Fig. 3. Absorption spectra of the pyridoxal-P hexylamine adduct in chloroform. (I) 1.0×10^{-4} M pyridoxal-P and 10^{-3} M hexylamine; (II) 1.0×10^{-4} M pyridoxal-P, 10^{-3} M hexylamine, and 0.40 M acetic acid; (III) 1.0×10^{-4} M pyridoxal-P, 10^{-3} M hexylamine, 0.40 M acetic acid, and 0.55 M butanol, spectrum taken 30 min after mixing with butanol

carbinolamine between the carbonyl group and the amino group is formed (Scheme 1B, step 1). On mixing pyridoxal-P with an excess of hexylamine in chloroform the solution will contain an excess of proton acceptors (free amine) and not of the proton donors which are necessary to catalyze step 2 (Scheme 1B) of the reaction. Therefore, under the conditions of Shaltiel and Cortijo (13), the reaction should stop at carbinol amine form of step 1. This carbinol amine form should have, as one can conclude from its absorption spectrum, the chemical structure of a pyridoxamine derivative (Scheme 1A, #3). The nitrogen of the amino group is very near to the oxygen of the phenolic group, so both of these ionized groups form an ion pair, the only form of contrary charged ionized groups which can exist in a solvent like chloroform.

Bridges et al. (4) report that pyridoxamine with a protonated amino group and a dissociated phenolic group shows no fluorescence emission. This would explain why the carbinolamine form in chloroform without acid is non fluorescent. Addition of acetic acid as a proton donor catalyzes then the forming of the Schiff base (step 2) as is seen in the absorbance peak at 425 nm. The carbinolamine form has to be activated for this by the proton donor. The most probable activation site seems to be the hydroxyl group of the carbinolamine (Scheme 1, Part A, and Part B, step 2). The constant fluorescence emission of the activated carbinolamine form indicates that this activated state and its reaction product, the Schiff base, are in a dynamic equilibrium. On radiation at 335 nm the proton activated carbinolamine form is excited, and the equilibrium is shifted towards the Schiff base. Shifting of equilibria as reflected by changes of dissociation constants are well known in the case of excitation of phenols (4, 14, 15).

That an equilibrium exists between the Schiff base and carbinol amine forms can be demonstrated by adding a nucleophilic compound, such as the alcohol, butanol, to the mixture with 0.4 M acetic acid. After adding the butanol the Schiff base maximum at 425 decreases, the 335 nm peak is decreased slightly, but the shoulder at 290 nm increases to a peak (Fig. 3). In a reference experiment with diethyl ether instead of butanol the absorbance spectrum shows no change except for volume dilution. The carbinolamine form with an undissociated phenolic group, which comes into existence on adding acetic acid to the chloroform solution, absorbs at ~290 nm as does pyridoxal-P in acetic solution (Fig. 3). These two forms, if excited at this wavelength, emit at two different places. Pyridoxal-P emits at 430 nm, the activated carbinol amine form at 535 nm (Fig. 2A).

If the hypotheses of Johnson et al. (6) and Shaltiel and Cortijo (13) are valid, then the results of exciting the pyridoxal-P - n-hexylamine adduct at ~330 nm should be explainable. They propose, as mentioned above, the existence of a Schiff base form with the proton bound to the phenolic oxygen. But this hypothesis is unable to explain why the quantum yield at neutrality and in chloroform is so low, and why it is increased on adding the acid. On the contrary, according to their proposal, addition of acid should cause the protonation of the basic imine nitrogen, which would prevent the proton transfer and lower the quantum yield. Additionally, the absorbance at 290 nm appearing on adding the acid, and its fluorescence emission at 515 nm, can only be explained with the carbinolamine form. The absorbance at about 290 nm is not reported for any of the proposed Schiff base forms so far studied (8, 9, 10). It is always reported to be due to pyridoxal or pyridoxamine derivatives with an undissociated phenolic group (16). As we show below, difference spectra of phosphorylase b also show a prominent absorbance maximum at 295 nm due to the presence of pyridoxal-P.

Very similar results to that obtained with chloroform as a solvent were also observed with pyridoxal-P and hexylamine in dioxane and methanol, and will be reported in detail elsewhere.

Another model compound employed was the adduct of pyridoxal-P and lysine ethylester. The absorption spectrum (Fig. 4) is quite similar to the thiazolidine product of cysteine and pyridoxal-P (Fig. 4) and, above 300 nm, nearly identical with the absorption spectrum of phosphorylase (17). Also, the fluorescence emission spectrum of the lysine ethylester - pyridoxal-P adduct is very similar to the enzyme. The main fluorescence emission peak appears at 535 nm (quantum yield 0.01), while another smaller peak at 400 nm is also observed (Fig. 1).

This lysine ester has two NH_2 groups, of which one can form the primary carbinolamine form with the pyridoxal-P. The remaining amino group can form, in exchange for the OH- group, an 8-membered cyclic carbinolamine. The unusual fluorescence behaviour is explained by the fact that an 8-membered ring would not be as stable as a 5-membered thiazolidine ring and would be split on excitation with light, forming a Schiff base.

If one examines an absorption spectrum of phosphorylase b, one always sees a small shoulder between 290 and 295 nm (see, for example, (17)). To determine if this should belong to the pyridoxal-P in the enzyme we took a difference spectrum of phosphorylase b vs. its apoenzyme (Fig. 4). There one sees two peaks, one at 333 nm and a second one at 295 nm. The 295 nm absorbance is normally buried in the tryptophan and tyrosine absorptions of the protein so that only the small shoulder can be seen. These peaks correspond, respectively, to the unprotonated and protonated phenolic group of the pyridoxamine derivatives

(4, 16, 18), as can be seen in comparison to the pyridoxal-P - cysteine and pyridoxal-P - L-lysine ethylester adducts (Fig. 4). The presence of a carbinolamine form with an undissociated phenolic group would be compatible with this result. Unfortunately, the maximum at 295 nm cannot be investigated directly with fluorescence measurements because the very high tryptophan fluorescence excitation overlaps that of the coenzyme with an undissociated phenolic group.

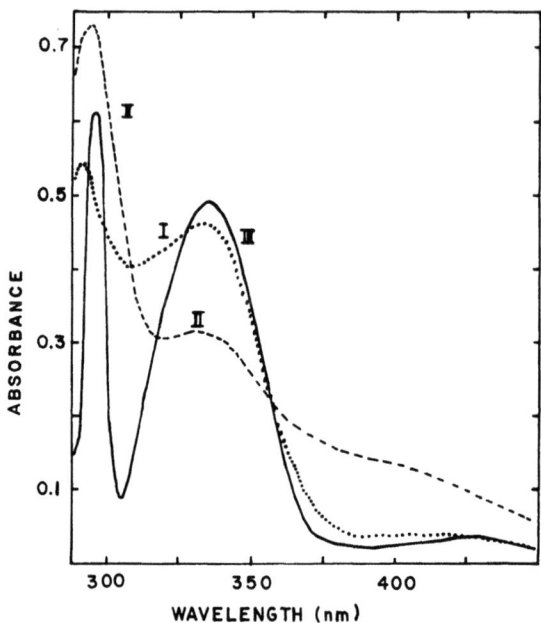

Fig. 4. Absorption spectrum of the pyridoxal-P - cysteine and the pyridoxal-P - L-lysine ethylester adduct and the difference spectrum of phosphorylase b vs. apophosphorylase b. (I) Absorption spectrum of a mixture of 2.3×10^{-4} M pyridoxal-P and 2.3×10^{-3} M L-lysine ethylester in 95% dioxane; (II) absorption spectrum of a mixture of 10^{-4} M pyridoxal-P and 4×10^{-4} M cysteine in 95% ethanol; (III) difference spectrum of phosphorylase b vs. apophosphorylase b. Reference cell: 9.8 mg/ml apophosphorylase b in 5×10^{-3} M glycerophosphate, 5×10^{-3} M mercaptoethanol, and 1.5×10^{-3} M EDTA, pH 6.8. Sample cell: 9.8 mg/ml phosphorylase b in the same buffer as the reference cell

While our models are only tools and not exact replicas of the binding of the cofactor in phosphorylase, they do permit us to make some relevant suggestions. The carbinolamine structure originally proposed by Kent et al. (1), and discussed in the introduction, would appear to be the most appropriate. In the enzyme, even in the non-excited state, an equilibrium exists between this carbinolamine and the Schiff base, but, as shown by the small absorbance of the enzyme at 415 nm, compared to that at 330 nm, the equilibrium is much in favor of the binding of the (unknown) nucleophilic group. In the excited state, however, this equilibrium is shifted to the Schiff base form by splitting off the nucleophile, so that a fluorescence emission typical of a Schiff base occurs from a compound with an absorbance typical of a carbinolamine (Scheme 2).

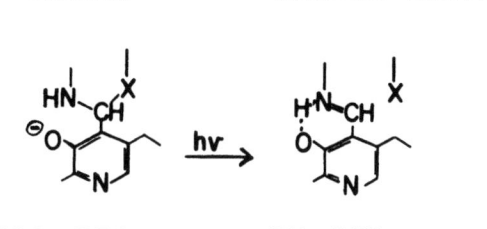

Scheme 2.
Postulated effect of light absorption on the structure and fluorescence of pyridoxal-P in phosphorylase

It will be noted in Fig. 1 that AMP and P_i decrease (quench) the fluorescence emission of phosphorylase b at 535 nm. This phenomenon has been exploited to study the interaction of various ligands with phosphorylase b and the results are summarized in Table I.

Table I

Ligand Interactions with Phosphorylase b
As Measured by Effects on PLP Fluorescence

Variable Ligand	Ligands Added (mMolar)	Effect on Fluorescence Emission at 535 nm	Apparent Dissociation Constant (mMolar)	Slope of Hill Plot
P_i	-	Quenches	70	1.08
P_i	AMP(1)	Quenches	7	2.0→0.93
P_i	AMP(1) + ATP(3)	Quenches	36	0.91
P_i	AMP(1) + G-6-P(2)	Quenches	36	0.8
P_i	ATP(0.5)	Quenches	8	2.1→0.93
P_i	G-6-P(3.8)	Quenches	30	1.0
G-1-P	-	Quenches	10	1.16
G-1-P	AMP(1)	Quenches	2.5	1.7→0.98
AMP	-	Quenches	0.37	1.32
AMP	P_i(10)	Quenches	0.14	2→0.7
AMP	ATP(0.66)	Enhances	10.5	1.0
ATP	-	Enhances	2.6	0.8
ATP	P_i(13)	Enhances	1.5	2
G-6-P	-	None	-	-
G-6-P	AMP(1) + P_i(10)	Enhances	2	1.5
G-6-P	AMP(1) + G-1-P(4.6)	Enhances	0.85	2

It may be seen that the interaction of AMP and P_i agree with the positive heterotropic interactions observed with kinetics or physicochemical methods, while ATP has the expected negative heterotropic effect. However, we find the unexpected result that ATP and, to a lesser extent, G-6-P, have a positive heterotropic effect on P_i binding in the absence of AMP. Furthermore, both the allosteric inhibitors cause, under certain conditions, an enhancement of fluorescence. One might suggest still another conformational state, in addition to the three discussed earlier by Helmreich and ourselves. This one would be induced by the allosteric inhibitors and would bind the substrates readily in a non-productive complex. Should one call it a T' state? Details of these experiments, as well as others in which sulfhydryl group reactivity leads to the same conclusions, will be published elsewhere.

We have recently grown large single crystals of phosphorylase <u>a</u> and <u>b</u> in 1.0 to 1.2 M sulfate or phosphate. The crystals are 1 to 2 mm in overall dimensions and their fluorescence, due to the pyridoxal-P moiety, is apparent to the naked eye and can be used to photograph them. An example of such a crystal is shown in Fig. 5.

<u>Fig. 5</u>. Fluorescent crystal of phosphorylase <u>a</u> grown in 1 M $(NH_4)_2SO_4$, pH 7.5, at 16°. The crystal was illuminated with the 334 and 365 nm lines of a mercury lamp while a barrier filter permitted only light of wavelength greater than 500 nm to reach the film. The crystal is 1.5 mm long

A study of these single crystals in the Turner Spectrofluorometer indicates considerable differences from the enzymes in solution. The size of the excitation spectral peak at 395 (monitored at 530) compared to the peak at 340 nm indicates that up to 25% of the pyridoxal-

P may be present as a Schiff base. This is illustrated in Fig. 6. This proportion is difficult to estimate because a scatter peak, due to the optical properties of the crystal, occurs at about 410 nm. However, large crystals do appear orange in color.

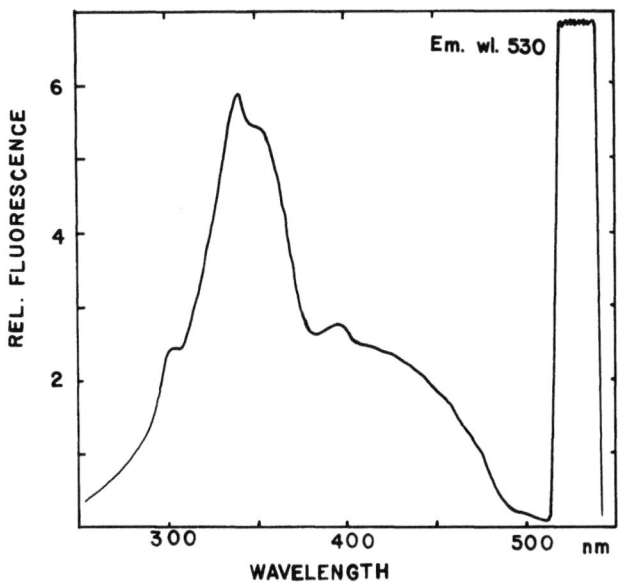

Fig. 6. Fluorescence excitation spectrum of a single crystal of phosphorylase a, grown as described for Fig. 5

Emission spectra of the crystals, using 340 nm as the exciting wavelength, are less subject to artifacts. An example for phosphorylase a is shown in Fig. 7 and it is noteworthy that a large proportion of the emission occurs at 418 nm, the remainder at 535 nm.

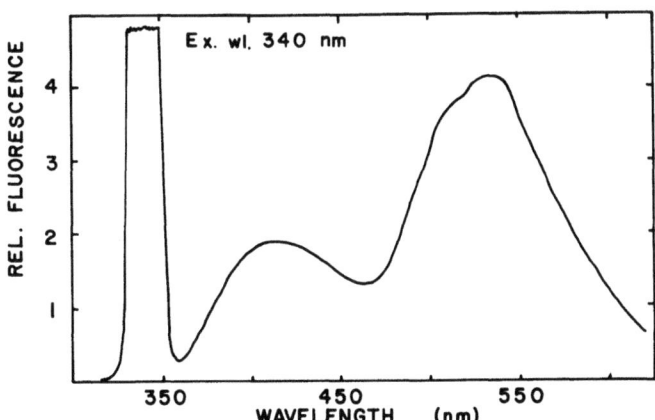

Fig. 7. Fluorescence emission spectrum of a single crystal of phosphorylase a, grown as described for Fig. 5

The emission spectra for phosphorylase b shows almost half the emission at 420 nm, the remainder at 535 nm (Fig. 8). The contrast with the enzymes in solution is obvious and suggests that there is a less facile equilibrium between the carbinolamine and Schiff base form when the protein is frozen into its crystal structure.

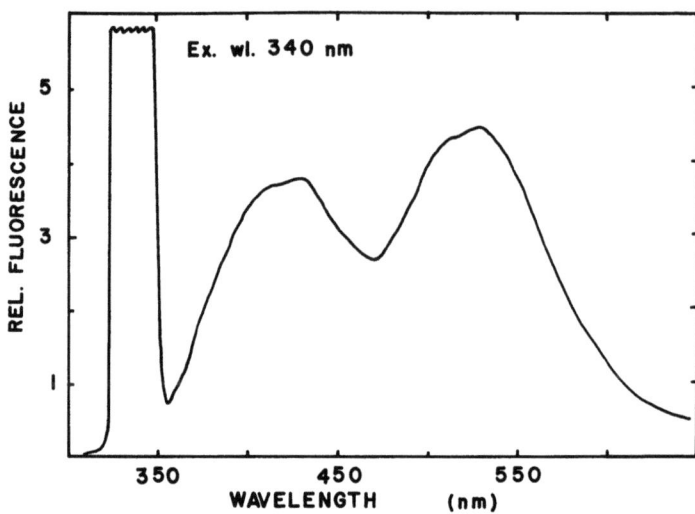

Fig. 8. Fluorescence emission spectrum of a single crystal of phosphorylase b, grown in 1.15 M phosphate, pH 6.8, at 16°

Crystals of phosphorylase a, b and of the CH_3Hg derivative were grown by the Zeppezauer technique (19) and photomicrographs of these crystals are shown in Fig. 9. Crystals from both $(NH_4)_2SO_4$ and K_2HPO_4 dialysis solutions have identical habits and diffraction patterns, and have an elongated rhombic shape (diamond) with well developed 010 faces. The quality of the crystals was a distinctly unpredictable parameter, but the best preparations had crystals of phosphorylase a measuring up to 1.0 mm x 0.8 mm x 0.4 mm with the shortest dimension parallel to the [010] axis. The methyl mercury derivative of phosphorylase a crystallized with a similar habit but with rounded edges and corners (Fig. 9(b)). These crystals also tended to be more tabular, with the b dimension of the crystals much thinner than that of the native enzyme.

The crystals of phosphorylase b tended to be less well developed than those of phosphorylase a. They had a distinctly different habit (Fig. 9(c)) than those of phosphorylase a but still retained a well developed 010 face and were diamond shaped. Many of the preparations of the phosphorylase b crystals showed interpenetrating growth and had a tendency for overgrowth on the larger crystals. These preparations contained few crystals that were of use in an X-ray study. The well-formed single crystals of the b form tended to be too small to give satisfactory diffraction patterns in our hands but this could be overcome with the use of a high intensity X-ray source such as a rotating anode generator.

All X-ray diffraction studies were carried out with the crystals mounted in thin-walled glass capillaries on a Jarrell-Ash precession camera (crystal to film distance = 75 mm). The X-ray source was a Hilger-Watts multifocus generator with a focal spot of 1.4 mm x 0.1 mm, operated at 40 kV and 7 mA.

The crystallographic data for the native phosphorylase a crystals may be summarized as follows: crystal system, monoclinic; space group, $P2_1$; unit cell dimensions as measured from precession photographs, a = 118.6 Å, b = 188.6 Å, c = 87.9 Å and β = 108.6°. These data correspond well with the values that were reported for crystals of phosphorylase b in the presence of AMP of: a = 120 Å, b = 190 Å, and c = 89 Å, β = 109°, space group $P2_1$ (20). This report implied one molecule of phosphorylase b per asymmetric unit of the above space group.

Our cell dimensions give a unit cell volume of 1.860×10^6 Å^3 or an asymmetric unit of 0.930×10^6 Å^3. The density of the crystals of the native enzyme was measured by the density-gradient technique (21) using a gradient column of bromobenzene and toluene. The column was calibrated using aqueous NaBr solutions and the most sensitive column employed had a range of 1.25 g cm^{-3} to 1.35 g cm^{-3} over a column length of approximately 10 cm. The mean crystal density was measured as 1.299 ±0.006 g cm^{-3} (the total range of 8 determinations was 1.294 to 1.304 g cm^{-3}).

The protein content of the crystals was also determined by weighing several crystals, that had had the mother liquor carefully wiped from their surfaces, on a Cahn microbalance. The weight loss due to evaporation of crystalline mother liquor was estimated by determining the weight at timed intervals over a period of 10-15 minutes and extrapolating back to zero time. These crystals were then dissolved in a measured volume of distilled water. The protein content of these solutions was determined from the A_{280}/A_{260} ratio of the solutions and an $E_{1cm}^{1\%}$ of 13.2 for phosphorylase (22). The mean of three such determinations was 52% protein in the crystal (range 49.7% - 54.4%). This stage of the analysis certainly contains the greatest source of error as it is very critical in determining the initial weight of the crystals as to how much mother liquor is left clinging to their surfaces.

The currently accepted molecular weight for the phosphorylase a tetramers is 370,000 daltons as determined from measurements of the

Fig. 9.

(a) Photomicrograph of phosphorylase <u>a</u> crystal grown from $(NH_4)_2SO_4$ as described in the text, taken in polarized light. The crystal is 1.05 mm long.

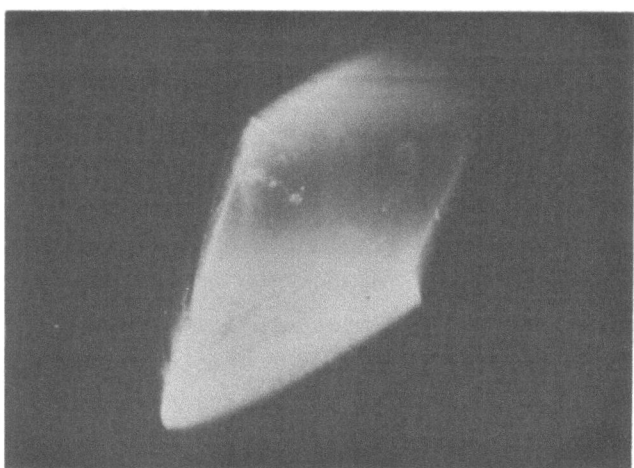

(b) Photomicrograph of a CH_3Hg derivative crystal of phosphorylase <u>a</u> grown from 1 M $(NH_4)_2SO_4$, taken with crossed polarizers. The large tabular face is 010. The crystal is 1.1 mm long. The CH_3Hg derivative crystallized from a 1% solution of phosphorylase <u>a</u> containing 2.5 moles of CH_3HgCl per monomer of protein.

(c) Photomicrograph of phosphorylase <u>b</u> crystals grown from K_2HPO_4, taken with crossed polarizers. The large crystal is 0.33 mm long

sedimentation and diffusion coefficients and by sedimentation equilibrium in the ultracentrifuge (23). With this value of the molecular weight and the above crystal data a unit cell content of 2.04 molecules for the crystals is calculated. This value agrees very closely to the integral value of 2 for the unit cell content of space group $P2_1$ and is well within the experimental error associated with the data. Thus our data indicate by an independent technique that the molecular weight of phosphorylase *a* tetramers is 370,000 daltons.

Diffraction patterns of the h0l and hk0 reciprocal lattice zones of the native and methyl mercury derivative are included in Fig. 10. The diffraction patterns of the native crystals are sharp and extend to a spacing of 3 Å resolution. The corresponding h0l zone of the CH_3Hg derivative is less well defined because of the very large crystal used. There are some significant intensity differences on these photographs, although a difference Patterson projection (UOW) is not immediately interpretable (24).

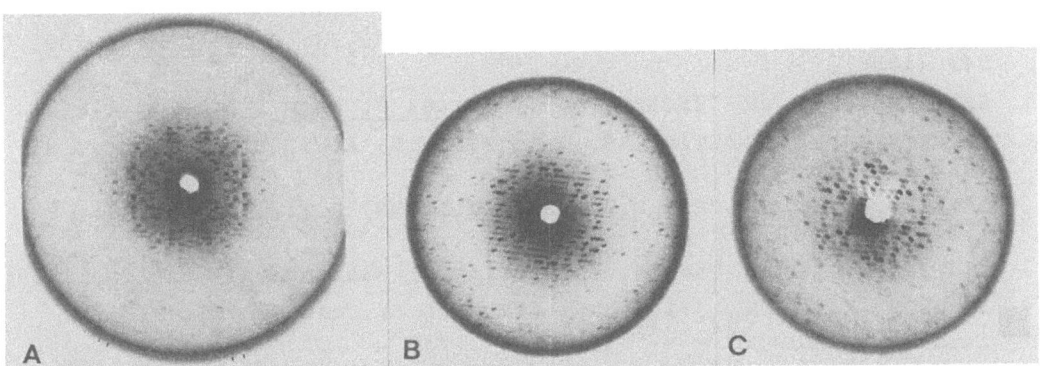

Fig. 10. (a) hk0 and (b) h0l diffraction patterns of the crystals of native phosphorylase *a*. Crystal to film distance 75 mm; exposure times h0l, 20 hours, hk0 50 hours. Precession angle 12° for h0l zone, 15° for hk0 zone.
(c) h0l diffraction pattern of a crystal of CH_3Hg derivative of phosphorylase *a*, exposure time 20 hours; precession angle 12°; crystal to film distance 7.5 cm

The significance of the above study is the fact that the molecules do not use a crystallographic symmetry element when crystallizing. That is, there is definitely a phosphorylase *a* tetramer in the asymmetric unit and the identity of subunits cannot be inferred from this particular crystalline modification.

Clearly the greatest immediate problem associated with the X-ray diffraction studies of a molecule of this size is the logistics of the data collection. The native crystals are relatively stable for up to

40-50 hours in the X-ray beam but the methyl mercury derivative crystals are much less stable. The most useful technique appears to be the photographic technique coupled with the use of a high speed scanner for the intensity measurement and work on this is in progress. Once a 6 Å native set of intensity data has been collected and measured the rotation function of Rossman and Blow (25) should provide considerable information regarding the subunit structure of phosphorylase a tetramers in this crystalline modification.

Acknowledgements

The authors wish to acknowledge the valuable technical assistance of Miss K. Hayakawa and Mrs. S. Shechosky. This work was supported by grants MA-3406 and MT-1414 from the Medical Research Council of Canada.

References

1. KENT, A.B., KREBS, E.G., and FISCHER, E.H., J. Biol. Chem., 232, 549 (1958).
2. SHALTIEL, S., and FISCHER, E.H., Israel J. Chem., 5, 127p (1967).
3. HEDRICK, J.L., SHALTIEL, S., and FISCHER, E.H., Biochemistry, 8, 2422 (1969).
4. BRIDGES, J.W., DAVIES, D.S., and WILLIAMS, R.T., Biochem. J., 98, 451 (1966).
5. KENT, A.B., Ph.D. Thesis, University of Washington, Seattle (1959).
6. JOHNSON, F.J., TU, J.I., BARTLETT, M.L.S., and GRAVES, D.J., J. Biol. Chem., 245, 5560 (1970).
7. SNELL, E.E., and Di MARI, S.J., in P.D. Boyer (Editor), The Enzymes, Vol. 2, 3rd ed., Academic Press, New York, 1970, p. 337.
8. HEINERT, D., and MARTELL, A.E., J. Am. Chem. Soc., 85, 183 (1963a).
9. HEINERT, D., and MARTELL, A.E., J. Am. Chem. Soc., 85, 188 (1963b).
10. MATSUSHIMA, Y., and MARTELL, A.E., J. Am. Chem. Soc., 89, 1322 (1967).
11. METZLER, D.E., J. Am. Chem. Soc., 79, 485 (1957).
12. BENTLEY, G.A., Ph.D. Thesis, University of Auckland, New Zealand (1971).
13. SHALTIEL, S., and CORTIJO, M., Biochem. Biophys. Res. Commun., 41, 594 (1970).
14. FORSTER, T., Z. Elektrochemie, 54, 531 (1950).
15. WELLER, A., Z. Elektrochemie, 56, 662 (1952).
16. METZLER, D.E., and SNELL, E.E., J. Am. Chem. Soc., 77, 2431 (1955).
17. SCHLISELFELD, L.H., DAVIS, C.H., and KREBS, E.G., Biochemistry, 9, 4959 (1970).
18. WILLIAMS, V.R., and NEILANDS, J.B., Arch. Biochem. Biophys., 53, 56 (1954).

19. ZEPPEZAUER, M., EDLUND, H., and ZEPPEZAUER, E.S., Arch. Biochem. and Biophys., 126, 564 (1968).
20. MATHEWS, F. SCOTT, Federation Proceedings, 26, 831 (1967).
21. LOW, B.W., and RICHARDS, F.M., J. Am. Chem. Soc., 74, 1660 (1952).
22. BUC, M.H., and BUC, H., "Symposium on regulation of enzyme activity and allosteric interactions", Oslo, July 1967, Academic Press, New York, 1968.
23. SEERY, V.L., FISHER, E.H., and TELLER, D.C., Biochemistry, 6, 3315 (1967).
24. JOHNSON, L., personal communication (1971).
25. ROSSMAN, M.G., and BLOW, D.M., Acta Cryst., 15, 24 (1962).

Probing and Mapping the Pyridoxal 5'-Phosphate Site of Glycogen Phosphorylase

Shmuel Shaltiel, Manuel Cortijo, and Yeshayahu Zaidenzaig

The Weizmann Institute of Science, Rehovot/Israel

Glycogen phosphorylase catalyses the phosphorolytic degradation of glycogen with the formation of glucose-1-phosphate (Fig. 1).

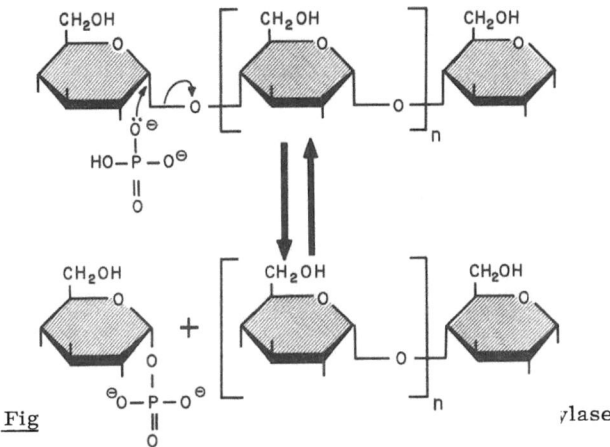

Fig ylase

From the work of Cori and his group it is known that this enzyme exists in an inactive dimer form (phosphorylase b) which can be activated (Fig. 2) either by the allosteric effector AMP (1-4) or by phosphorylation of the enzyme at unique serine residues (5-9). This phosphorylation "locks" the enzyme in its active conformation (phosphorylase a).

Each of the protomers of phosphorylase (M.W. 92500 (11,12) contains one molecule of PLP[*](13-15) which is tightly bound to the protein and appears to be at a key position in the enzyme. This is indicated by the following observations:

---

[*]Abbreviations: DMF, N,N'-dimethylformamide; DNP, 2,4-dinitrophenyl; FDNB, 1-fluoro-2,4-dinitrobenzene; F_2DNB, 1,5-difluoro-2,4-dinitrobenzene; PLP, pyridoxal 5'-phosphate; PMP, pyridoxamine 5'-phosphate.

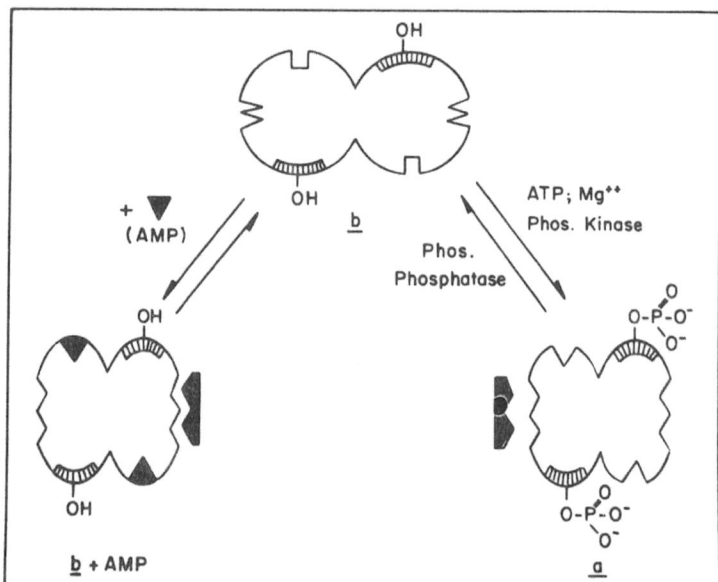

Fig. 2. Active and inactive forms of glycogen phosphorylase. ▼ , AMP; ◄►, substrate. In the rabbit muscle enzyme the phosphorylation is accompanied by the formation of tetramers. However, this aggregation is not indispensable for activity (10)

a) The presence of PLP in the enzyme is indispensable for its catalytic activity. Removal of the cofactor even under very mild, fully reversible conditions (15), results in total loss of the catalytic activity of the enzyme.

b) Stoichiometric amounts of PLP were found in all glycogen phosphorylases isolated so far - from yeast to man (16), indicating that the vitamin is an essential feature of the enzyme which was preserved during evolution.

c) The presence of PLP has a pronounced effect on the stability of the enzyme (17) and on its antigenic properties (18).

d) The vitamin is recognized by the apoenzyme with high specificity, to the extent that it cannot be replaced by any other of the naturally occurring forms of vitamin B_6 (19). Even among the synthetic analogues of PLP, only few are capable of endowing the enzyme with catalytic activity (19, 20).

e) The PLP site in phosphorylase strongly interacts not only with the catalytic site in the enzyme but also with the known regulatory sites of phosphorylase, namely, the site that accommodates the allosteric activator AMP or the site that contains the serine residue which becomes phosphorylated upon conversion of phosphorylase b to a. This interaction is best illustrated by the fact that in the presence of AMP or upon phosphorylation of the enzyme with phosphorylase b kinase, removal of PLP by the use of deforming agents becomes blocked (15, 21).

f) The process by which PLP can be made to leave or re-enter the enzyme exhibits very high specific requirements (Fig. 3). It will occur with L-cysteine (as the PLP reagent), but not with D-cysteine (22). Furthermore, there appears to be a specific L-cysteine binding site in phosphorylase.

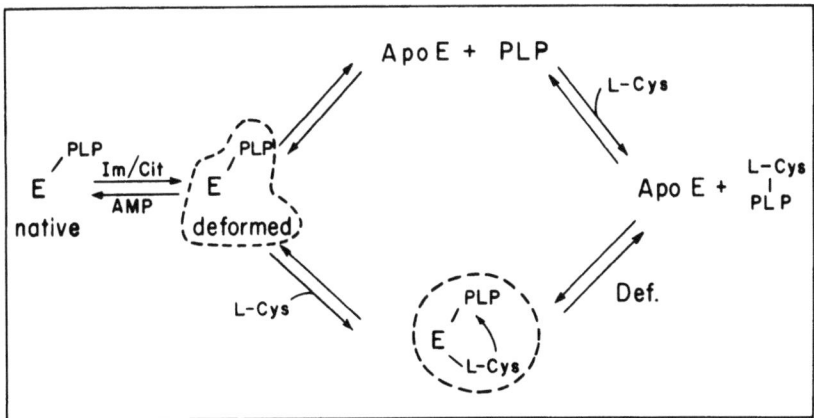

Fig. 3. Mechanism of removal of PLP from phosphorylase by the use of deforming agents. Encircled species were shown to be intermediates in the process of resolution of the enzyme (15, 21, 22)

All the above-mentioned observations raise the possibility that PLP may not be merely a structural building block in phosphorylase, but rather play a key role in the enzyme. It is possible, for example, that PLP is part of the catalytic site of the enzyme or occupies another site which may be involved in the transfer of a regulatory signal from or to phosphorylase. In view of the important function of PLP in phosphorylase, we attempted to probe and map its microenvironment and to establish its mode of binding to the protein.

Absorption and Fluorescence Properties of Glycogen Phosphorylase

Above 260 nm, glycogen phosphorylase has three absorption bands (23) with maxima at 280, 333 and 425 nm (at pH 7.0, $A_{280}:A_{333}:A_{425} = 100:5.3:0.4$). Upon excitation of the enzyme at these wavelengths, two types of fluorescence can be obtained (24, 25): one with a maximum at 335 nm (quantum yield 0.12), caused by the protein moiety, and the other with a maximum at 535 nm (quantum yield 0.012) caused by the cofactor PLP (Fig. 4). Taking an excitation spectrum of the green fluorescence, one obtains maxima not only at 425 and 333 nm but also at 280 nm (Fig. 5). This is due to the occurrence of energy transfer from the protein moiety to the cofactor. This transfer of electronic excitation energy occurs most probably via the resonance transfer mechanism described by Förster (26), since its efficiency (33%) is not a function of protein concentration, and since there is a good overlap between the protein fluorescence spectrum and the absorption spectrum of the PLP residue when bound to the protein. The occurrence of this transfer of energy also accounts for the difference in quantum yield between the protein fluorescence of the holoenzyme (0.12) and the apoenzyme (0.18).

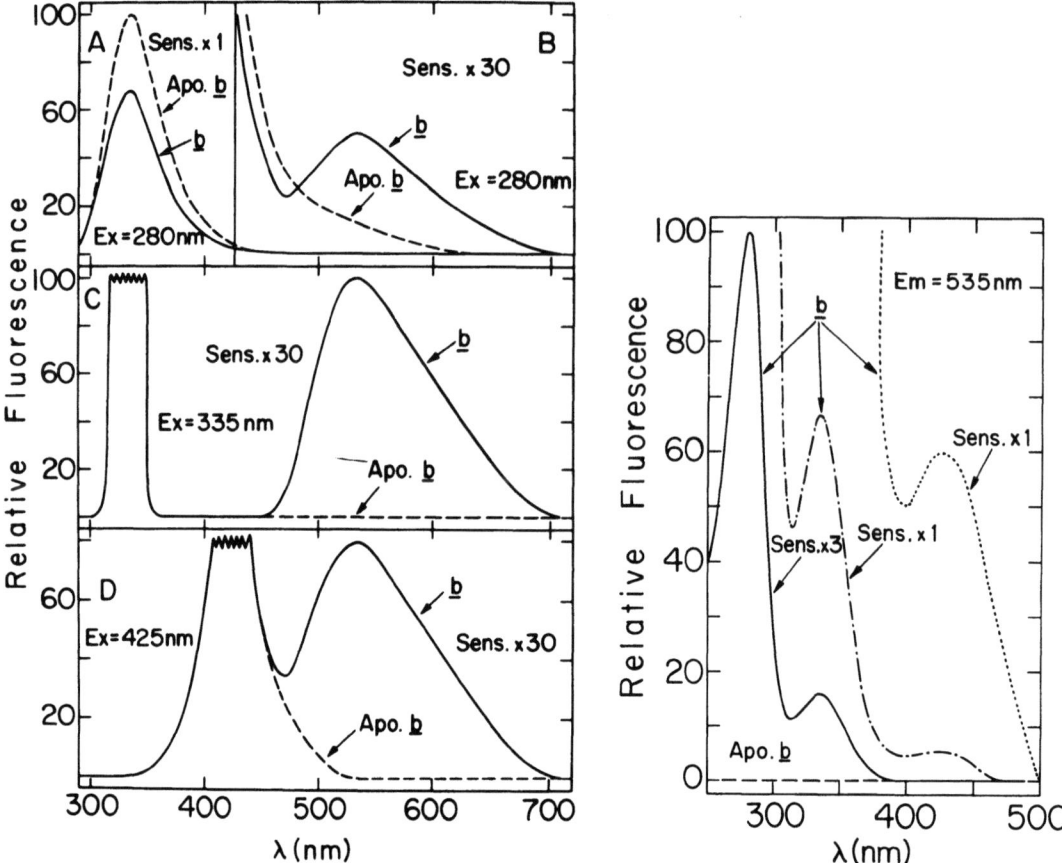

Fig. 4. Emission spectra of glycogen phosphorylase b (——) and its apoenzyme (---) at pH 7.0 and 30°. The spectra were taken with excitations and instrument sensitivities as indicated in the figure. The optical densities at the excitation wavelengths were: O.D.$_{280}$ = 0.09 for A and B; O.D.$_{335}$ = =0.05 for C; and O.D.$_{425}$ = 0.04 for D

Fig. 5. Excitation spectra of the 535 nm fluorescence of phosphorylase b (pH 7.0, 30°). The spectra were taken at three protein concentrations: 0.08 mg/ml (——) (O.D.$_{280}$ = 0.09); 0.76 mg/ml (—·—) (O.D.$_{335}$ = 0.05); 7.6 mg/ml (······) (O.D.$_{425}$ = 0.04). Under these conditions the apoenzyme (---) had no fluorescence. Instrument sensitivity - as indicated for each curve

The protein and PLP fluorescence of phosphorylase can be used to monitor structural transitions occurring in the enzyme as a function of pH. Figure 6 illustrates the effect of pH on the excitation and emission spectra of phosphorylase, when emission and excitation wavelengths are kept constant at 535 and 335 nm, respectively. Lowering the pH below 5.8 brings about quenching of the 535-nm fluorescence (excitation at 335 nm). Around pH 5.1 the quenching is accompanied by a blue shift of the emission maximum towards 500 nm and a concomitant rise in the excitation band around 425 nm (Fig. 6, A and B). Changes in the fluorescence characteristics of the enzyme are also observed upon raising the pH of the solution. Up to pH 9.5 there is a gradual quenching of the fluorescence. At higher pH values there is also a red shift in the emission maximum towards 550 nm with an appearance of an excitation maximum at 395 nm (Fig. 6, C and D).

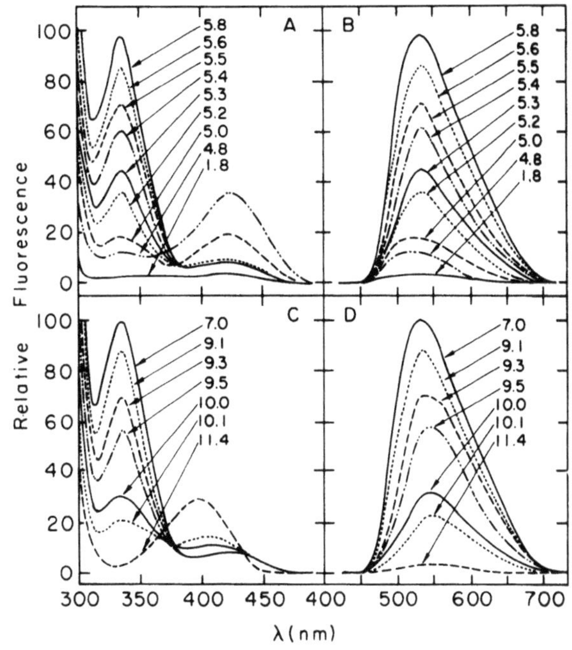

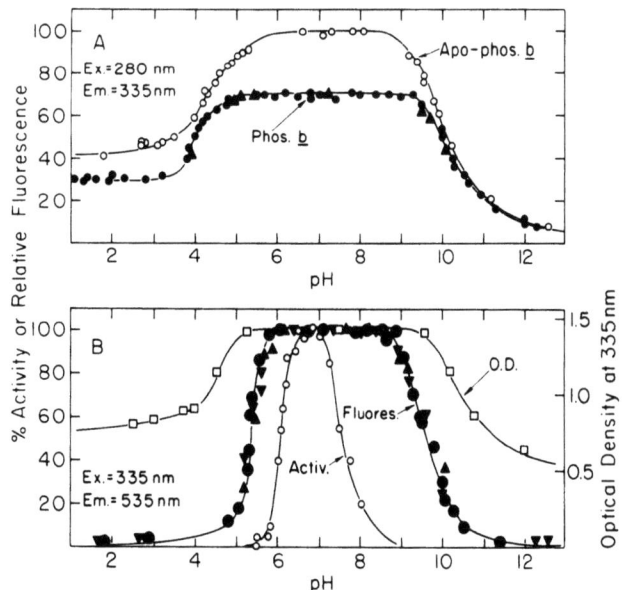

Fig. 6. Effect of pH on the PLP fluorescence of phosphorylase b. A and C are excitation spectra keeping the emission at 535 nm. B and D are emission spectra keeping the excitation at 335 nm. Enzyme concentration was 1.1 mg/ml

Fig. 7. Comparison of the effect of the pH on the fluorescence, absorbance and catalytic activity of phosphorylase b. A, protein fluorescence of the apoenzyme (—O—) and the holoenzyme (—●—). B, comparison of the PLP fluorescence of the enzyme (—●—) with its absorption at 335 nm (—□—) and its catalytic activity (—O—). For experimental details see reference 25

Figure 7A compares the effect of pH on the protein fluorescence (ex. at 280 nm, em. at 335 nm) of phosphorylase and its apoenzyme. Between pH 5.0 and 9.5 there is no change either in the intensity or in the excitation and emission maxima of the fluorescence of the holoenzyme. This plateau of the curve coincides with the well known "stability range" of phosphorylase (23). In the case of apophosphorylase the plateau of the pH dependence curve is restricted to a narrower range (pH 5.5 to 8.5).

The pH dependence of the PLP fluorescence of phosphorylase is depicted in Fig. 7B. With this fluorescence it is possible to show an occurrence of transitions in the structure of the enzyme around pH 5.4 and 9.6. It should be noted that these transitions do not coincide with those obtained by following the PLP absorption of the enzyme at 335 nm, or with the pH-activity curve of the enzyme. Around the transition points of the PLP fluorescence there are considerable changes in the state of aggregation of phosphorylase (27, 28). It is therefore possible that either the changes in the aggregation state per se, or conformational changes associated with them, affect the microenvironment of the PLP residue and cause the observed changes in the PLP fluorescence. The difference between the PLP absorption and fluorescence profiles may well be due to the fact that chromophores in the excited state are much more sensitive to changes in their environment than when in the ground state (29).

The Microenvironment of PLP in Phosphorylase

On the basis of the resemblance between the absorption properties of the PLP site in phosphorylase and of conjugates between PLP and amino acids, it was proposed (23) that at neutral pH, PLP is bound to the protein through a substituted aldamine structure (I, Fig. 8) involving an ϵ-amino group of a lysine residue and another group (X) which was not identified. It was also suggested (23) that in acidic or alkaline pH, PLP is detached from group X and becomes bound to the protein through a Schiff base structure (II, Fig. 8). These suggestions were supported by the finding (30) that at pH 4 PLP can be reduced onto the protein with $NaBH_4$ to yield structure III (Fig. 8) and that such reduction does not occur at neutral pH.

The finding that PLP endows phosphorylase with a characteristic green fluorescence (24, 25) provided additional parameters for investigating the microenvironment of this cofactor and its mode of binding to the protein. A systematic study was conducted (31) in an attempt to find a model PLP conjugate that would simulate both the absorption and the fluorescence properties of the PLP site in glycogen phosphorylase.

In a neutral aqueous solution there are several conjugates of PLP and amino acids which resemble the enzyme in having an absorption maximum around 330 nm (Fig. 9A). These include conjugates between PLP and cysteine (or cysteamine, Fig. 9B), histidine and tryptophan (32-35). Yet in a non-aqueous solvent such as DMF, CCl_4 or ethanol (Fig. 9C) a simple Schiff

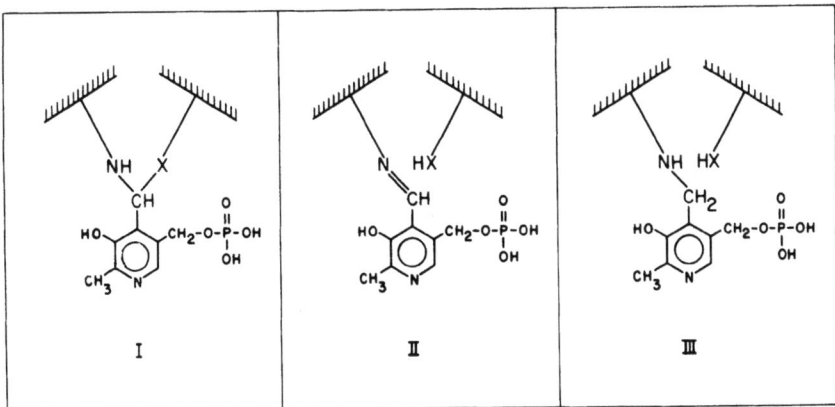

Fig. 8. Mode of binding of PLP to phosphorylase according to Kent et al. (23). I, at neutral pH; II, at acidic or basic pH; III, after reduction with $NaBH_4$. The ionization state of the functional groups on PLP is not indicated here

base conjugate between PLP and n-hexylamine can also simulate the PLP site in phosphorylase: it has a major absorption maximum at 335 nm and a weaker band between 415 and 425 nm. In fact, the relative intensities of these two bands depend on the nature of the solvent. For

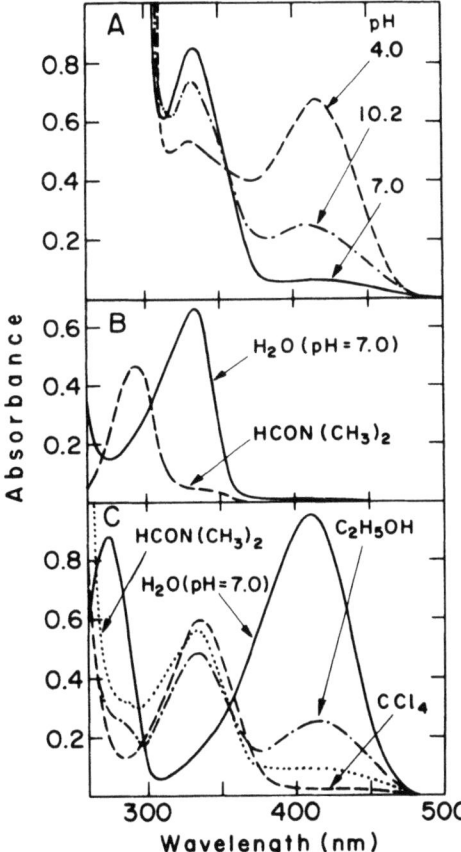

Fig. 9. Absorption spectra of glycogen phosphorylase b and of PLP derivatives. A. Aqueous solutions of the enzyme (15 mg/ml) at the indicated pH values. B. Thiazolidine derivatives of PLP (10^{-4} M) obtained by reaction with an excess of cysteamine (10^{-2} M) in the indicated solvent. C. Schiff base derivatives of PLP (1.1×10^{-4} M) obtained by reaction with an excess of n-hexylamine (10^{-2} M in the indicated media

example, in dioxane-water mixtures (Fig. 10) the Schiff base has a major peak at 335 nm when the solvent contains 5% water but it has a major 415 nm peak when the solvent contains more than 35% water. Moreover, the transition that occurs upon changing the polarity of the solvent has an isobestic point at 359 nm (Fig. 10), very similar to the isobestic point (at 357 nm) that was reported (23) for the titration of the enzyme from pH 7 to 4.

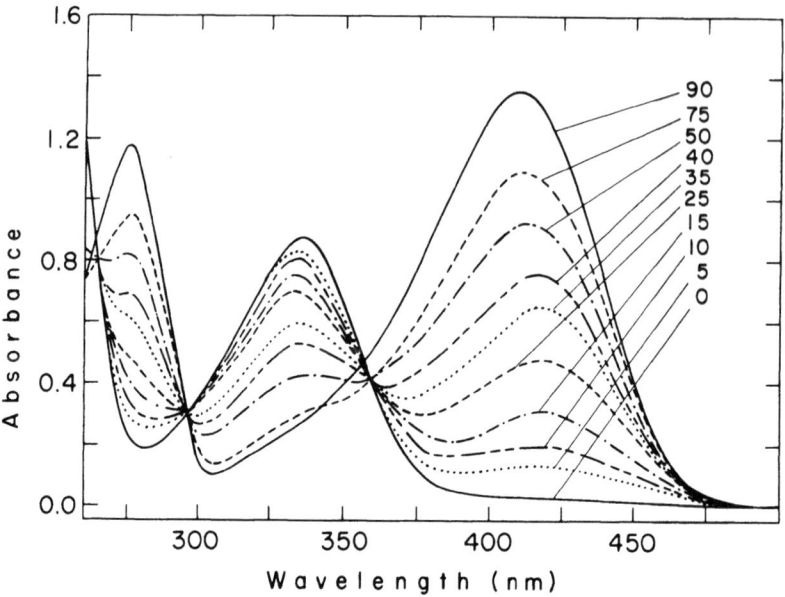

Fig. 10. Absorption spectra of a Schiff base derivative of PLP in dioxane-water mixtures. PLP (1.6×10^{-3} M) was reacted with an excess of n-hexylamine (0.1 M) in absolute dioxane and diluted ten-fold into dioxane-water mixtures. The final solutions had the indicated percentage of water (by volume)

As for the fluorescence properties of phosphorylase, it was found (31) that they can also be simulated by a Schiff base derivative of PLP in non-aqueous solvents. As seen in Fig. 11, the excitation and emission spectra of the enzyme at neutral pH are practically identical with the spectra of a PLP-Schiff base derivative in chloroform, but not in water (pH 7). On the other hand, PLP conjugates which have an absorption maximum at 330 nm in water do not have fluorescence properties similar to the enzyme. Figure 12 illustrates the fluorescence spectrum of a conjugate between PLP and cysteamine. When excited at 330 nm it does not give the 535 nm fluorescence in either an aqueous or a non-aqueous medium. Several other PLP conjugates were screened (with cysteine, homocysteine, histidine, histamine and tryptophan). All of these have absorption maxima around 330 nm in water (pH 7) but fluoresce between 390 and 400 nm.

An appropriate model for the PLP site should account not only for the absorption and fluorescence properties of the enzyme, but also for the finding reported by Fischer and his

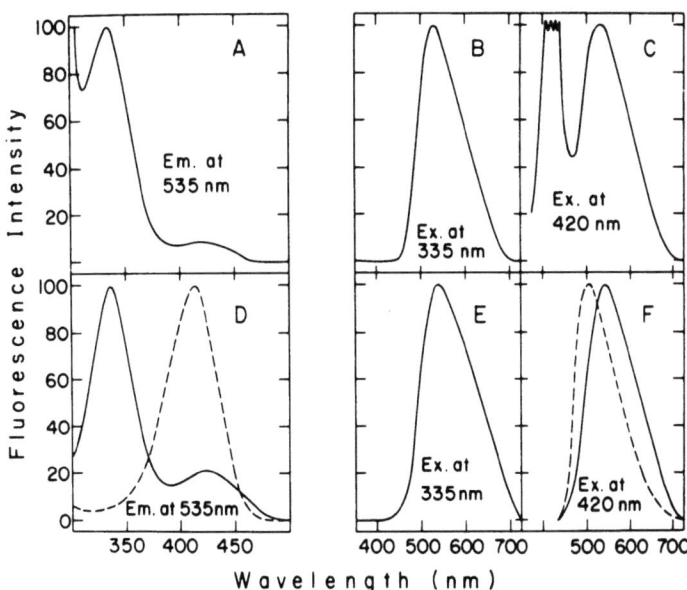

Fig. 11. Fluorescence properties of glycogen phosphorylase b and of a Schiff base derivative of PLP. The enzyme (1.3 mg/ml) was dissolved in a buffer composed of sodium glycerophosphate (0.05 M), 2-mercaptoethanol (0.05 M) and EDTA (10^{-3} M), pH 7. The Schiff base was prepared by reaction of PLP (10^{-5} M) with n-hexylamine (10^{-2} M) in either water, pH 7 (- - - -) or in chloroform (———). A, excitation spectrum of the enzyme; B and C, emission spectra of the enzyme; D, excitation spectrum of the Schiff base; E and F, emission spectra of the Schiff base

associates (30) that at neutral pH, PLP cannot be fixed onto the protein by reduction with $NaBH_4$. Now, Schiff base model compounds in organic solvents (e.g. DMF) can be reduced with $NaBH_4$. However, it was found that the rate of reduction in such a medium ($t_{1/2}$ = 245 sec, Fig. 13) is much slower than the rate of decomposition of $NaBH_4$ in water ($t_{1/2}$ < 12 sec (36)). It is therefore possible that $NaBH_4$ fails to reduce aqueous solutions of phosphorylase at pH 7

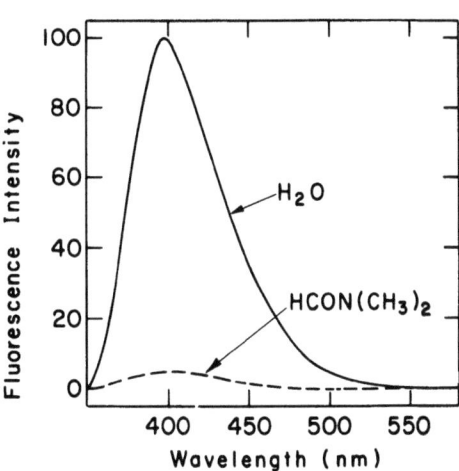

Fig. 12. Fluorescence of a thiazolidine derivative of PLP (excitation at 330 nm). This conjugate was prepared by reaction of PLP (10^{-5} M) with an excess of cysteamine (10^{-3} M) either in water, pH 7 (———) or in DMF (- - - - -)

since the reducing agent decomposes much faster than it can react with the bound PLP in its hydrophobic microenvironment. The low solubility of the BH_4^- ions in this hydrophobic site may in itself be a dominant factor in preventing the penetration of these ions into the PLP site and the subsequent reduction of the cofactor onto the protein.

Kent et al. (23) have observed that upon treatment of phosphorylase with urea or with detergents at neutral pH, there is an increase in the 415 nm-absorption of the enzyme and it is possible then to reduce the cofactor onto the protein with $NaBH_4$ (23, 37, 38). This observation can also be accounted for by the suggested hydrophobic microenvironment of the PLP site in the native conformation of phosphorylase at pH 7. Urea and detergents, being denaturants, apparently expose the PLP site to a more polar environment, thus facilitating the approach of the BH_4^- ions and the rapid reduction of the Schiff base.

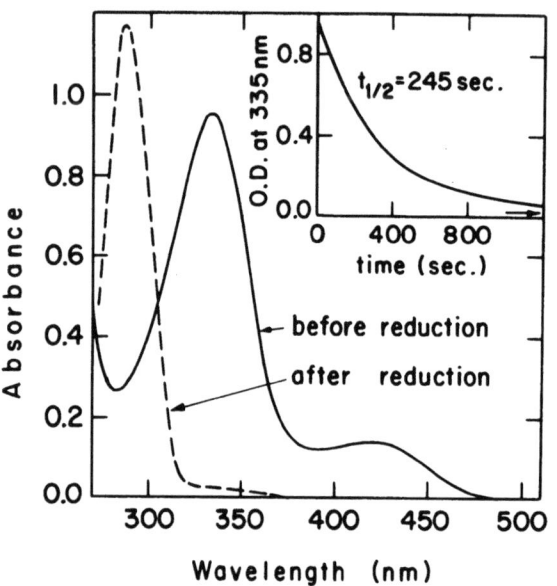

Fig. 13. Reduction of a Schiff base derivative of PLP in DMF. A solution of PLP (3×10^{-4}M) and n-hexylamine (4.5×10^{-2}M) was reduced by addition of 50 moles $NaBH_4$ per mole of PLP. The spectrum of the reaction mixture was taken before and after the addition of $NaBH_4$. <u>Inset</u>: Rate of reduction of the Schiff base followed by the drop in absorption at 335 nm

Optical Properties of $NaBH_4$-Reduced Phosphorylase

Reduction of glycogen phosphorylase with $NaBH_4$ results in an enzyme which is still catalytically active (up to 60%) and is essentially indistinguishable from the native enzyme in many of its physicochemical properties (30, 37, 38). It was therefore of interest to investigate the absorption and fluorescence properties of the modified chromophore (a PMP residue, see structure III in Fig. 8) for probing the microenvironment of the cofactor (39, 40).

Figure 14 depicts a spectrophotometric titration of the $NaBH_4$-reduced enzyme. It is clearly seen that at neutral pH it has a small absorption band around 330 nm (at pH 7, $\epsilon_M \sim 950$) which increases abruptly upon acidifying the enzyme solution to pH 5.5. In contrast, the absorption of ϵ-pyridoxal-5'-phosphate-lysine increases gradually already below pH 9 (Fig. 15).

It was found that the transition occurring with the enzyme could be simulated with free PMP by increasing the water content of DMF-water mixtures (Fig. 16), i.e. by transferring PMP to a less hydrophobic medium.

$NaBH_4$-reduced phosphorylase b fluoresces at 392 nm when excited at 330 nm. At pH 7 and 30°, the quantum yield of this fluorescence is very low (0.01) compared with that of free PMP in water (0.12, see reference 42). In the case of the reduced enzyme, both the wavelength of the emission maximum and its intensity are subject to considerable changes with pH (Fig. 17). The enhancement of the PMP fluorescence of the enzyme upon lowering the pH of the medium from 7 to 5.5 is of particular interest. This enhancement is not due to the increase in absorption of the enzyme at 330 nm (Fig. 14) as it is reflected also in the quantum yield (from 0.015 at pH 7 to 0.11 at pH 5.6). Once again it was found that this behaviour could be simulated with PMP by transferring it from a non-aqueous to an aqueous medium (Fig. 18). The quantum yield of PMP in DMF was found to be 0.004 as compared with 0.12 in water (pH 7 and 30°).

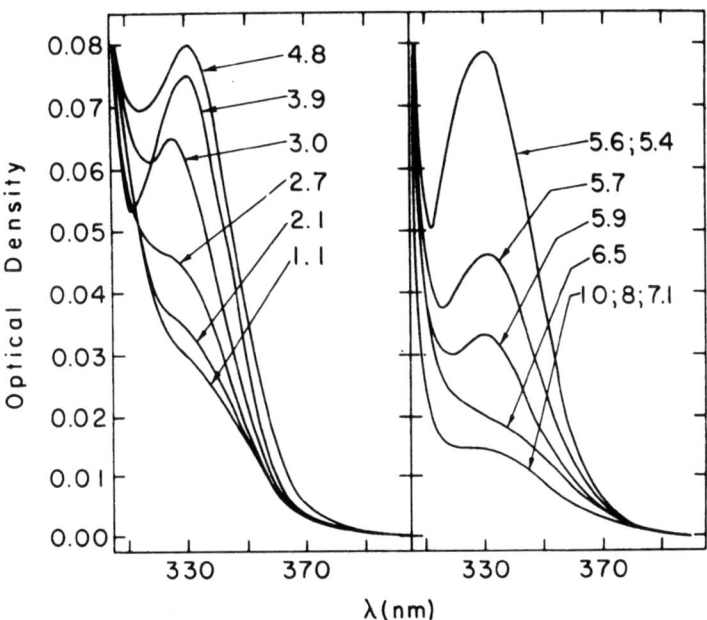

Fig. 14. Effect of pH on the absorption spectrum of $NaBH_4$-reduced phosphorylase b.
Enzyme concentration — 1.17 mg/ml and the pH - as indicated for each curve

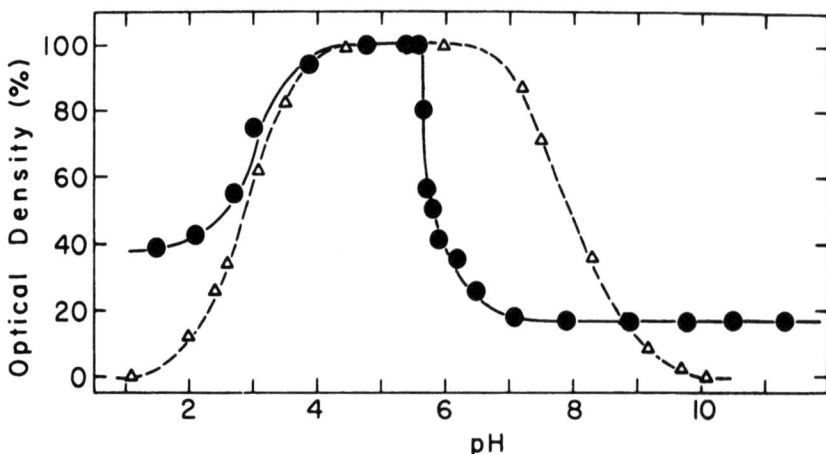

Fig. 15. Spectrophotometric titration of NaBH$_4$-reduced phosphorylase b̲ as compared with pyridoxyl-5'-phosphate-lysine. The absorption values of the enzyme at 330 nm (——●——) were taken from Fig. 14, and those of ε-pyridoxyl-5'-phosphate-lysine (——△——) were taken from reference 41. The two curves were normalized by assigning a value of 100% to the absorption values at pH 5

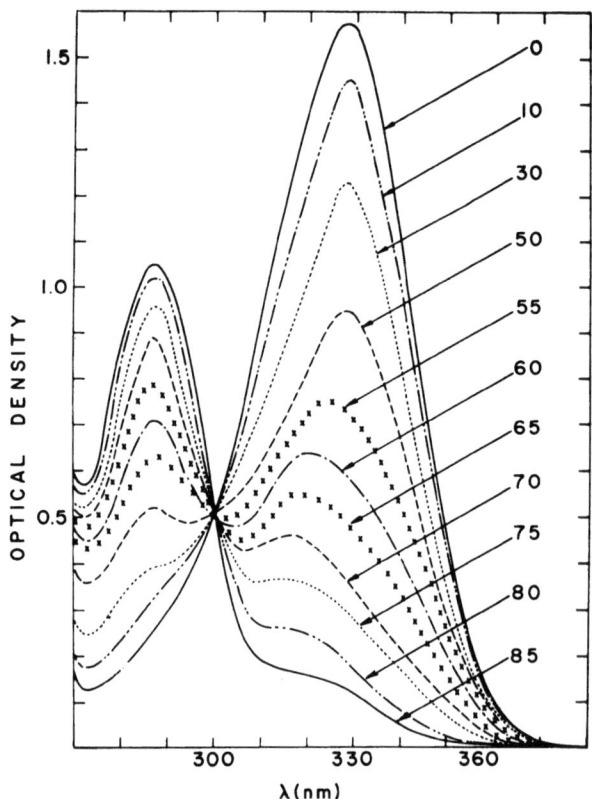

Fig. 16. Absorption spectra of PMP in DMF-water mixtures. PMP concentration was 1.88×10^{-4}M and the solvent mixtures contained the indicated percentage of DMF (by volume)

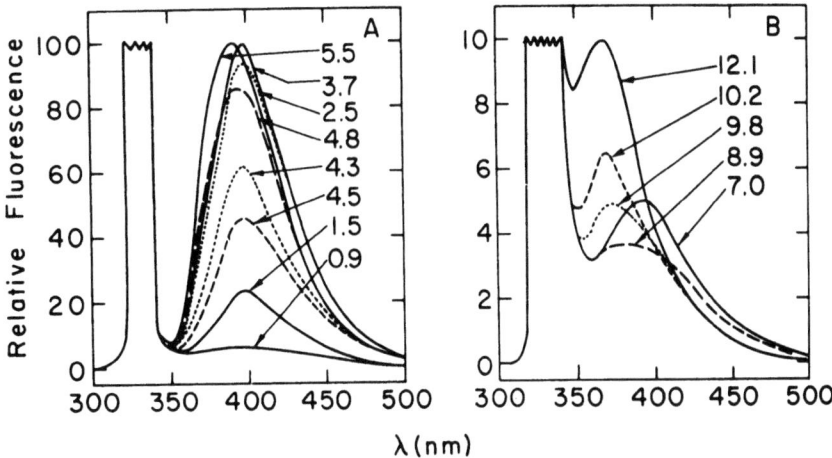

Fig. 17. Effect of pH on the PMP fluorescence of $NaBH_4$-reduced phosphorylase b. Enzyme concentration was 1.33 mg/ml and the pH - as indicated. The fluorescence was measured with excitation at 330 nm in all cases. Note the difference in scale between A (acidic region) and B (basic region)

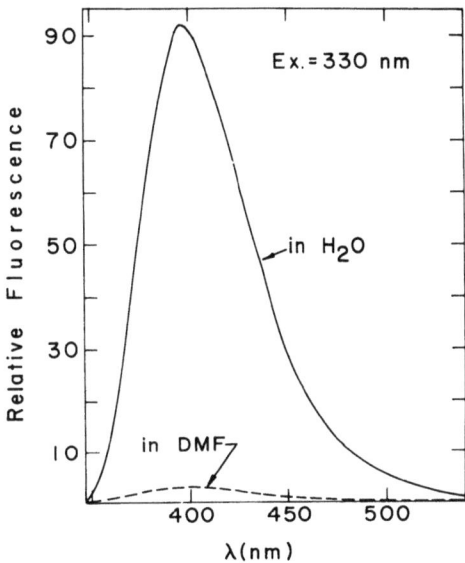

Fig. 18. Emission spectrum of PMP in water (pH 6.5) and in DMF. The optical density at the excitation wavelength (330 nm) was 0.08 in both cases

Mode of Binding of PLP in Glycogen Phosphorylase

The evidence presented above clearly demonstrates that a Schiff base derivative of PLP in an organic solvent adequately simulates the PLP site of glycogen phosphorylase (at pH 7) in both its absorption and fluorescence properties. We therefore suggested (31) that at neutral pH, PLP in phophorylase is embedded in a hydrophobic microenvironment and bound to an ϵ-amino group of a lysine residue in the protein through a hydrogen-bonded Schiff base structure (Fig. 19).

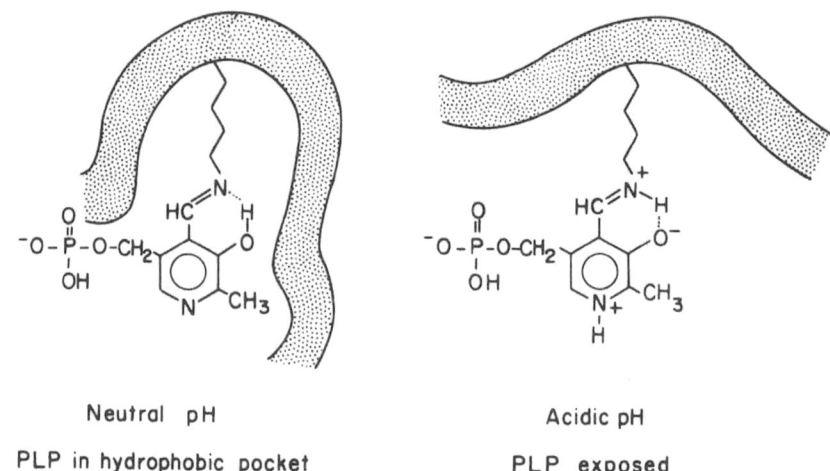

Fig. 19. Structures proposed for the PLP site in glycogen phosphorylase (31)

By assigning the above structures to the PLP site it is possible to account also for the chemical reactivity of this site towards $NaBH_4$, and for the absorption and fluorescence properties of the PMP residue formed after reduction of the enzyme with $NaBH_4$. The changes that occur in the optical and chemical properties of the PLP site in native phosphorylase (or in the $NaBH_4$-reduced enzyme) could be attributed to a conformational change or a change in the state of aggregation which results in exposure of the cofactor to a more polar environment.

The structures proposed above (Fig. 19) are based on the formulae assigned to PLP Schiff bases in non-aqueous (43, 44) and aqueous (45, 46) media. It should be emphasized, however, that each of these structures is representative of two tautomeric forms, enol-imine and keto-enamine, the existence of which was previously demonstrated by spectroscopic studies in the visible and infrared (43, 47) as well as by nuclear magnetic resonance data (48).

Assigning these tautomeric structures to the PLP site of phosphorylase at neutral pH makes it possible to account for the fluorescence properties of the enzyme (24, 25), especially for its large Stokes shift. Such Stokes shifts have been previously observed with a Schiff base derivative of 5-chlorosalicylaldehyde (49). They were attributed to the occurrence of an intramolecular proton transfer at the excited state (50, 51) from the phenolic hydroxyl to the nitrogen atom forming the Schiff base.

According to the structures proposed here*, PLP is bound to phosphorylase <u>covalently</u> only through the azomethine (—CH=N—) double bond. This does not exclude the possibility

---

*On the basis of absorption and circular dichroism studies with native and $NaBH_4$-reduced phosphorylase, Graves and his coworkers (52) independently raised the possibility that PLP in phosphorylase may be embedded in a hydrophobic environment and bound to the protein through a Schiff base structure.

that other, non-covalent bonds (hydrogen bonds, ionic or hydrophobic interactions, etc.) also participate in binding PLP to the protein. In fact, the high degree of specificity with which PLP is recognized by apophosphorylase strongly suggests that such interactions may well exist (19).

Mapping the PLP Site in Phosphorylase

Mapping the environment of the PLP site in glycogen phosphorylase was attempted by labeling the functional groups which become exposed upon removal of PLP from phosphorylase (53,54).

When apophosphorylase b is reacted with increasing amounts of FDNB, it loses its potential catalytic activity**. As seen in Fig. 20, 1.5 moles of FDNB per mole of enzyme protomer suffice to cause an inactivation of about 85%. Under the same conditions, native phosphorylase b, or an apoenzyme which was reconstituted with PLP prior to labeling, retain essentially all their activity. The low percentage of inactivation with the holoenzyme is reflected also in the small amount of label which becomes covalently attached to the protein (Fig. 21). It seems therefore that removal of PLP from phosphorylase exposes functional groups which are preferentially labeled with FDNB.

As expected, the pH of the medium strongly affects the rate of inactivation. Assuming that this is due to the ionization of a functional group, then this group would have an apparent pK of 7.5 (Fig. 22). The spectral properties of the labeled protein (absorption maximum at 335 nm, pH 7.5) indicate that the labeled functional group is a sulfhydryl (55, 56). Indeed, after acid hydrolysis (5.7 N HCl; 110°; 22 hours) followed by paper electrophoresis and chroma-

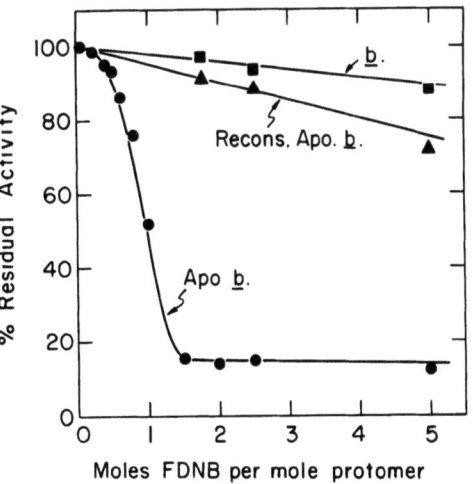

Fig. 20. Inactivation of phosphorylase b (—■—), apophosphorylase b (—●—) and reconstituted apophosphorylase b (—▲—) with FDNB. Each of the proteins was reacted with the indicated amount of FDNB (pH 7.5; 60 minutes; 22°) and then assayed. In the case of the apoenzyme, the potential catalytic activity was measured after reconstitution with 5 moles of PLP per mole of enzyme protomer (pH 7.5; 10 minutes; 37°)

**The potential catalytic activity of apophosphorylase is measured after reconstitution with PLP (17).

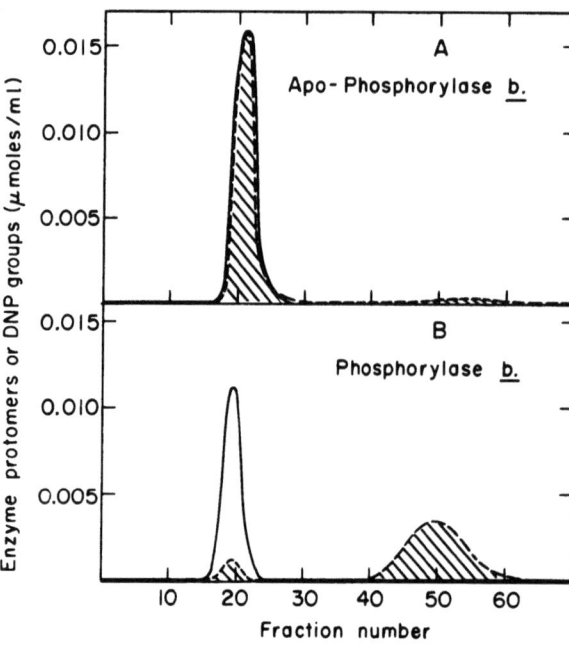

Fig. 21. Covalent labeling of apophosporylase b (A) and phosphorylase b (B). Each of the enzymes was allowed to react with one mole of $[^{14}C]$ FDNB per mole of enzyme protomer at pH 8 and 22°. After one hour they were subjected to gel filtration on a Sephadex G-25 column (50x1.5 cm). Fractions of 2 ml were collected and their radioactivity as well as their optical density were used to calculate the concentration of enzyme protomers (——) and of DNP groups (- - - -) in the various fractions

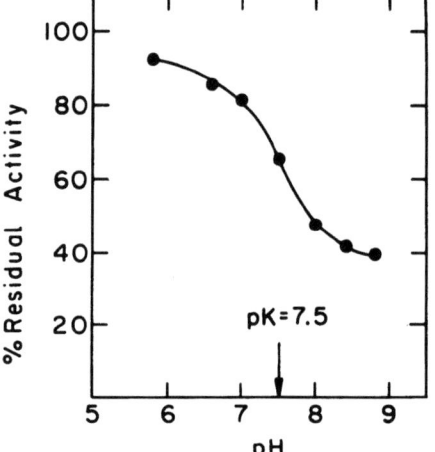

Fig. 22. Effect of pH on the inactivation of apophosphorylase b. The apoenzyme was reacted with one mole of FDNB per mole of enzyme protomer at 22° and the indicated pH. The reaction was allowed to proceed for 5 minutes and then aliquots were removed, reconstituted with PLP under standard conditions (17) and assayed. Control samples were incubated without FDNB and assayed as standards of activity after exposure to the various pH values

tography, essentially all the radioactivity is located in a spot that migrates identically with a marker of S-DNP-Cys (Fig. 23).

It was previously shown by Madsen (57-59) that upon treatment of phosphorylase with an excess of p-chloromercuribenzoate the enzyme undergoes dissociation into monomers. Since the state of aggregation of apophosphorylase is very sensitive to its environment (17), the possibility was considered that the inactivation of apophosphorylase upon stoichiometric dinitrophenylation is due merely to a change in the state of aggregation of the protein. However, it was found that under the conditions used here for inactivation, this does not appear to be the case. As seen in Fig. 24, the apoenzyme and its dinitrophenylated derivative have very similar (though not identical) sedimentation patterns.

In spite of the fact that the bond formed upon dinitrophenylation of apophosphorylase b is not cleaved during exhaustive hydrolysis of the protein, it is ruptured by thiols (55). This

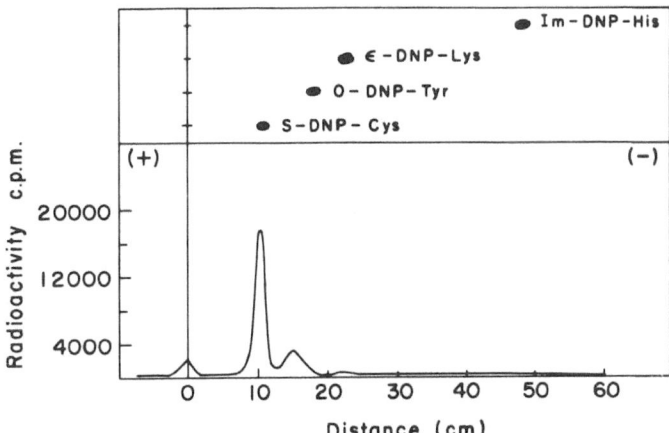

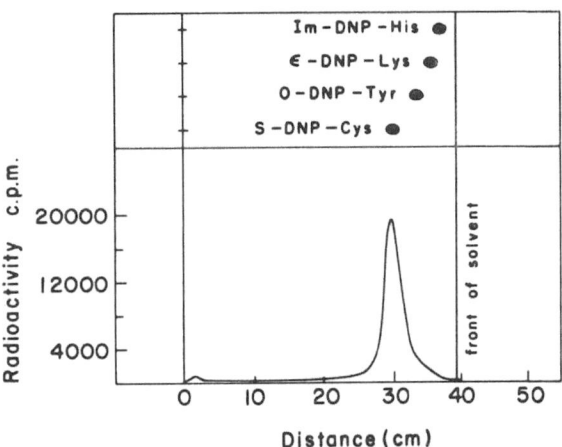

Fig. 23. High voltage electrophoresis of hydrolyzed [^{14}C] DNP-apophosphylase b. The protein was dinitrophenylated with 1.2 moles [^{14}C] FDNB per mole of enzyme protomer (pH 8, 60 minutes, 22°) and hydrolyzed with 5.7 N HCl (22 hours, 110°). Electrophoresis was carried out at pH 1.9 (3000 V, 3 hours). The paper was cut into 1cm strips and their radioactivity was monitored. Lower panel: Paper chromatography of the same protein hydrolysate using the lower layer of butanol : acetic acid : water (4 : 1 : 5) as the developing solvent. Radioactivity monitored as described above

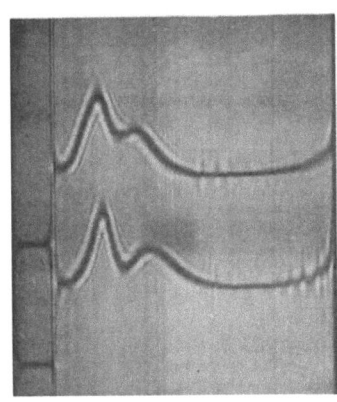

Fig. 24. Sedimentation velocity patterns for apophosphorylase b and its dinitrophenylated derivative (prepared as described in the legend to Fig. 21). Top: apophosphorylase b. Bottom: DNP-apophosphorylase b. Protein concentration 8.6 mg/ml, pH 7.5, 12°. The picture was taken 24 minutes after attaining the speed of 56100 rpm

nucleophilic displacement, which occurs at pH 7.0 and room temperature, involves a resoration of potential catalytic activity of the apoenzyme (up to 80%, Fig. 25). It was not possible yet to find conditions under which all the label could be removed with concomitant restoration of full catalytic activity (Figs. 25 and 26).

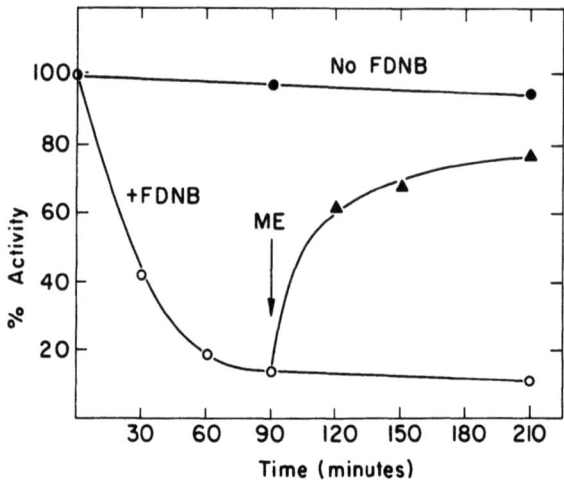

Fig. 25. Inactivation of apophosphorylase b with FDNB, and subsequent reactivation by thiolysis. The apoenzyme was reacted with 1.5 moles of FDNB per mole of enzyme protomer at pH 8 (22°), and its inactivation was followed with time (— ○ —). After 90 minutes, thiolysis was initiated (— ▲ —) by addition of 2-mercaptoethanol to a final concentration of 0.1 M. The control (— ● —) was kept under identical conditions except for FDNB and 2-mercaptoethanol which were not added

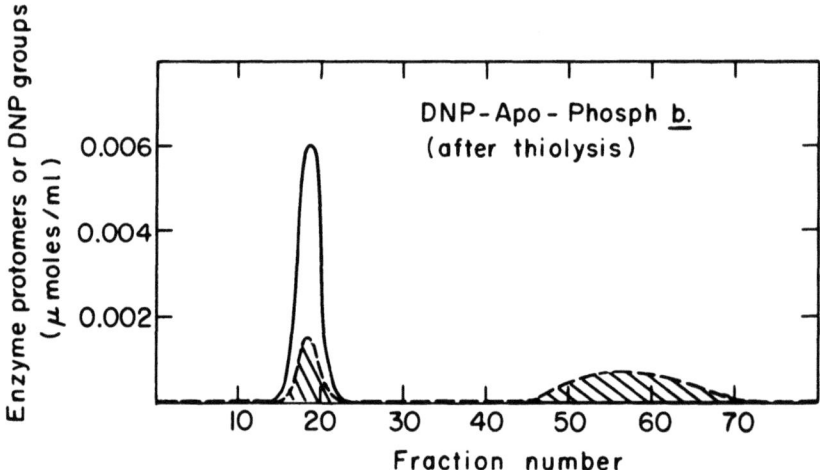

Fig. 26. DNP-apophosphorylase b was prepared as described in the legend to Fig. 21 and then thiolyzed with 2-mercaptoethanol (0.1 M) at pH 6.5 and 30°. Thiolysis was allowed to proceed for 3 hours and then the reaction mixture was subjected to gel filtration on a Sephadex G-25 column (50x1.5 cm). Fractions of 2 ml were collected and their radioactivity as well as their optical density were used to calculate the concentration of enzyme protomers (———) and of DNP groups (- - - -) in the various fractions

Fluorescence studies indicate that stoichiometric dinitrophenylation of apophosphorylase prevents the re-entry of PLP to its native site in the enzyme (53, 54). As seen in Fig. 27, when PLP is added to a dinitrophenylated apoenzyme preparation which had lost about 80% of its potential catalytic activity it gives rise to about 25% of the green fluorescence that would appear if the apoenzyme had not been dinitrophenylated prior to reconstitution with PLP. If, however, the dinitrophenylated apoenzyme is thiolyzed before reconstitution with PLP to restore 70% of its potential catalytic activity, it is possible to show that 65% of its PLP sites are vacant and accessible for reconstitution with the cofactor (54).

One of the advantages of dinitrophenylation as a method for labeling sulfhydryl groups is that in spite of the fact that the label can be easily removed under mild conditions by thiolysis, its binding is stable enough to withstand the manipulations required for fragmentation and sequence studies (60), so that the exact location of the labeled amino acid can be identified.

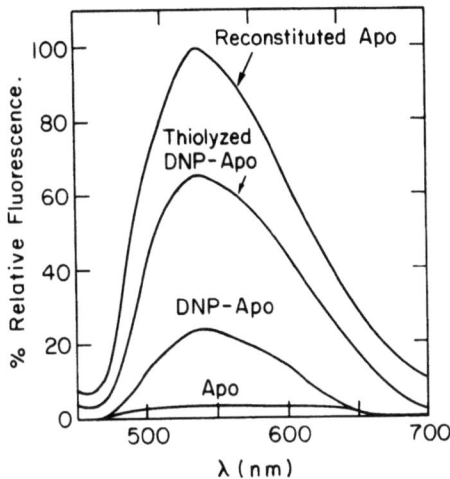

Fig. 27. Effect of dinitrophenylation and thiolysis on the fluorescence properties of apophosphorylase b after reconstitution with PLP. Dinitrophenylation was performed with one mole of FDNB per mole of enzyme protomer (pH 7.5, 60 minutes, 22°). Protein concentration was in all cases 1.9 mg/ml. Fluorescence spectra (excitation at 335 nm) were taken after reconstitution with PLP, except for the control apoenzyme (bottom curve) which was not reconstituted

Apophosphorylase b (2 μMoles of enzyme protomers) was dinitrophenylated with 2 μMoles of [^{14}C] FDNB at pH 7.5. The reaction was allowed to proceed for 1 hour at 22°, dialysed exhaustively against 0.01 N HCl, and digested with 10 mg pepsin at 30°. After 22 hours, the reaction mixture was subjected to gel filtration on a Sephadex column (Fig. 28). The fractions from the two radioactive peaks (S_1 and S_2) were pooled separately. They contained 90% of the total counts applied on the column, of which 32.5% were in the S_1 pool and 57.5% in S_2. Each of these fractions was subjected to further purification on Dowex 50X8 (Fig. 29). The pool S_1 yielded only one readioactive peak on Dowex (S_1D) which was further purified by preparative paper electrophoresis and chromatography (54). The pool of fractions S_2 yielded two major radioactive peaks when applied on the Dowex column. Each of these peptides (S_2D_1 and S_2D_2) was further purified by preparative paper chromatography (54). Table I summarizes the sequence of the three radioactive peptides as well as their yield.

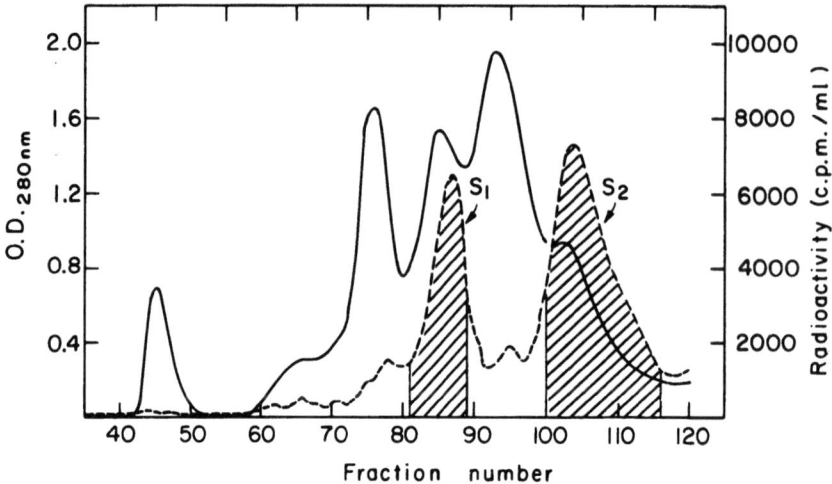

Fig. 28. Elution pattern of the peptic digest of DNP-apophosphorylase b. The column (Sephadex G-25 fine, 180x1.5 cm) was equilibrated at 22° with 2% acetic acid. Fractions of 3.2 ml were collected and their absorbancy at 280 nm (————) as well as their radioactivity (— — —) were monitored

When apophosphorylase is reacted with 1.5 moles of FDNB per mole of enzyme protomer there is a 90% loss of potential activity. However, when the reaction is performed with only 1 mole FDNB per mole of protomer, the loss of activity is 50-55%, although 0.9-1.0 DNP group becomes covalently attached to each enzyme protomer. This already indicates that some of the label may be attached to a site where it does not inhibit the potential activity of the apoenzyme. Therefore, one is faced with the problem of deciding which of the DNP-labeled peptides are responsible for the loss of potential activity and presumably originate from the PLP site.

Several pieces of evidence indicate that two of the three labeled peptides isolated (Asx-Ala-Cys-Asp and Ala-Cys) are apparently derived from the PLP binding site:

a) In a labeled apoenzyme sample which was found to have lost 55% of its potential activity, the extent of labeling on these two peptides together (the S_2 fraction, see Fig. 28) was 57.5%.

b) By thiolysis of the dinitrophenylated apoenzyme with various thiols it was possible to obtain partially labeled preparations, in which the residual label was distributed in different ratios among the above cysteine residues (61). The loss of potential catalytic activity in these preparations was found to be roughly proportional to the extent of labeling in peptides S_2D_1 and

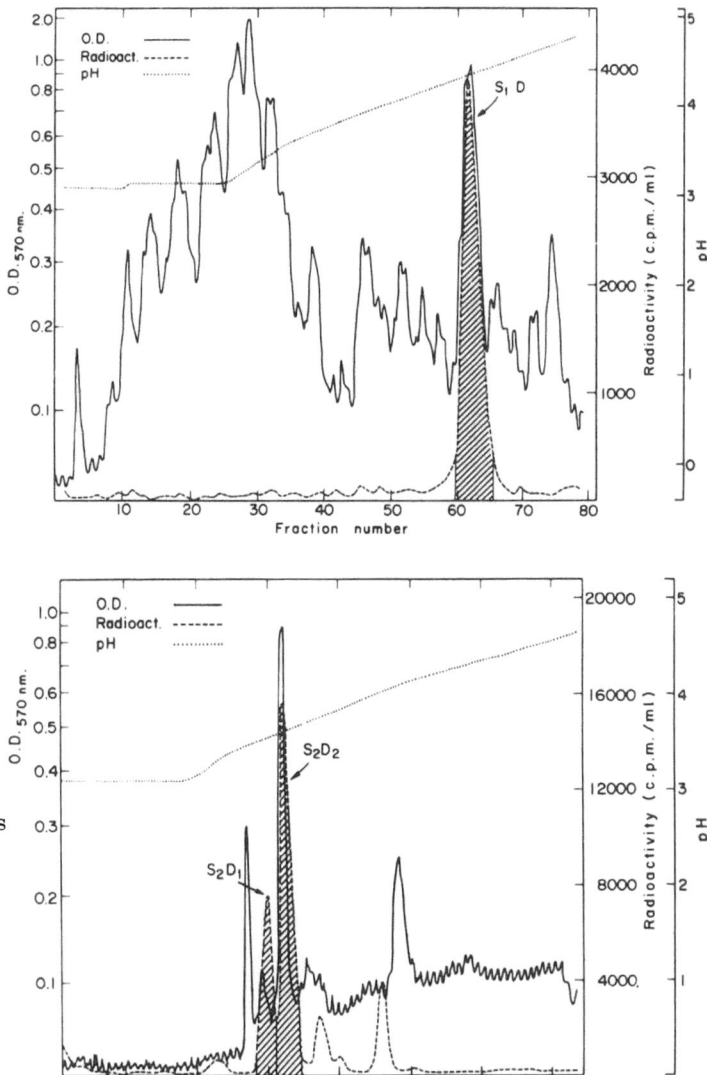

Fig. 29. Fractionation of S_1 (above) and S_2 (below) on Dowex 50X8. The column (35 x 0.9 cm) was water jacketed and eluted at 50°, first with 0.2 M pyridine acetate, pH 3.1 and then with a linear gradient using 100 ml of this buffer in the mixing chamber and 100 ml of 2 M pyridine acetate (pH 5.1) in the reservoir. The flow rate was 0.5 ml/min; 0.05 ml/min were removed for alkaline hydrolysis and subsequent reaction with ninhydrin. Fractions of 2.5 ml were collected and their ninhydrin color (O. $D_{570\,nm}$) as well as their radioactivity and pH were measured.

——— absorbancy at 570 nm
- - - radioactivity
.... pH

S_2D_2. An example of such preferential thiolysis is illustrated in Fig. 30. In this case (thiolysis with 2-mercaptoethanol) the extent of residual label in fraction S_2 was 24% (12% in each of the peptides S_2D_1 and S_2D_2) which fits the loss of activity of this sample (23%). Fluorescence studies with the same preparation showed that 26% of the PLP sites were inaccessible to the cofactor.

c) PLP affords protection from labeling to the cysteine residues in the peptides S_2D_1 and S_2D_2 but not to the cysteine residue in the peptide S_1D. Dinitrophenylation of the holoenzyme or the $NaBH_4$-reduced enzyme (under the conditions used above for the apoenzyme) brings about labeling of the peptide S_1D only (62) with essentially no loss of enzymatic activity (Fig. 31).

Table I. Sequence of the Labeled Peptides from Apophosphorylase b

Peptide	R_f [a]	Sequence [b]	Yield [c]
S_1D	0.35	Asx-Glx-Lys-Ileu-Cys(DNP)-Gly-Gly	18.6
S_2D_1	0.50	Asx-Ala-Cys(DNP)-Asp	22.0
S_2D_2	0.63	Ala-Cys(DNP)	39.0

[a] In 1-butanol : acetic acid : water (4 : 1 : 4, upper layer).

[b] The amino acid composition (―――) was determined by quantitative amino acid analysis except for S-DNP-cysteine which was determined from the radioactivity of the sample. The symbol ―→ indicates a determination by the Dansyl-Edman procedure and ⇒ represents an N-terminal determination by the FDNB method. Carboxypeptidase A was used for C-terminal analysis (←―).

[c] Represents the percentage of radioactivity in S_1D, S_2D_2 and S_2D_2 out of the total radioactivity found in the fractions obtained from the two Dowex columns (Fig. 29).

d) If, during dinitrophenylation, the cofactor site is exposed by the use of deforming agents (15), it becomes possible to introduce the DNP label also at the peptides S_2D_1 and S_2D_2 (Fig. 32), with proportional loss of activity.

The evidence summarized above shows that upon removal of PLP from phosphorylase or upon exposure of the PLP site by deforming agents, two (out of 9) cysteine peptides become available for preferential labeling with FDNB. Since each of these peptides contain the sequence Ala-Cys, the possibility was considered (54) that this dipeptide is a fraction of the tetrapeptide (Asx-Ala-Cys-Asp), and that both contain the same cysteine residue in the peptide chain. It was mentioned, however (54), that in the case of phosphorylase this need not be the case. Madsen and his collaborators who determined the amino acid sequence around all nine cysteine residues in phosphorylase, had previously shown (63) that there are two different peptides containing the sequence Ala-Cys, one peptide Asn-Ala-Cys-Asp and the other Ala-Cys-Ala-Phe.

In an attempt to establish whether the labeled dipeptide which we isolated (Ala-Cys) originates from the sequence Asn-Ala-Cys-Asp or from the sequence Ala-Cys-Ala-Phe, we attempted to isolate the labeled peptides after shorter digestions with pepsin (64), in the hope

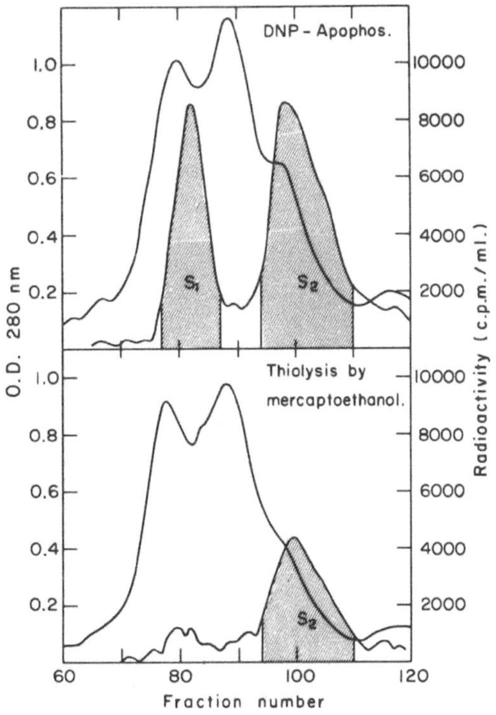

Fig. 30. Elution pattern of peptic digests of DNP-apophosphorylase b before and after thiolysis with 2-mercaptoethanol. Dinitrophenylation was performed with 1 mole [^{14}C] FDNB per mole of enzyme protomer (pH 7.5, 22°, 60 minutes). Thiolysis was initiated by addition of 2-mercaptoethanol (final concentration 0.1 M) and carried out for 3 hours at pH 7.5 and 30°. Radioactive peaks are marked as shaded areas. Peptic digestion and gel filtration on Sephadex - as described for Fig. 28

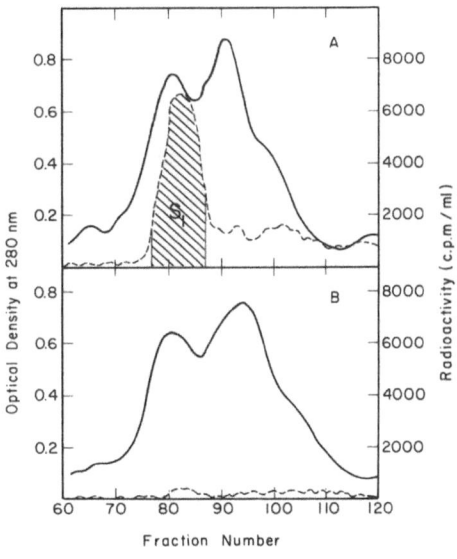

Fig. 31. Elution pattern of a peptic digest from dinitrophenylated NaBH$_4$-reduced phosphorylase b, before (A) and after (B) thiolysis with 2-mercaptoethanol. The enzyme was dinitrophenylated with one mole of [^{14}C] FDNB per mole of enzyme protomer (pH 7.5, 22°, 60 minutes) and 0.43 moles of the label became covalently attached to the protein with loss of ca 5% of its catalytic activity. Thiolysis was performed with 2-mercaptoethanol (final concentration: 0.05 M) at pH 7.5 and 22° (3 hours). Peptic digestion and gel filtration on Sephadex - as described in the legend to Fig. 28. (————), optical density and (– – – –), radioactivity

that it will be possible to obtain larger peptides. It was found that if the peptic digestion is shortened to 4 hours, there is some increase in the yield of the tetrapeptide Asx-Ala-Cys-Asp, indicating that some of the peptide Ala-Cys does originate from further digestion of this tetrapeptide. However, if the peptic digestion was limited to one hour only - it was possible to isolate a tripeptide Ala-Cys-Ala, which could have originated only from the sequence Ala-Cys-Ala-Phe. It appears, therefore, that there are two cysteine residues which are made available for preferential labeling upon removal of PLP from phosphorylase. These cysteines might therefore originate from the PLP binding site.

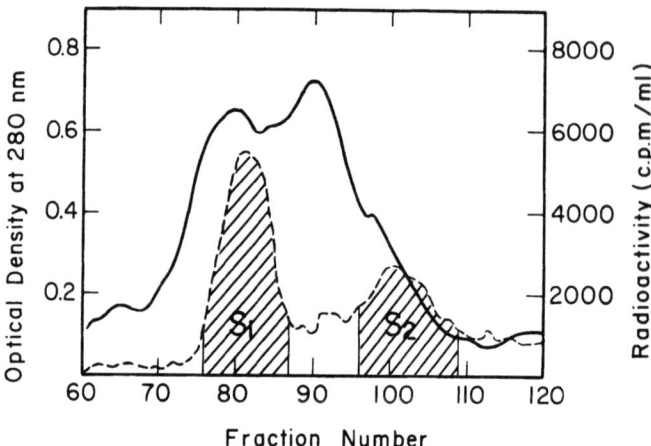

Fig. 32. Dinitrophenylation of $NaBH_4$-reduced phosphorylase b in the presence of a deforming agent (0.1 M citrate, pH 7). Experimental details as described in the legend to Fig. 31. The extent of labeling in this case was 0.58 moles of DNP groups per mole of enzyme protomer. (———), optical density and (– – – –), radioactivity

When we first suggested the existence of a cysteine at the PLP site of phosphorylase (53), we emphasized that this cysteine might originate either from the PLP site itself or from a site which becomes exposed upon removal of PLP from the enzyme. In dealing with an enzyme like phosphorylase, which is composed of protomers each of which contains several sites strongly interacting with one another, the assignment of a functional group to a specific site cannot be unambiguous, at least on the basis of chemical evidence alone. Figure 33 illustrates four possible locations for a cysteine residue, which could give rise to the results described here.

The possibility that the sulfhydryls are buried in the contact region between the two protomers (No. 2, Fig. 33) arises from the fact that upon removal of PLP from phosphorylase it becomes an associating-dissociating system (17), a process which could expose the contact region between the protomers as a result of the formation of monomers. This possibility does not seem likely, since dinitrophenylation of the apoenzyme under the conditions described here does not affect much the aggregation state of the apoenzyme (Fig. 24). As far as the possibility No. 3 is concerned, the evidence presented here regarding the mode of binding of PLP to phosphorylase suggests that the cofactor is bound covalently to the enzyme only through the ε-amino group of a lysine, so that the involvement of a sulfhydryl in covalent binding of PLP seems remote. As for the possibilities No. 1 and 4 - our evidence at this stage does not make it possible to discriminate between them.

LOCATION OF LABELED SULFHYDRYLS (POSSIBILITIES).

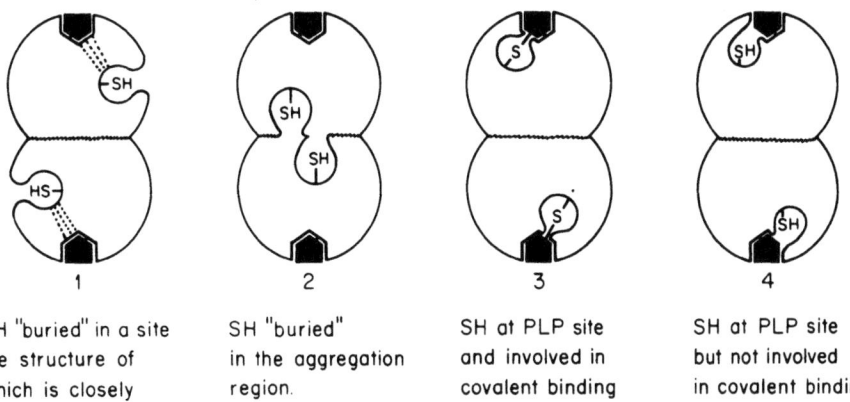

Fig. 33. Darkened areas indicate the location of PLP

The reactivity of sulfhydryl groups in phosphorylase b (the holoenzyme) and the consequences of their labeling on the activity of the enzyme have been studied by several investigators (63, 65-70). It was shown, for example, that alkylation of the peptides Ala-Cys-Ala-Phe and Asn-Ala-Cys-Phe results in dissociation of the protein and loss of catalytic activity (70). These two peptides are preferentially labeled upon dinitrophenylation of the apoenzyme (53, 54) and their labeling prevents the re-entry of PLP to its native site. In a previous communication we have reported (53) that the apoenzyme could be protected from the loss of its potential catalytic activity not only by PLP but also by AMP, which brings about aggregation of the apoenzyme (17). The question is whether these two peptides are associated primarily with the aggregation state of the protein or with the PLP binding site. Now, in a study concerned with the removal of PLP from phosphorylase, we have shown (15, 21) that in general, compounds and reaction conditions which promote dissociation of the enzyme also expose the PLP site. Since it is currently assumed (71) that upon aggregation the structure of the protomers within the aggregated state becomes constrained, it is possible that the PLP site becomes exposed upon dissociation of phosphorylase as a result of the loosening in the structure of the monomers.

The results summarized above imply that the two cysteine residues in the peptides Ala-Cys-Ala-Phe and Asn-Ala-Cys-Asp may be part of the PLP site. We have therefore attempted to find out whether the sulfhydryl groups of these two cysteines may be vicinal in the 3-dimensional structure of the protein. For this purpose we used the bifunctional reagent F_2DNB (72) which is very similar to FDNB in size, shape and reactivity (Fig. 34).

Upon anchoring of such a bifunctional reagent to a sulfhydryl group in a protein (73), the other reactive carbon atom may either react with water or become attached to one of the nucleophiles in the vicinity (Fig. 35). Preliminary results with this reagent (64) indicate that upon treatment of apophosphorylase with F_2DNB, a crosslink is established between the two cysteines mentioned above (Fig. 36).

On the basis of the above evidence, we would like to propose a plausible structure for the PLP site, both when buried and exposed. Figure 37 depicts these structures, illustrating the microenvironment and mode of binding of PLP in phosphorylase, and placing the sulfhydryl groups of the sequences Ala-Cys-Ala-Phe and Asn-Ala-Cys-Asp in the vicinity of the cofactor.

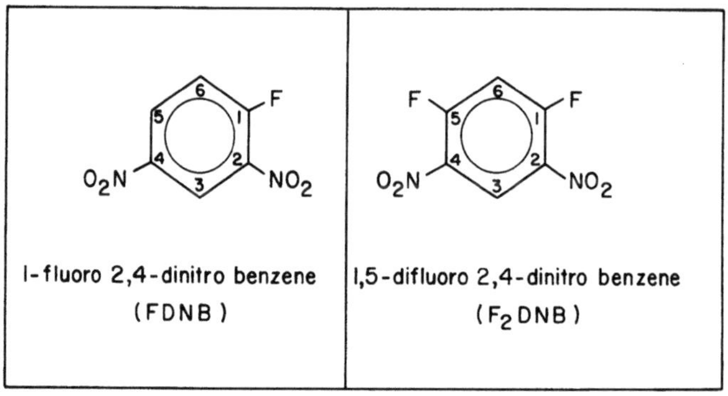

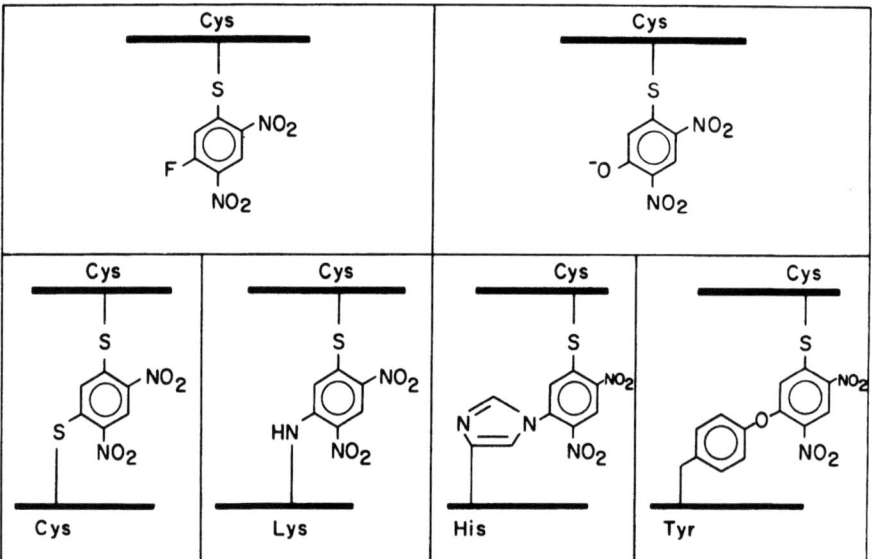

Fig. 34

Fig. 35. Possible reactions of the bifunctional reagent F_2DNB once anchored on a cysteine residue of a protein. The other reactive carbon atom (C-5) may either react with water or establish crosslinks with other amino acid residues

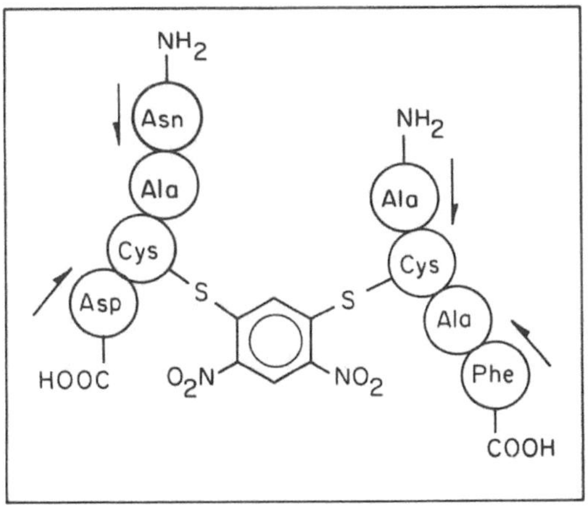

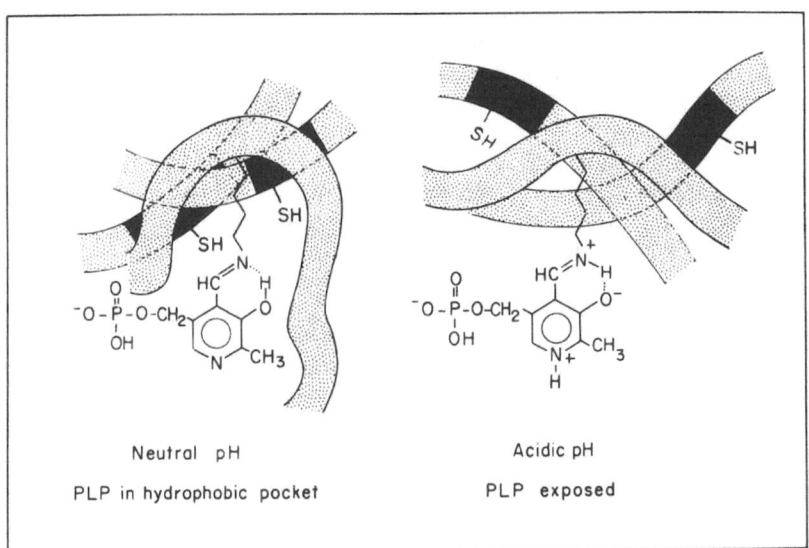

Fig. 36

Fig. 37

It should be kept in mind that at present it is difficult to discriminate between the possibility that these sulfhydryls are at the PLP site and the possibility that they are at another site which becomes exposed due to a conformational change resulting from the removal of PLP. Further studies will be needed to confirm or disprove these suggested structures

REFERENCES

1. GREEN, A.A. and CORI, G.T., J.Biol.Chem., 151, 21 (1943).
2. CORI, C.F. and CORI, G.T., J.Biol.Chem., 158, 341 (1945).
3. CORI, C.F. and CORI, G.T., Proc.Soc.Exp.Biol.Med., 34, 702 (1936).
4. CORI, G.T., COLOWICK, S.P. and CORI, C.F., J.Biol.Chem., 123, 375 (1938).
5. CORI, G.T., J.Biol.Chem., 158, 333 (1945).
6. SUTHERLAND, E.W., Ann.N.Y.Acad.Sci., 54, 693 (1951).
7. FISCHER, E.H. and KREBS, E.G., J.Biol.Chem., 216, 121 (1955).
8. KREBS, E.G., KENT, A.B. and FISCHER, E.H., J.Biol.Chem., 231, 73 (1958).
9. NOLAN, C., NOVOA, W.B., KREBS, E.G. and FISCHER, E.H., Biochemistry, 3, 542 (1964).
10. WANG, J.H. and GRAVES, D.J., Biochemistry, 3, 1437 (1964).
11. SEERY, V.L., FISCHER, E.H. and TELLER, D.C., Biochemistry, 6, 3315 (1967).
12. DE VINCENZI, D.L. and HEDRICK, J.L., Biochemistry, 6, 3489 (1967).
13. BARANOWSKI, T., ILLINGWORTH, B., BROWN, D.H. and CORI, C.F., Biochim. Biophys.Acta, 25, 16 (1957).
14. ILLINGWORTH, B., JANSZ, H.S., BROWN, D.H. and CORI, C.F., Proc.Natl.Acad. Sci. U.S., 44, 1180 (1958).
15. SHALTIEL, S., HEDRICK, J.L. and FISCHER, E.H., Biochemistry, 5, 2108 (1966).
16. FISCHER, E.H., POCKER, A. and SAARI, J.C., Essays in Biochemistry, 6, 23 (1970).
17. HEDRICK, J.L., SHALTIEL, S. and FISCHER, E.H., Biochemistry, 5. 2117 (1966).
18. SHALTIEL, S., Israel J.Chem. 6, 104p (1968).
19. SHALTIEL, S., HEDRICK, J.L., POCKER, A. and FISCHER, E.H., Biochemistry, 8, 5189 (1969).
20. EHRLICH, J., FELDMANN, K., HELMREICH, E. and PFEUFFER, T., in preparation.
21. HEDRICK, J.L., SHALTIEL, S. and FISCHER, E.H., Biochemistry, 8, 2422 (1969).
22. SHALTIEL, S., HEDRICK, J.L. and FISCHER, E.H., Biochemistry, 8, 2429 (1969).
23. KENT, A.B., KREBS, E.G. and FISCHER, E.H., J.Biol.Chem., 232, 549 (1958).
24. SHALTIEL, S. and FISCHER, E.H., Israel J.Chem., 5. 127p (1967).
25. CORTIJO, M., STEINBERG, I.Z. and SHALTIEL, S., J.Biol.Chem., 246, 933 (1971).
26. FÖRSTER, T., Faraday Soc.Discuss., 27, 7 (1959).
27. LIVANOVA, N.B., MOROZKIN, A.D., PIKHELGAS, V.Y. and SHPIKITER, V.O., Biokhimiya, 31, 194 (1966).
28. HEDRICK, J.L., SMITH, A.J. and BRUENING, G.E., Biochemistry, 8, 4012 (1969).
29. VAN DUUREN, B.L., Chem.Rev., 63, 325 (1963).
30. FISCHER, E.H., KENT, A.B., SNYDER, E.R. and KREBS, E.G., J.Am.Chem.Soc., 80, 2906 (1958).
31. SHALTIEL, S. and CORTIJO, M., Biochem.Biophys.Res.Commun., 41, 594 (1970).
32. HEYL, D., HARRIS, S.A. and FOLKERS, K., J.Am.Chem.Soc., 70, 3429 (1948).
33. BUEL, M.V. and HANSEN, R.E., J.Am.Chem.Soc., 82, 6042 (1960).
34. DEMPSEY, W.B. and CHRISTENSEN, H.N., J.Biol.Chem., 237, 1113 (1962).
35. SNELL, E.E. and DI MARI, S.J., The Enzymes (3rd ed.), 2, 335 (1970).
36. GARDINER, J.A. and COLLAT, J.W., J.Am.Chem.Soc., 87, 1692 (1965).
37. STRAUSBAUCH, P.H., KENT, A.B., HEDRICK, J.L. and FISCHER, E.H., Methods in Enzymol. 11, 671 (1967).
38. KENT, A.B., Ph.D. Dissertation, Univ. of Washington, Seattle, Wash. (1959).
39. CORTIJO, M. and SHALTIEL, S., Biochem.Biophys.Res.Commun., 39, 212 (1970).
40. CORTIJO, M. and SHALTIEL, S., in preparation.
41. FORREY, A.W., Ph.D. Dissertation, Univ. of Washington, Seattle, Wash. (1963).
42. CHEN, R.F., Science, 150, 1593 (1965).
43. HEINERT, D. and MARTELL, A.E., J.Am.Chem.Soc., 85, 183 (1963).
44. MATSUSHIMA, M. and MARTELL, A.E., J.Am.Chem.Soc., 89, 1322 (1967).
45. METZLER, D.E., J.Am.Chem.Soc., 79, 485 (1957).
46. CHRISTENSEN, H.N., J.Am.Chem.Soc., 80, 99 (1958).

47. HEINERT, D. and MARTELL, A. E., J. Am. Chem. Soc., 84, 3257 (1962).
48. DUDEK, G. O. and DUDEK, E. P., Chem. Commun., 464 (1965).
49. COHEN, M. D. and SCHMIDT, G. M. J., J. Phys. Chem., 66, 2442 (1962).
50. WELLER, A., Z. Elektrochem., 60, 1144 (1960).
51. WELLER, A., in "Progress in Reaction Kinetics" (G. Porter, ed.), Vol. 1, p. 189. Macmillan, New York (1961).
52. JOHNSON, G. F., TU, J. I., SHONKA-BARTLETT, M. L. and GRAVES, D. J., J. Biol. Chem., 245, 5560 (1970).
53. ZAIDENZAIG, Y. and SHALTIEL, S., Israel J. Chem., 7, 128p (1969).
54. SHALTIEL, S. and ZAIDENZAIG, Y., Biochem. Biophys. Res. Commun., 39, 1003 (1970).
55. SHALTIEL, S., Biochem. Biophys. Res. Commun., 29, 178 (1967).
56. SHALTIEL, S. and SORIA, M., Biochemistry, 8, 4411 (1969).
57. MADSEN, N. B. and CORI, C. F., J. Biol. Chem., 223, 1055 (1956).
58. MADSEN, N. B., J. Biol. Chem., 223, 1067 (1956).
59. MADSEN, N. B. and GURD, F. R. N., J. Biol. Chem., 223, 1075 (1956).
60. SHALTIEL, S., and TAUBER-FINKELSTEIN, M., FEBS Letters, 8, 345 (1970).
61. ZAIDENZAIG, Y. and SHALTIEL, S., Proc. Ann. Meeting of the Israeli Biochem. Soc. p. 28 (1971).
62. SHALTIEL, S. and ZAIDENZAIG, Y., Abs. of the 7th FEBS Meeting, Varna, No. 164 (1971).
63. ZARKADAS, C. G., SMILLIE, L. B. and MADSEN, N. B., J. Mol. Biol., 38, 245 (1968).
64. ZAIDENZAIG, Y. and SHALTIEL, S., in preparation.
65. DAMJANOVICH, S. and KLEPPE, K., Biochim. Biophys. Acta, 122, 145 (1966).
66. GOLD, A. M., Biochemistry, 7, 2106 (1968).
67. PHILIP, G. and GRAVES, D. J., Biochemistry, 7, 2093 (1968).
68. BATTELL, M. L., ZARKADAS, C. G., SMILLIE, L. B. and MADSEN, N. B., J. Biol. Chem., 243, 6202 (1968).
69. KASTENSCHMIDT, L. L., KASTENSCHMIDT, J. and HELMRICH, E., Biochemistry, 7, 3590 (1968).
70. AVRAMOVIC-ZIKIC, O., SMILLIE, L. B. and MADSEN, N. B., J. Biol. Chem., 245, 1558 (1971).
71. MONOD, J., WYMAN, J. and CHANGEUX, J. P., J. Mol. Biol., 12, 88 (1965).
72. ZAHN, H. and STUERLE, H., Biochem. Z., 331, 29 (1958).
73. SHALTIEL, S. and TAUBER-FINKELSTEIN, M., Biochem. Biophys. Res. Commun., 44, 484 (1971).

Association-dissociation Properties of $NaBH_4$-reduced Phosphorylase b

D. J. Graves, Jan-I. Tu, R. A. Anderson, T. M. Martensen, and B. J. White

Department of Biochemistry and Biophysics,
Iowa State University of Science and Technology, Ames, Iowa/USA

Molecular weight studies show that rabbit muscle glycogen phosphorylase is composed of monomeric units with a weight of $9.25 - 10 \times 10^4$ daltons (1,2). The enzyme exists as dimers and tetramers, and it has been shown that both phosphorylase b and a undergo dissociation-association reactions which affect enzymic activity (3,4). The formation of the monomer was first demonstrated by Madsen and Cori (5) by reaction of phosphorylase with PMB. Other methods have been used but all of these are rather stringent, and it has not been established whether the monomeric form has catalytic activity. Recent studies with $NaBH_4$-reduced phosphorylase showed that monomer formation can occur upon enzyme dilution (6). This fact and other studies of its subunit structure (7) led us to study the association-dissociation properties of $NaBH_4$-reduced phosphorylase* to determine whether the monomeric form of this enzyme is catalytically active.

Ultracentrifugal studies showed that $NaBH_4$-reduced phosphorylase b is sensitive to low concentrations of formamide whereas native phosphorylase b is not. This is illustrated in Fig. 1 (insert); the reduced enzyme after one hour of incubation with formamide has an $S_{20,w}$ of 5.6 S and native enzyme has an $S_{20,w}$ of 8.4 S. These constants correspond to values previously determined for monomeric and dimeric forms of the enzyme and thus suggest that dissociation of the reduced enzyme has occurred. The effect on enzyme activity is also seen in Fig. 1. $NaBH_4$-reduced enzyme upon incubation with 6.7% formamide shows a time dependent change of enzymic activity, and after several hours, it has approximately 30% of the activity of enzyme not incubated with formamide. Native phosphorylase b does not show a time dependent inactivation, but it is 70% less active when measured in the presence of this concentration of formamide.

---

* This is a form of enzyme which results from $NaBH_4$ reduction of an imine between pyridoxal phosphate and the ε-amino group of a protein lysyl residue.

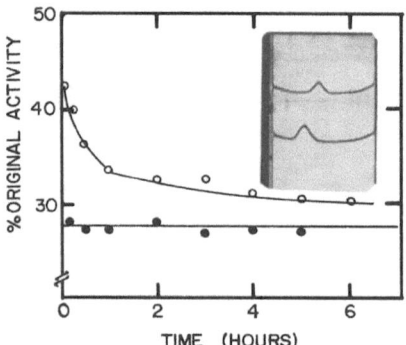

Fig. 1. Effect of formamide on activity and ultracentrifugal properties of native and reduced phosphorylase b. Native phosphorylase b (●) and NaBH$_4$-reduced phosphorylase b (O) were incubated in 80 mM glycerol-P—60 mM β-mercaptoethanol and 6.7% formamide (pH 6.2) at 24°. At various times, aliquots were removed and tested for enzymic activity with 0.4% glycogen, 30 mM glucose-1-P, 0.1 mM AMP, and 6.7% formamide at pH 6.2. Activity of phosphorylase in the absence of formamide is taken as 100%. For native and NaBH$_4$-reduced phosphorylase, specific activities are 9.8 and 5.3 μmoles/min/mg, respectively. For ultracentrifugation, enzymes were incubated in the above buffer for 1½ hours. Upper curve, native phosphorylase b (3.4 mg/ml); lower curve, NaBH$_4$-reduced phosphorylase b (4 mg/ml). Pictures were taken at 42 minutes after attainment of full speed at 60,000 rpm. Direction of the sedimentation was from left to right

The possible relationship between change of enzyme activity and alteration of subunit structure was further examined by testing the effects of different concentrations of formamide on NaBH$_4$-reduced phosphorylase. Figure 2 illustrates the time courses of inactivation and shows that the extent of change of enzymic activity depends upon formamide concentration.

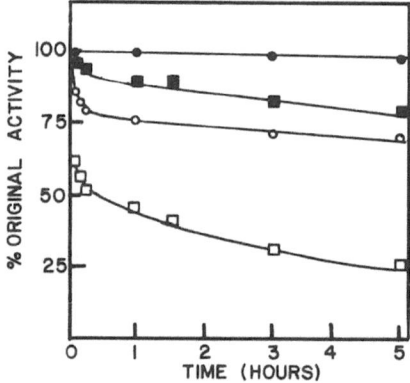

Fig. 2. Rate of inactivation of NaBH$_4$-reduced phosphorylase b with different concentrations of formamide. NaBH$_4$-reduced phosphorylase b was preincubated at 24°C in 80 mM glycerol-P—60 mM β-mercaptoethanol (pH 6.2) and (●) 0, (■) 2.5%, (O) 5% or (□) 6.7% formamide. Assay was done as in Fig. 1 with different concentrations of formamide

Ultracentrifugal measurements (Fig. 3) likewise show that change in the sedimentation pattern is a function of concentration. In the absence of formamide, reduced phosphorylase b at pH 6.2 (picture A) sediments mainly as a dimer ($S_{20,w}$ = 8.4 S). A small amount of a slower sedimenting component is also present in (A) indicating that dissociation of $NaBH_4$-reduced phosphorylase is occurring to some extent. In the presence of 2.5% formamide (B), two distinct species are present with $S_{20,w}$ of 5.6 S and an 8.4 S. At 5% formamide, one broad peak is seen with an $S_{20,w}$ of 6.55. At 6.7% formamide (C) only the 5.6 S species is present.

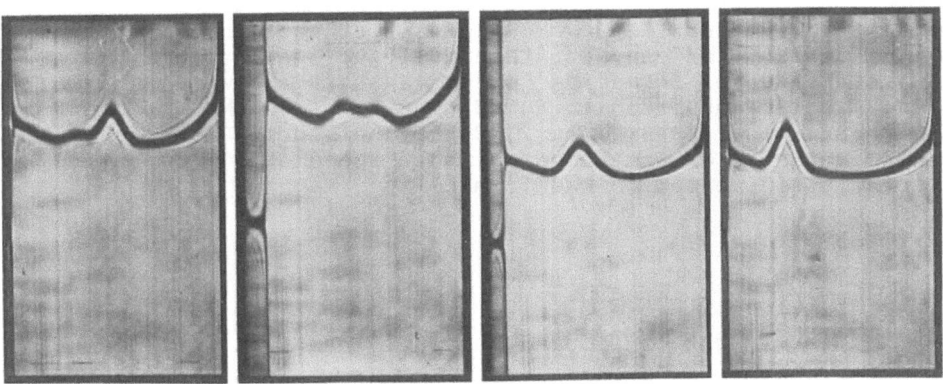

Fig. 3. Dependence of ultracentrifugal patterns of $NaBH_4$-reduced phosphorylase b upon formamide concentration. Enzyme was preincubated at 24°C for three hours in 80 mM glycerol-P—60 mM β-mercaptoethanol with variable concentrations of formamide at pH 6.2 (A) 0, (B) 2.5%, (C) 5% and (D) 6.7% formamide. Pictures were taken at 40 minutes after attainment of full speed at 60,000 rpm. Direction of the sedimentation was from left to right

These data (Fig. 2 and 3), therefore, indicate that the extent to which activity is altered is related to the degree to which the subunit structure of $NaBH_4$-reduced phosphorylase is modified.

To test whether the effect of formamide on reduced phosphorylase b is reversible, enzyme was first incubated for one hour with 6.7% formamide and then dialyzed to remove it. Figure 4 shows schlieren patterns of enzyme after one hour of incubation with 6.7% formamide (lower curve) and that obtained after dialysis (upper curve). The $S_{20,w}$ of these forms, respectively, are 5.6 S and 8.4 S showing that the effect of formamide on the subunit assembly was reversible. Before dialysis, the enzyme in formamide had 30% of the activity of enzyme measured in the absence of formamide. After its removal by dialysis, activity was restored completely to the control value.

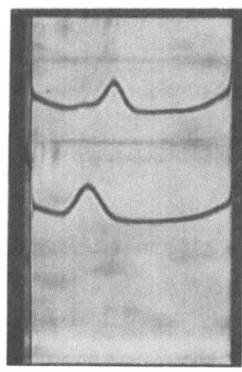

Fig. 4. Effect of formamide on the ultracentrifugal patterns of NaBH$_4$-reduced phosphorylase b. Upper curve: enzyme (4 mg/ml) was incubated in 80 mM glycero-P—60 mM β-mercaptoethanol and 6.7% formamide (pH 6.2) at 23°C for one hour then dialyzed against same buffer except in the absence of formamide for three hours. Lower curve: enzyme (4 mg/ml) was incubated in 80 mM glycero-P—60 mM β-mercaptoethanol 6.7% formamide (pH 6.2) at 23°C for one hour. Pictures were taken at 42 minutes after attainment of full speed at 60,000 rpm. Direction of the sedimentation was from left to right

The molecular weight of reduced phosphorylase b in formamide was determined by the sedimentation equilibrium method of Van Holde. Figure 5 shows that a constant slope is found for the data indicating no obvious polydispersity is present in the sample. The molecular weight was calculated to be 103,000 ± 2,200 showing that enzyme dissociation of reduced phosphorylase b to a monomer is complete in 6.7% formamide.

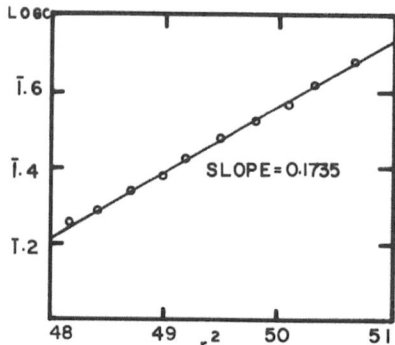

Fig. 5. Sedimentation equilibrium of NaBH$_4$-reduced phosphorylase b in formamide. Enzyme (2 mg/ml) was preincubated at 17°C for four hours in 80 mM glycerol-P—1 mM β-mercaptoethanol and 6.7% formamide at pH 6.2. Enzyme was diluted to 0.51 mg/ml with same buffer just before the ultracentrifugation. The experiment was carried out at 11,000 rpm for 20 hours

The fact that reduced phosphorylase b is fully dissociated in 6.7% formamide and possesses 30% of enzymic activity in the test system described in Fig. 1 can be interpreted to mean 1) the monomeric form of the enzyme is active or 2) it is inactive but reassociates to some degree in the assay to give an active dimer. In order to determine whether substrates and the activator, AMP, do cause association of reduced phosphorylase b, their effects were tested by ultracentrifugation. Glycogen could not be used in these experiments because it would bind to phosphorylase and cause rapid sedimentation due to the high molecular weight of glycogen. Amylodextrin is not a good primer because of its lower molecular weight and structure, but it was useful for these experiments because it allowed a direct comparison to be made between ultracentrifugal experiments and enzyme assays. Figure 6 shows schlieren patterns obtained with substrates and activator. Part A shows patterns obtained in the presence of 30 mM glucose-1-P and 1 mM AMP. The sedimentation constants ($S_{20,w}$) were 5.4 S and 5.25 S, respectively, in their absence and in their presence, showing that at these concentrations no association of reduced phosphorylase b occurs. In part B, the combined effects of amylodextrin, AMP, and glucose-1-P were tested. The sedimentation constant of the main peak has an $S_{20,w}$ of 5.35 S. The light component seen in the

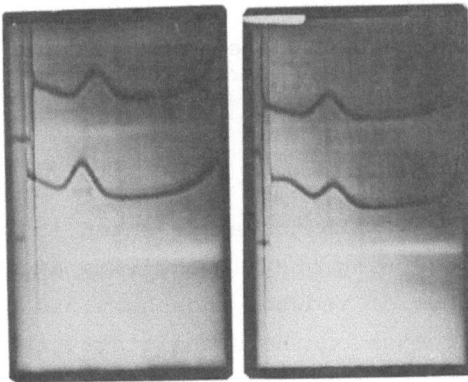

Fig. 6. Effect of formamide on ultracentrifugal patterns of $NaBH_4$-reduced phosphorylase b. Enzyme (3.5 mg/ml) was preincubated in 80 mM glycerol-P—60 mM β-mercaptoethanol and 6.7% formamide (pH 6.2) at 24°C for three hours. (A) Upper curve; no addition of substrate and activator; lower curve: 30 mM glucose-1-P and 1 mM AMP. (B) Upper curve: no addition; lower curve: 0.4% amylodextrin (a α 1 → 4 linkage polysaccharide of 40 glucosyl units with 3% branch chains), 30 mM glucose-1-P and 1 mM AMP. Activator and substrates were added in (A) and (B) just before centrifugation. Pictures were taken at 52 minutes after attainment of full speed at 60,000 rpm. Direction of the sedimentation was from left to right

picture is due to free amylodextrin. Thus, it appears that little or no reassociation of the monomeric form of reduced phosphorylase b is occurring under these experimental conditions.

With these concentrations of substrates and activator (part B), enzymic activities were measured at 24°C with 3.6 mg/ml of reduced phosphorylase b. Before incubation, a specific activity of 2.6 μmoles/min/mg was obtained and after exposure to 6.7% formamide, specific activity dropped approximately 83% after one hour and was essentially constant for several hours (Fig. 7).

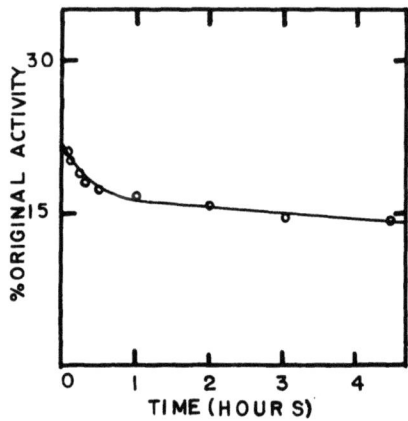

Fig. 7. Effect of formamide on enzymic activity of $NaBH_4$-reduced phosphorylase b. Enzyme was incubated in 80 mM glycero-1-P—60 mM β-mercaptoethanol and 6.7% formamide (pH 6.2). At various times, aliquots were removed and tested for enzymic activity with 0.4% amylodextrin, 30 mM glucose-1-P, 1 mM AMP and 6.7% formamide

These results taken along with those presented in Fig. 6 indicate that the form of enzyme possessing enzymic activity after incubation with formamide is the monomer of reduced phosphorylase b.

The effect of glycogen on the subunit structure of $NaBH_4$-reduced phosphorylase cannot be directly tested but certain inferences can be drawn from its effects on enzymic activity. If glycogen promotes the association of the monomeric form of phosphorylase b to a dimer, it might be expected that glycogen would influence the rate and/or extent of inactivation of reduced phosphorylase in formamide. For Fig. 8, enzyme was incubated in 6.7% formamide in the absence and presence of 0.4% glycogen. At various intervals, aliquots were removed and tested for enzymic activity. No difference in the rate or the extent of activity loss is apparent suggesting that glycogen does not affect the dimer-monomer equilibrium.

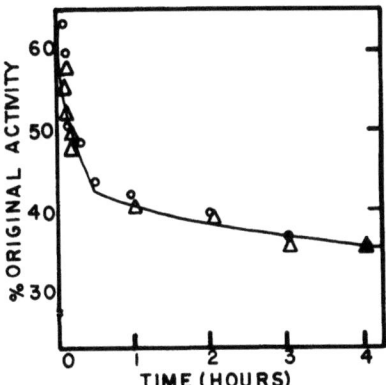

Fig. 8. Effect of glycogen on the change of enzymic activity of NaBH$_4$-reduced phosphorylase b in formamide. Enzyme (4.8 mg/ml) was incubated at 24°C for one hour in the presence (O) and absence (Δ) of 0.4% glycogen. At this time, formamide was added to 6.7%, and aliquots were removed at times indicated and tested for enzymic activity with 0.4% glycogen, 30 mM glucose-1-P, and 1 mM AMP. Activity of both samples in the absence of formamide is taken as 100%. Specific activity equals 11.1 μmoles/min/mg

Figure 9 shows the specific activity of enzyme in formamide as a function of enzyme concentration. The plot shows that specific activity is independent of protein concentration. If a nonlinear plot had been obtained, this would indicate the presence of two enzyme forms which differ in catalytic activity. The result, shown in the figure could also be interpreted to mean that all the enzyme is present as a dimer or as a monomer.

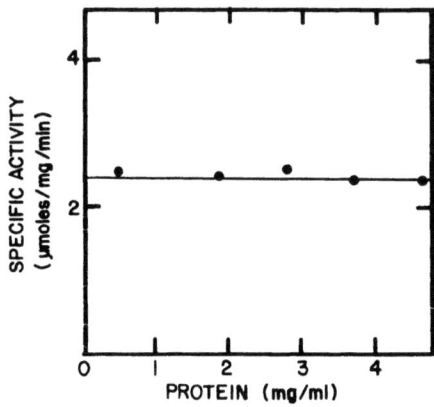

Fig. 9. Effect of protein concentration on specific activity of NaBH$_4$-reduced phosphorylase b in formamide. Enzyme was preincubated in 80 mM glycerol-P—60 mM β-mercaptoethanol and 6.7% formamide (pH 6.2) at 24°C for one hour. Aliquots were removed and tested for enzymic activity in 0.4% glycogen, 30 mM glucose-1-P, 0.1 mM AMP and 6.7% formamide. Reactions were terminated by addition of 0.18 ml perchloric acid (4 M) and then neutralized with potassium carbonate, centrifuged, and the Pi produced was determined colorimetrically

Kinetics were done with respect to AMP, as it is known that cooperativity seen with respect to AMP is due to site-site interactions present in the dimer. Even if the enzyme was present as a dimer, these site-site interactions could be changed by formamide to the extent that no interaction could occur. Thus, if the kinetics were of the Michaelis-Menten type, two explanations are possible. If cooperative, this would certainly indicate the presence of a normal type dimer. The data illustrated in Fig. 10 show that cooperativity is abolished after $NaBH_4$-reduced phosphorylase b is incubated with formamide.

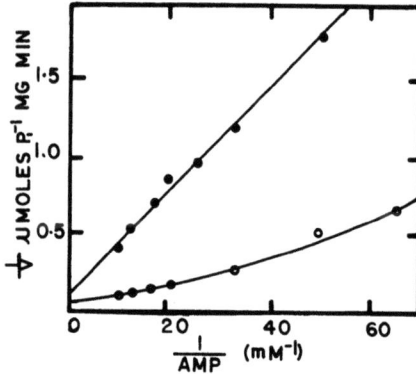

Fig. 10. Effect of formamide on the reciprocal plot for AMP activation of $NaBH_4$-reduced phosphorylase b. Enzyme (4 mg/ml) was preincubated at 24°C for one hour in 80 mM glycerol-P—60 mM β-mercaptoethanol (pH 6.2) in the presence and absence of 6.7% formamide. Reactions were initiated by addition of enzyme to 0.4% glycogen, 30 mM glucose-1-P and variable concentrations of AMP. Closed circles (●) assay in the presence of 6.7% formamide; open circles (O) no additions

Kinetics were also done with respect to AMP with native phosphorylase b as its activity is changed by formamide but no dissociation occurs. These results (Fig. 11) show that homotropic interactions still occur with respect to AMP and do not seem to be changed by this concentration of formamide.

As formamide does not desensitize the homotropic interactions of native phosphorylase b for AMP and the Michaelis-Menten kinetics seen with respect to AMP in Fig. 10 further indicate the presence of an active monomer.

In conclusion, $NaBH_4$-reduced phosphorylase b is easily and reversibly dissociated to a monomeric form in low concentrations of formamide. The extent of activity change appears to be related to change in the subunit structure of the enzyme. Comparison of ultracentrifugal experiments with enzymic assays under identical conditions provides strong support for the hypothesis of an active monomer. Experiments done with glycogen as a primer are also indicative that the enzyme form active after incubation with formamide is a monomer.

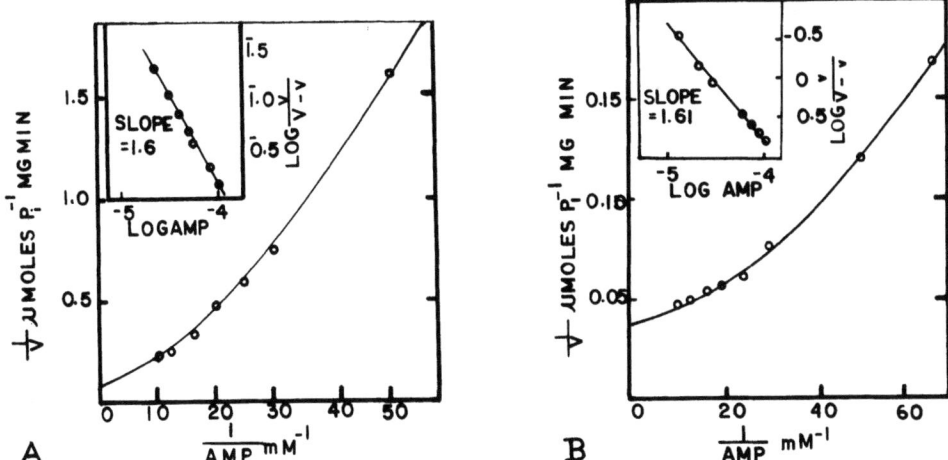

Fig. 11. Effect of formamide on the reciprocal plots for AMP activation of native phosphorylase b. Enzyme (4 mg/ml) was incubated in 80 mM glycerol-P—60 mM β-mercaptoethanol (pH 6.2) in the presence and absence of 6.7% formamide. Reactions were initiated by addition of enzyme to 0.4% glycogen, 30 mM glucose-1-P and variable concentration of AMP. (A) Assay in the presence of 6.7% formamide; (B) no additions

The biological significance of a functional monomer can be examined. It is not expected that conditions reported here for dissociation are related to a physiological process, but with these conditions, kinetic studies can be made for example with phosphorylase b kinase. From a comparison of results obtained with monomeric, dimeric, and tetrameric forms, it should be possible to establish the biological importance of the subunit structure of phosphorylase in phosphorylase activation.

References

1. SEERY, V.L., FISCHER, E.H. and TELLER, D.C., *Biochemistry*, **9**, 3591 (1970).
2. COHEN, P., DUEWER, T. and FISCHER, E.H., *Biochemistry*, **10**, 2683 (1971).
3. WANG, J.H., SHONKA, M.L. and GRAVES, D.J., *Biochemistry*, **4**, 2296 (1965).
4. WANG, J.H., KWOK, S.C., WIRCH, E. and SUZUKI, I., *Biochem. Biophys. Res. Commun.*, **40**, 1340 (1970).
5. MADSEN, N.B. and CORI, C.F., *J. Biol. Chem.*, **233**, 1055 (1956).
6. DeVINCENZI, D.L. and HEDRICK, J.L., *Biochemistry*, **9**, 2048 (1970).
7. JOHNSON, G.F., TU, J.-I, BARTLETT, M.L.S. and GRAVES, D.J., *J. Biol. Chem.*, **245**, 5560 (1970).

The Mechanism of Action of Cyclic AMP in the Activation of Phosphorylase Kinase[1]

E. G. Krebs, J. A. Beavo, C. O. Brostrom[2], J. D. Corbin[3], T. Hayakawa[4], and D. A. Walsh

Department of Biological Chemistry, School of Medicine, University of California, Davis, California/USA

Rabbit skeletal muscle phosphorylase kinase exists in two kinetically different forms (1). One of these forms, referred to as nonactivated phosphorylase kinase, has very little activity at pH 6.8 and is only partially active at higher pH values. The other form of the enzyme, activated phosphorylase kinase, is highly active at pH 6.8 and at pH 8.2. Conversion of nonactivated phosphorylase kinase to the activated form can be brought about in vitro by limited proteolysis (2, 3) or as a result of phosphorylation by ATP. It is doubtful whether the first type of activation is of physiological significance, but it seems very likely that activation by ATP does constitute a meaningful regulatory device. Phosphorylation and activation of the muscle kinase is accelerated by cyclic AMP (1, 4, 5); moreover, it has been possible to correlate the activation of the enzyme with changes in the cyclic AMP concentration produced by epinephrine administration in vivo (6). A further indication that phosphorylation and activation of phosphorylase kinase actually occurs in muscle is the fact that this tissue contains a phosphatase capable of reversing the process (7). Many of the properties noted for the skeletal muscle enzyme have also been observed with heart muscle phosphorylase kinase (8-10).

Nonactivated and activated phosphorylase kinase both exhibit an absolute requirement for Ca^{2+} (2, 11, 12), and it can be demonstrated that this metal binds to the purified protein (12). There is excellent evidence from a number of laboratories that the reversible binding of

---

[1] Supported by Grant AM 12842 from the U.S. Public Health Service and a Grant-in-Aid from the American Heart Association.
[2] Present address: Department of Biochemistry, Rutgers Medical School, New Brunswick, New Jersey 08903.
[3] Present address: Department of Physiology, School of Medicine, Vanderbilt University, Nashville, Tennessee 37203
[4] Present address: Department of Physiological Chemistry, Roche Institute of Molecular Biology, Nutley, N. J. 07110.

Ca^{2+} to the kinase constitutes an important regulatory device in addition to regulation brought about through activation by phosphorylation (2, 11-16). Phosphorylase kinase also binds to glycogen (5) and this latter substance is believed to be an important factor in regulating the activity of the enzyme (5, 17).

Physicochemical Properties of Rabbit Skeletal Muscle Phosphorylase Kinase — Phosphorylase kinase from muscle is a large molecule having a molecular weight of 1.33×10^6. It is made up of three types of subunits, A, B and C, with molecular weights of 118,000, 108,000 and 41,000 respectively (18). From an estimate of the relative amounts of each type of subunit made by scanning electrophoresis gels, a formula of $A_4B_4C_8$ can be ascribed to the enzyme. When phosphorylase kinase is activated by ATP, the B subunit is the first one to be phosphorylated. This is followed by phosphorylation of the A subunit after a brief lag period. The C subunit does not appear to be phosphorylated (19).

Catalysis of the Phosphorylase Kinase Activation Reaction —

The phosphorylation and activation of skeletal muscle phosphorylase kinase is catalyzed in vitro by two different enzymes. One of these is phosphorylase kinase itself (5) and the other is a cyclic AMP-dependent phosphorylase kinase kinase (20, 21). The latter enzyme is generally referred to as a cyclic AMP-dependent protein kinase (20) since its specificity is broader than would be implied by calling it a kinase kinase. The two activation reactions are shown in Eq. 1 and 2:

$$\text{Nonactivated Phosphorylase Kinase} \xrightarrow[\text{ATP, Mg}^{2+}, \text{Ca}^{2+}]{\text{Activated Phosphorylase Kinase}} \text{Activated (phosphorylated) Phosphorylase Kinase} \quad (1)$$

$$\text{Nonactivated Phosphorylase Kinase} \xrightarrow[\text{ATP, Mg}^{2+}, \text{cyclic AMP}]{\text{Protein Kinase}} \text{Activated (phosphorylated) Phosphorylase Kinase} \quad (2)$$

The autocatalytic component of the activation reaction (Eq. 1) is inhibited by chelating agents that bind Ca^{2+}. The reaction of Equation 2 is inhibited by a heat-stable protein of unknown physiological significance which has been isolated from skeletal muscle (22). Highly purified phosphorylase kinase from rabbit muscle still contains traces of the cyclic AMP-dependent protein kinase.

Functions of Cyclic AMP-Dependent Protein Kinase — In addition to catalyzing the phosphorylation and activation of phosphorylase kinase, cyclic AMP-dependent protein kinases are now known to be involved in the inactivation of glycogen synthetase (23, 24) and in the activation of the hormone-sensitive lipase of fat cells (25, 26). Protein kinases are widely distributed in the biological kingdom (27, 28), and it has been postulated that all actions of cyclic AMP may be mediated by protein phosphorylation. Whether or not this viewpoint is too sweeping in scope remains to be determined, but it does seem probable that a number of manifestations of the cyclic nucleotide in animals may be due to the action of protein kinases.

The Mechanism of Action of Cyclic AMP at the Molecular Level — Cyclic AMP-dependent protein kinases contain two types of subunits. One of these subunits, designated as R, is a regulatory subunit which binds cyclic AMP. The other subunit, designated as C, is a catalytic subunit. In the presence of cyclic AMP, the enzymatically inert RC complex dissociates to yield the active catalytic subunit according to Eq. 3:

$$RC + \text{cyclic AMP} \rightleftharpoons R \cdot \text{cyclic AMP} + C \qquad (3)$$

Evidence for the mechanism shown in Eq. 3 has been obtained in several different laboratories (29-34). It is possible, of course, that a ternary complex consisting of RC plus cyclic AMP may also be active, but the existence of such a complex has not been detected with the rabbit muscle enzyme. Majumder and Turkington have studied a cyclic AMP-dependent protein kinase from mammary gland which does not appear to dissociate in the presence of the cyclic nucleotide (35).

Factors other than cyclic AMP may conceivably be of importance with reference to the dissociation and activation of the protein kinase. Reimann et al. (36) observed that preincubation of this enzyme with the histone substrate prior to starting the reaction by addition of Mg^{2+}-ATP resulted in a loss of the cyclic AMP requirement. That this behavior was probably due to dissociation of the RC complex by its interaction with the basic protein was reported recently by Miyamoto et al. (37). Similar observations have been made independently in this laboratory.

Unsolved Problems of Current Interest — A number of problems relating to the role and functions of cyclic AMP-dependent protein kinases in cellular regulation are of current research interest. One of these is the question of specificity. What makes one protein able to serve as a substrate for a protein kinase while another does not? What determines that a specific seryl residue in protein may be phosphorylatable? What is the significance of isozyme forms of the cyclic AMP-dependent protein kinase within a given tissue (36, 38)? Is there an excess of R over C in various tissues? Does R regulate any enzymes other than the protein kinase? Finally, in phosphorylation-dephosphorylation sequences is active regulation confined to the protein kinase reactions or are phosphoprotein phosphatases also subject to active control?

References

1. Krebs, E. G., Graves, D. J., and Fischer, E. H., J. Biol. Chem., 234, 2867 (1959).
2. Meyer, W. L., Fischer, E. H., and Krebs, E. G., Biochemistry, 3, 1033 (1964).
3. Huston, R. B. and Krebs, E. G., Biochemistry, 7, 2116 (1968).
4. Krebs, E. G., Love, D. S., Bratvold, G. E., Trayser, K. A., Meyer, W. L., and Fischer, E. H., Biochemistry, 3, 1022 (1964).
5. DeLange, R. J., Kemp, R. G., Riley, W. D., Cooper, R. A., and Krebs, E. G., J. Biol. Chem., 243, 2200 (1968).
6. Posner, J. B., Stern, R., and Krebs, E. G., J. Biol. Chem., 240, 982 (1965).
7. Riley, W. D., DeLange, R. J., Bratvold, G. E., and Krebs, E. G., J. Biol. Chem. 243, 2209 (1968).
8. Hammermeister, K. E., Yunis, A. A., and Krebs, E. G., J. Biol. Chem., 240, 986 (1965).
9. Drummond, G. I., and Duncan, L., J. Biol. Chem., 241, 3097 (1966).
10. Drummond, G. I., Duncan, L., and Friesen, A. J. D., J. Biol. Chem., 240, 2778 (1965).
11. Ozawa, E., Hosoi, K., and Ebashi, S., J. Biochem., 61, 531 (1967).
12. Brostrom, C. O., Hunkeler, F. L., and Krebs, E. G., J. Biol. Chem., 246, 1961 (1971).
13. Heilmeyer, L. M. G., Meyer, F., Haschke, R. H., and Fischer, E. H., J. Biol. Chem. 245, 6649 (1970).
14. Villar-Palasi, C., and Wei, S. I., Proc. Nat. Acad. Sci.U.S., 67, 345, (1970).
15. Drummond, G. I., Harwood, J. P., and Powell, C. A., J. Biol. Chem., 244, 4235 (1969).
16. Mayer, S. E., Namm, D. H., and Hickenbottom, J. P., Adv. in Enz. Reg., 8, 205 (1970).
17. Meyer, F., Heilmeyer, L. M. G., Haschke, R. H., and Fischer, E. H., J. Biol. Chem., 245, 6642 (1970).
18. Hayakawa, T., Perkins, J. P., Walsh, D. A., and Krebs, E. G., Unpublished results.
19. Hayakawa, T., Perkins, J. P., and Krebs, E. G., Unpublished results.
20. Walsh, D. A., Perkins, J. P., and Krebs, E. G., J. Biol. Chem., 243, 3763 (1968).
21. Walsh, D. A., Perkins, J. P., Brostrom, C. O., Ho, E. S., and Krebs, E. G., J. Biol. Chem., 246, 1968 (1971).
22. Walsh, D. A., Ashby, C. D., Gonzales, C., Calkins, D., Fischer, E. H., and Krebs, E. G., J. Biol. Chem., 246, 1977 (1971).

23. Schlender, K. K., Wei, S. H., and Villar-Palasi, C., Biochim. Biophys. Acta, 191, 272 (1969).
24. Soderling, T. R., Hickenbottom, J. P., Reimann, E. M., Hunkeler, F. L., Walsh, D. A., and Krebs, E. G., J. Biol. Chem., 245, 6317 (1970).
25. Corbin, J. D., Reimann, E. M., Walsh, D. A., and Krebs, E. G., J. Biol. Chem., 245, 4849 (1970).
26. Huttunen, J. K., Steinberg, D., and Mayer, S. E., Proc. Nat. Acad. Sci. U.S., 67, 290 (1970).
27. Kuo, J. F., and Greengard, P., Proc. Nat. Acad. Sci. U.S., 64, 1349 (1969).
28. Kuo, J. F., and Greengard, P., J. Biol. Chem., 244, 3417 (1969).
29. Brostrom, M. A., Reimann, E. M., Walsh, D. A., and Krebs, E. G., Adv. in Enz. Reg., 8, 191 (1970).
30. Gill, G. N., and Garren, L. D., Biochem. Biophys. Res. Commun., 39, 335 (1970).
31. Tao, M., Salas, M. L., and Lipmann, F., Proc. Nat. Acad. Sci. U.S., 67, 408 (1970).
32. Kumon, A., Yammamura, H., and Nichizuka, Y., Biochem. Biophys. Res. Commun., 41, 1290 (1970).
33. Reimann, E. M., Brostrom, C. O., Corbin, J. D., King, C. A., and Krebs, E. G., Biochem. Biophys. Res. Commun., 42, 187 (1971).
34. Erlichman, J., Hirsch, A. H., and Rosen, O. M., Proc. Nat. Acad. Sci. U.S., 68, 731 (1971).
35. Majumder, G. C., and Turkington, R. W., J. Biol. Chem. 246, 2650 (1971).
36. Reimann, E. M., Walsh, D. A., and Krebs, E. G., J. Biol. Chem., 246, 1986 (1971).
37. Miyamoto, E., Petzold, G. L., Harris, J. S., and Greengard, P., Biochem. Biophys. Res. Commun., 44, 305 (1971).
38. Chen, L. J., and Walsh, D. A., Biochemistry, 10, 3614 (1971).

Discussion:

Fasold

We have also grown large crystals of phosphorylase. We started with the b form and grew these to about 0,8 mm length. The monoclinic crystals were investigated by Dr. Huber, Munich, who found a P 2_1 space group, a unit cell of 188,6 x 119,4 x 88,2 Å, this contains one tetramer. We grew phosphorylase a to about 1,5 mm, and Dr. Huber showed these crystals to be practically isomorphous with the b form. The crystals give a good pattern down to at least 3 Å resolution, and he has started the gathering of data of the native form now, while heavy atom derivates are being tried out. We have been interested in furnishing biochemical aids, such as a heavy atom-labelled AMP analogue.

Fischer

Do you have further informations as to how the heat stable inhibitor works, and secondly, have you tried to look at the phosphorylation of the calcium free kinase with and without cyclic AMP?

Krebs

With regard to the last question we have not done the experiment to see whether or not Ca^{++} affects the phosphorylatability of the kinase. We have tried to produce the heat stable inhibitor. These experiments have not worked, and we do not know whether there is a structural relationship between I and R, either one of them is capable of blocking C.

Cori

What is the mechanism of the inhibition?

Krebs

The inhibition is brought about by combination of I with C, we can demonstrate this as free catalytic subunit, we produce a high molecular weight complex. The dissociation constant of I and C is considerably greater than that of R and C and beyond there we really don't know whether they combine in the same way with C.

O. Wieland

You gave us examples of the physiological action of protein kinases on several enzymes. How about the physiological significance of other proteins, not enzymes, due to protein kinase action? In this connection I would like to refer to a recent paper by Marinetti, I believe, who showed that isolated plasma membranes can be phosphorylated by a cyclic AMP dependent protein kinase from liver.

Krebs

Of course this possibility exists. I think that many such instances will show up, because protein kinase is not entirely specific and when one used a very hot ATP^{32} and a very large amount of protein kinase then one would find counts in many different proteins.

Helmreich

You have just said the protein kinase is not very specific. But it has a very specific feature in so far that - as I remember - it phosphorylates phosphorylase b kinase, glycogen synthetase I, but does not touch phosphorylase b. So I think specificity must exist although we do not understand the underlaying features.

Holzer

This brings me to a question concerning the lipase. If I understood you correct the phosphorylation of lipase from fat cells has been achieved with protein kinase from muscle. My question is whether there is such a kinase in fat cells.

Krebs

Yes, in the published work the rabbit skeletal muscle enzyme has been used to activate the fat cell lipase. However Dr. Steinberg and Dr. Corbin have shown that the fat cell kinase will also work.

Shaltiel

The dephosphorylated peptide from phosphorylase a can it be phosphorylated by the kinase?

Krebs

This experiment has not been done.

Phosphorylase *a* as a Glucose Receptor

Willy Stalmans, Thierry de Barsy, Monique Laloux, Henri De Wulf, and Henri-Géry Hers

Laboratoire de Chimie Physiologique, Université de Louvain, Louvain/Belgique

The effect of glucose on the enzymatic inactivation of muscle phosphorylase *a* by phosphorylase phosphatase has been studied by several groups of workers with, however, discrepant results (1, 2, 3). An enhanced conversion of phosphorylase *a* into *b* in the presence of the hexose has also been demonstrated in liver preparations (4). The mechanism by which glucose may influence the phosphorylase phosphatase reaction was investigated by Holmes and Mansour (1) who did not reach a definite conclusion.

We have reinvestigated the effect of glucose on the enzymatic inactivation of muscle and liver phosphorylase *a*. Our main conclusion is that the concentration of glucose regulates glycogen metabolism in the liver through the binding of the hexose to its receptor, phosphorylase *a*. The binding of glucose to muscle phosphorylase *a* seems to have no physiological implication.

§ 1. The effect of glucose on the conversion of muscle phosphorylase *a* to *b*

Holmes and Mansour (1) have shown a stimulatory effect of glucose on the conversion of muscle phosphorylase *a* into *b* by the phosphatase present in a crude extract from diaphragm. Torres and Chelala (2), using a muscle extract that had been filtered on Sephadex G-25, were not able to demonstrate this effect of glucose; furthermore, many years ago Cori and Cori (3) reported that purified muscle phosphorylase phosphatase was insensitive to glucose. On the other hand, the conversion of muscle phosphorylase *a* into *b'* by trypsin is subject to inhibition by AMP (5) and to stimulation by glucose (6, 7).

The purpose of this study was to reach a better insight in the mechanism of the glucose effect on the phosphorylase phosphatase reaction, and to provide an explanation for the disagreement concerning the effect of the hexose on this conversion.

The interaction between AMP and glucose. When purified muscle phosphorylase *a* was incubated at 30° with a mouse muscle extract, the conversion of phosphorylase *a* into *b* was stimulated 2 to 3-fold by the addition of 0.5 % glucose (Fig. 1 A).

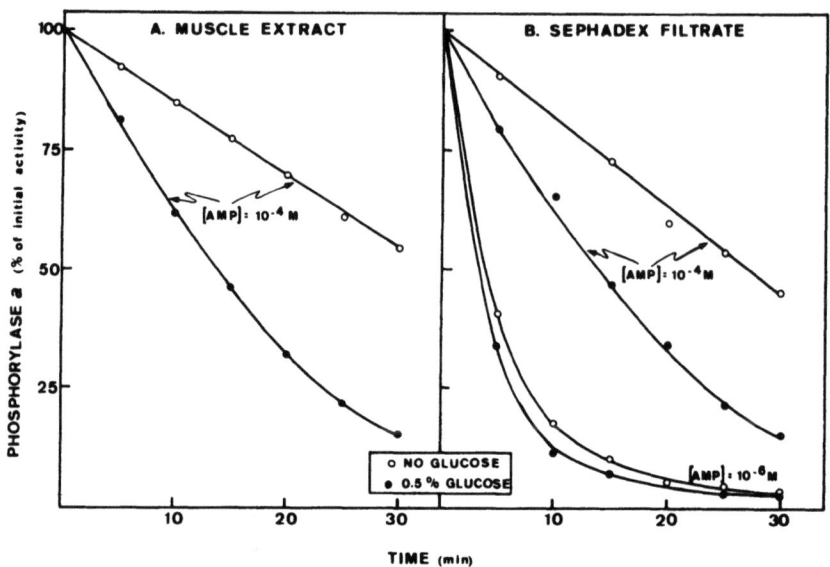

Fig. 1. The interaction of glucose and AMP in the conversion of phosphorylase *a* into *b* at 30°. Purified muscle phosphorylase *a* (0.2 mg/ml) was incubated with a muscle extract (0.7 mg protein/ml) or a Sephadex filtrate (0.5 mg/ml), in 0.1 M glycylglycine at pH 7.4. The phosphatase reaction was stopped by dilution with cold 0.1 M NaF; glucose and AMP were added in order to reach a uniform final concentration of either reagent. The samples were maintained for 10 min at 30° prior to the assay of phosphorylase *a*, performed at 30° in the presence of 0.5 mM caffeine (8)

A previous filtration of the extract on Sephadex G-25 not only produced a marked increase in the activity of phosphorylase phosphatase, but also abolished the effect of glucose (Fig. 1 B). It appeared that these changes were due to the removal of AMP by gel filtration : when 10^{-4} M AMP (i.e. the concentration present in the experiment of Fig. 1 A) was reintroduced, the rate of inactivation decreased and the sensitivity of the reaction to glucose was restored (Fig. 1 B). Similar results were obtained with a phosphorylase phosphatase, purified from rabbit muscle; the reaction was insensitive to glucose, unless AMP was added (not shown).

Our findings explain why the effect of glucose on the phosphorylase phosphatase reaction was not observed by Cori and Cori (3) or by Torres and Chelala (2), who performed their experiments at 27° and 30° respectively, using preparations of phosphorylase phosphatase that presumably contained very low amounts of AMP. The stimulation of this reaction by glucose could be observed by Holmes and Mansour (1) because their preparation contained an appreciable amount of AMP.

The influence of temperature. It has been reported that the conversion of muscle phosphorylase a into b' by trypsin is stimulated by glucose at 20°, but to a much smaller extent at 30° (7). Therefore it was of interest to investigate the effect of temperature on the stimulation of the phosphorylase phosphatase reaction by glucose. It is shown in Fig. 2A that at 13°, 0.5 % glucose stimulated the a to b conversion by the phosphatase present in a Sephadex filtrate, whereas (as already shown in Fig. 1B) at 30° the reaction was insensitive to glucose; similar results were obtained with the purified phosphorylase phosphatase (Fig. 2B). The effect of glucose at 13° was still greater when AMP was added (not shown).

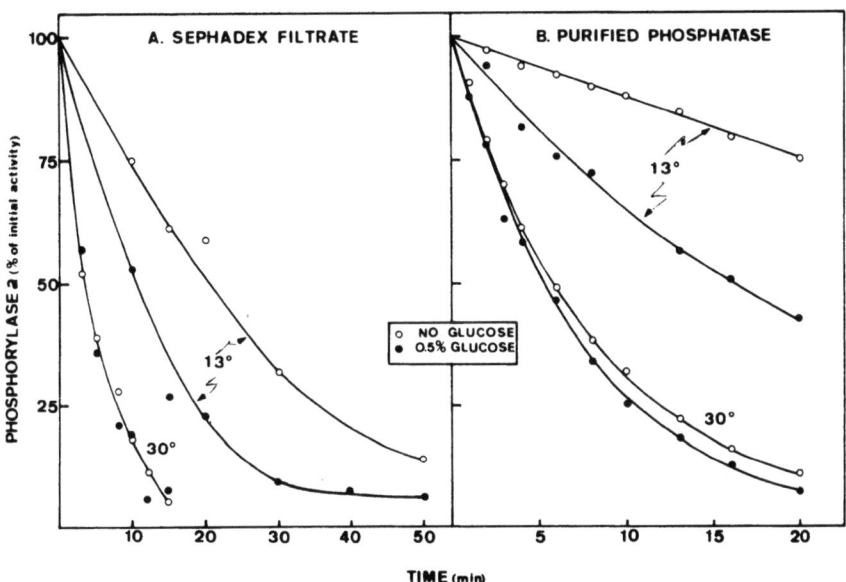

Fig. 2. The interaction of glucose and temperature in the conversion of phosphorylase a into b. A muscle Sephadex filtrate (0.6 mg protein/ml at 30°; 1.8 mg/ml at 13°) or purified (9) phosphorylase phosphatase (5 µg/ml at 30°; 10 µg/ml at 13°) were used. Other experimental conditions were as in Fig. 1. The final concentration of AMP was 10^{-6} M in part A, 5×10^{-8} M in part B

The conversion of phosphorylase a *into* b'. We have studied how AMP and temperature influence the effet of glucose on the attack of phosphorylase a by trypsin. At 30°, this reaction was stimulated by glucose only if AMP was present (Fig. 3A); at 13° (Fig. 3B) AMP was not required for the glucose effect. This behaviour is quite similar to that of the phosphorylase phosphatase reaction.

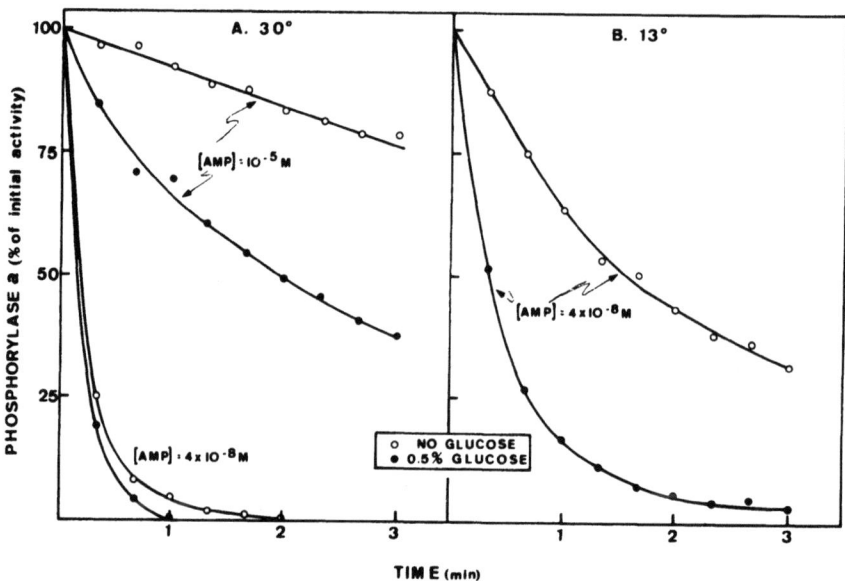

Fig. 3. The interaction of temperature, glucose and AMP in the conversion of phosphorylase a to b'. The experimental conditions were as in Fig. 1, except that the muscle preparation was replaced by trypsin (20 µg/ml), and that soybean trypsin inhibitor was used to stop the reaction

The mechanism of the glucose effect. The striking similarities between the glucose effect on the phosphorylase phosphatase reaction and on the conversion of phosphorylase a into b' by trypsin allow one to conclude that glucose interferes in these reactions by binding to the substrate, phosphorylase a. This mechanism could have been inferred from the several effects of glucose on phosphorylase a that have been described. It has been known for a long time that the hexose inhibits muscle phosphorylase a (10) and this inhibition has been interpreted as the result of the binding of glucose to the glucose-1-phosphate site of the enzyme (11). On the other hand, glucose displaces the equilibrium between tetrameric and dimeric phosphorylase a towards the dimer (12). Since raising the temperature to 30° has a similar effect (12), our results could be interpreted as indicating that dimeric phosphorylase a is a better substrate for the phosphatase and for trypsin than the tetramer. Such a conclusion has been reached by Graves et al. (7) for the attack of phosphorylase a by trypsin. It seems, however, that this interpretation cannot account for all our observations. In order to obtain more information about the relative quantities of dimer and tetramer present in our experiments, we have taken advantage of the dependence of the fluorescence of the MNS-phosphorylase a complex on the state of aggregation of phosphorylase a (13). Table 1 summarizes the results of these experiments, which were performed in the experimental conditions used for the study of the conversion of phosphorylase a to b or b'. It appears that the state of aggregation of the enzyme at 30° was not influenced by AMP or glucose; presumably the enzyme was entirely dimeric at this temperature. At 13°, the amount of tetramer decreased by adding either glucose or AMP. Thus, whereas both glucose and AMP tend to disaggregate phosphorylase a tetramers, their effects on the inactivation of phosphorylase a, by the phosphatase as well as by trypsin, are antagonistic. Therefore, it seems that at 13° glucose produces some conformational change in the enzyme, that leads to its dissociation and to an enhanced attack by the phosphatase and by trypsin, and that these phenomena are not causally related. This conclusion is further strengthened by the fact that glucose also stimulates the inactivation, whether by phosphatase or by trypsin, of liver phosphorylase a (see § 2), which seems not to form aggregates (14).

Table 1. The effect of temperature, glucose and AMP on the fluorescence of the 2-methylanilino-naphtalene-6-sulphonate (MNS)-phosphorylase a complex. The fluorescence of MNS (20 µM) in the presence of phosphorylase a (0.2 mg/ml) and of 0.1 M glycylglycine at pH 7.4 was measured against a blank without phosphorylase. In each experiment, the same mixture was brought to a different temperature or received an addition of glucose (final concentration : 0.5 %) or of AMP (final concentration: 10^{-5}M) in the order as indicated. The changes of volume due to these additions did not exceed 0.1 %. The results are expressed as percent of the initial fluorescence at 13°

Experimental Conditions		Relative Fluorescence			
Temperature	Addition	Expt. nr 1	Expt. nr 2	Expt. nr 3	Expt. nr 4
13°	-	100	100	100	100
30°	-	38	38	37	-
30°	Glucose	-	38	-	-
30°	AMP	-	-	37	-
13°	-	100	70	61	-
13°	AMP	-	69	-	54
30°	-	-	36	-	-
30°	Glucose	-	-	39	-

At 30°, phosphorylase a is predominantly dimeric (12) and the only effect of glucose on the enzymatic inactivation is to counteract the inhibitory action of AMP. This antagonistic effect of the two ligands might be related to their specific affinity for two conformational states of the enzyme (11). It has also been studied at 20° by Wang and Black (15), who proposed a more complex model.

Since no free glucose occurs inside the muscle cell (16), the observed stimulation of the a to b conversion by the hexose has presumably no physiological implication.

§ 2. The effect of glucose on the conversion of liver phosphorylase a to b

Stalmans et al.(4) have shown that the conversion of phosphorylase a to b, in a liver Sephadex filtrate or in a partially purified preparation from liver, is stimulated 5 to 10-fold by 0.5 % glucose. The stimulation is highly specific for this hexose and is concentration dependent with an apparent K_m of 0.13 %. It is additive to the effect of caffeine.

Since, in the case of the muscle system, the effect of glucose on the phosphorylase phosphatase reaction appears to be due to the binding of the hexose to phosphorylase a, it was necessary to investigate possible effects of glucose on liver phosphorylase a. The influence of different concentrations of glucose on the saturation kinetics of purified liver phosphorylase a with glucose-1-phosphate is represented as a double reciprocal plot in Fig. 4.

Glucose decreased the affinity of the enzyme for glucose-1-phosphate. Furthermore, it augmented the interaction between glucose-1-phosphate binding sites. The value of n, as estimated from Hill plots, increased from 1.1 in the absence of glucose to 1.3 in the presence of 0.5 % (35 mM) glucose. These results are similar to those previously obtained with muscle phosphorylase a (11). The K_m for glucose 1-phosphate in the absence of glucose was found to be 0.45 mM, a value that agrees well with the determinations by Maddaiah and Madsen (17), who used purified rabbit liver phosphorylase a in similar conditions of assay.

As shown in Fig. 5, glucose also markedly protected purified liver phosphorylase a against inactivation at 55°. Protection against thermal inactivation was also provided by AMP, as noted originally by Sutherland (18). It is shown in Fig. 6 that half maximal protection against heat inactivation was afforded by 0.13 to 0.18 % glucose.

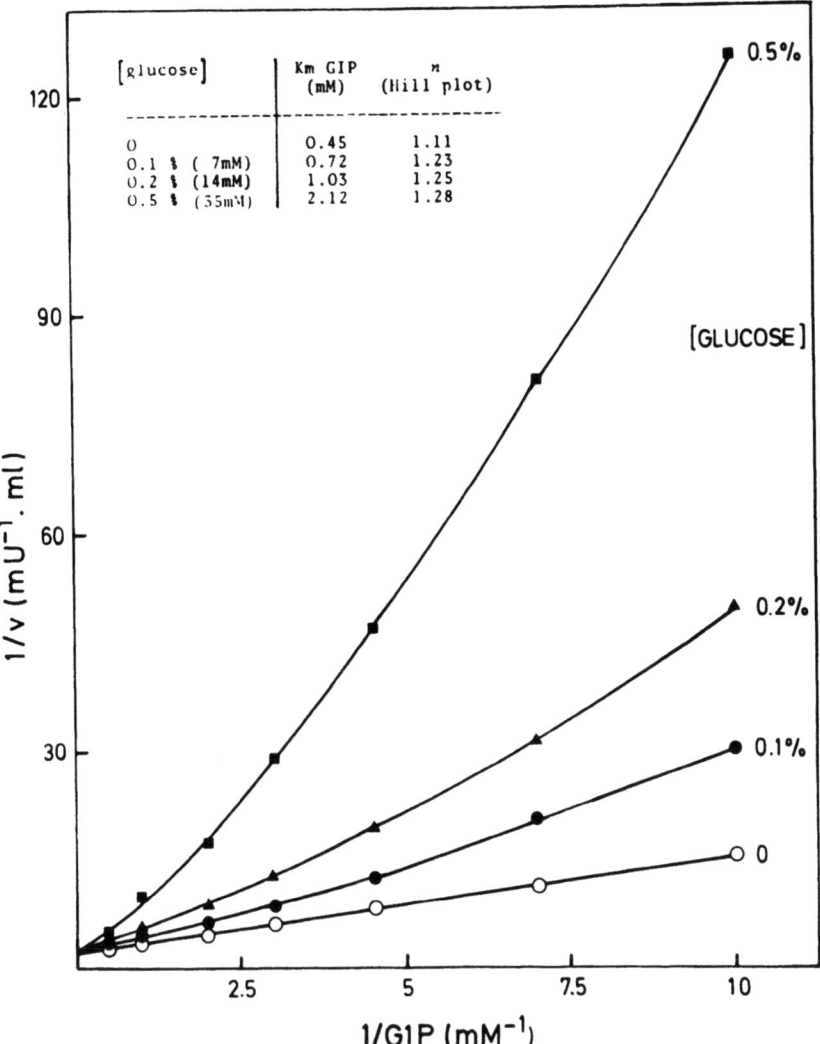

Fig. 4. Saturation kinetics of liver phosphorylase a with glucose 1-phosphate in the absence and presence of glucose. Initial velocity was measured at 20° by the incorporation of radioactivity from $[U-^{14}C]$ glucose 1-phosphate into glycogen. The assay mixture contained 1 % glycogen, 1 mM EDTA, 30 mM imidazole at pH 6.6, 0.1 to 2 mM glucose-1-P, glucose as indicated, and purified dog liver phosphorylase a. The final volume was 2 ml

Glucose also stimulated the inactivation of purified liver phosphorylase a by trypsin, whereas AMP had the reverse effect (Fig. 7). It is not possible, however, to discriminate in this experiment between a specific action of trypsin, similar to the conversion of muscle phosphorylase a to b', and a loss of activity due to unspecific proteolysis.

The several effects of glucose on liver phosphorylase a, that we have described here, suggest that the stimulatory effect of glucose on the inactivation of liver phosphorylase a by its phosphatase is due to the binding of the hexose to phosphorylase a.

§ 3. The control of glycogen metabolism in the liver by glucose : the sequence of events

We summarize in Fig. 8 our present knowledge of the various mechanisms that regulate glycogen synthesis or breakdown in the liver. This scheme is based on work that has been recently reviewed (19) and on the data presented in this paper. The primary effect of glucose is to bind to phosphorylase a, which is then slightly inhibited and, much more important, becomes more sensitive to the action of phosphorylase phosphatase. When phosphorylase a is converted to phosphorylase b, glycogenolysis is stopped. Furthermore, phosphorylase a is a potent

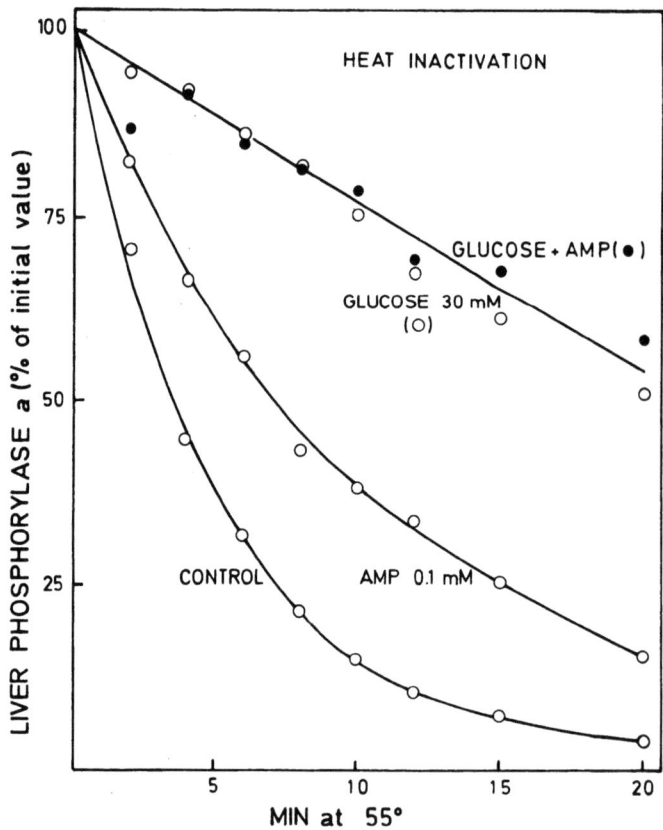

Fig. 5. The effect of glucose and of AMP on the thermal inactivation of liver phosphorylase a. Purified liver phosphorylase a (25 µg/ml) was incubated at 55° with or without glucose or AMP. Phosphorylase a was assayed at 20° in the presence of 0.5 mM caffeine (8)

inhibitor of the enzyme that activates glycogen synthetase (i.e. synthetase phosphatase); its disappearance allows the latter enzyme to become active and thus triggers glycogen synthesis.

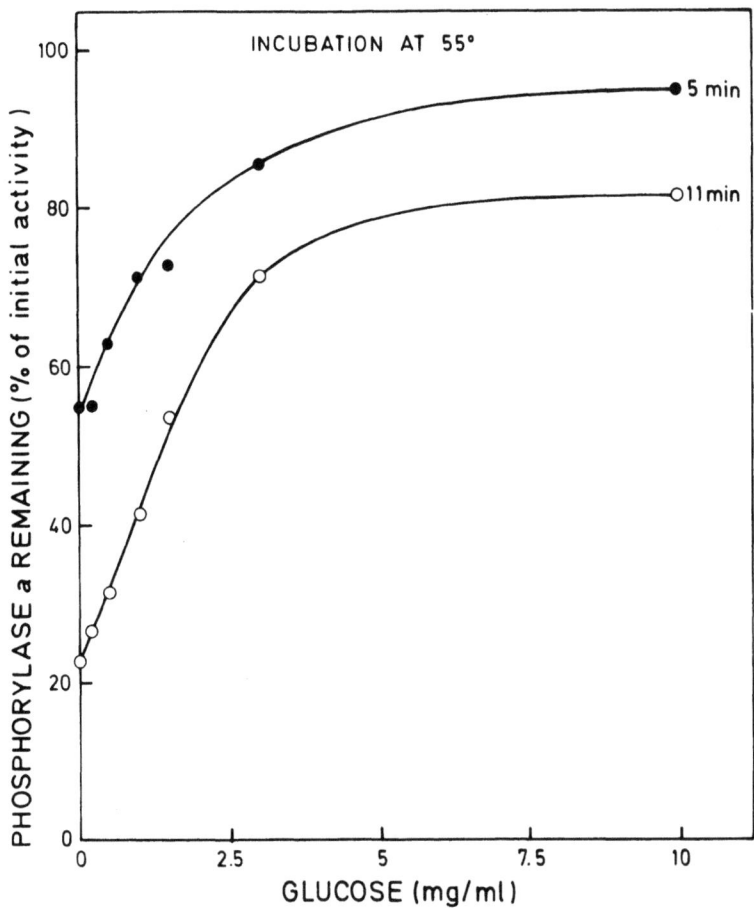

Fig. 6. The effect of the concentration of glucose on the thermal inactivation of liver phosphorylase a. Purified liver phosphorylase a (25 µg/ml) was maintained at 55° for 5 or 11 min in the presence of various concentrations of glucose

An important implication of this mechanism is that the conversion of phosphorylase a into b must precede the activation of glycogen synthetase. This sequence of events has actually been observed in the liver *in vitro* (8) but had not been studied *in vivo*. It is indeed difficult to know the actual level of active phosphorylase *in vivo*; as a rule, the values obtained are much higher than would be expected from the actual rate of glycogenolysis *in vivo*. This is presumably due to a rapid activation of the enzyme during the time that elapses between the death of the animal and the preparation of the homogenate. We have

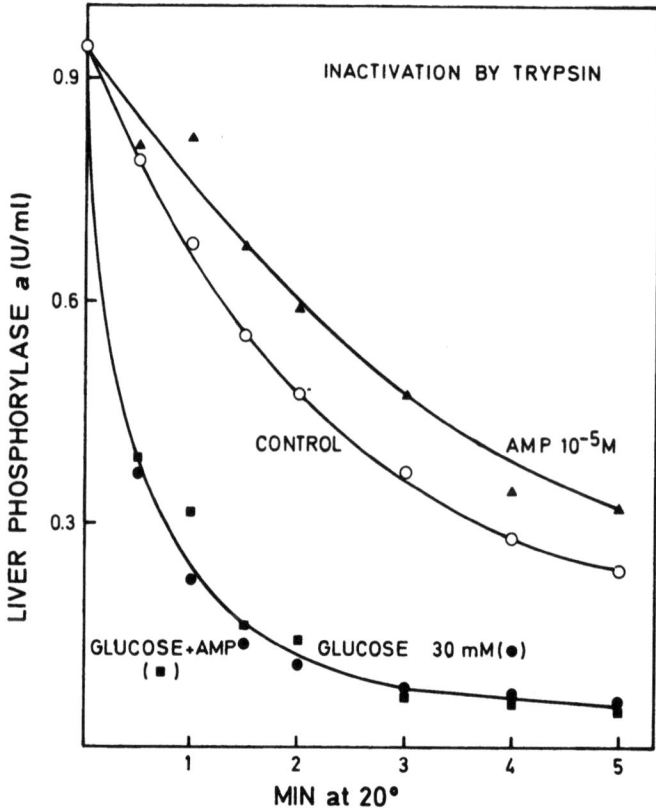

Fig. 7. The effect of glucose and of AMP on the inactivation of liver phosphorylase a by trypsin. Purified liver phosphorylase a (0.17 mg/ml) was incubated at 20° with trypsin (60 μg/ml) in 0.1 M glycylglycine at pH 7.4, and with other additions as indicated

recently been able to obtain values that are presumably closer to the *in vivo* activity, by using the technique outlined in the legend of Fig. 9. The first change that occurred after the intravenous administration of glucose to mice was the inactivation of phosphorylase (Fig. 9). The fact that, in contrast to what has been observed *in vitro*, this inactivation was not complete, is probably due to an artifactual reactivation of the enzyme. The *in vivo* inactivation was very rapid (within one minute), whereas the activity of glycogen synthetase remained low for another minute. In the second phase, the conversion of synthetase b into a occurred while the activity of phosphorylase a remained stable at a low level.

The fact that this sequence of events has been observed in mice indicates that the regulatory mechanism outlined in Fig. 8 operates *in vivo*. Phosphorylase a is the primary receptor for glucose and, by its interaction with synthetase phosphatase, also behaves as a second messenger in the control of glycogen synthesis.

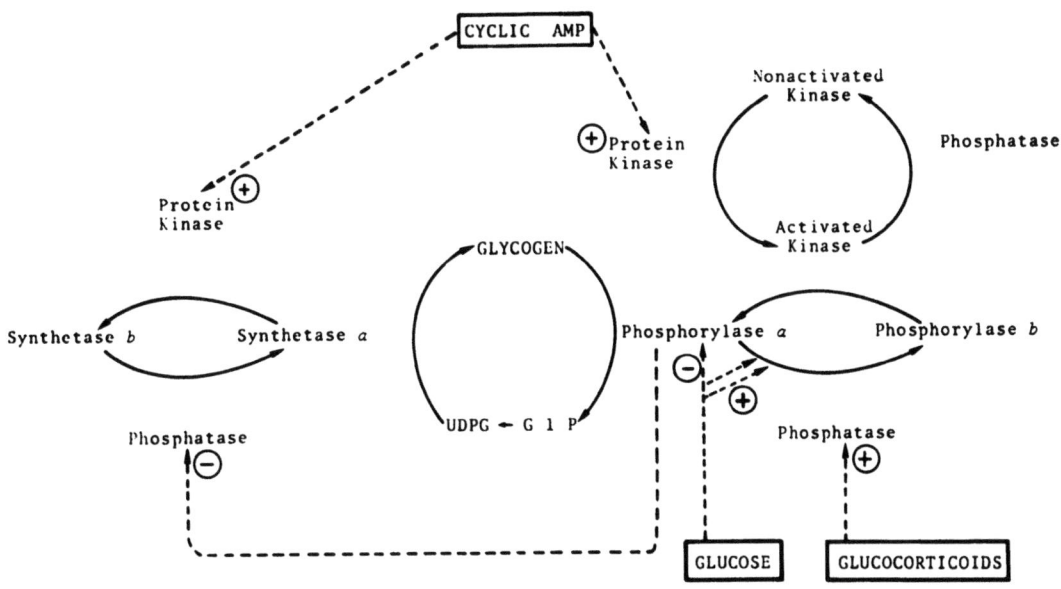

Fig. 8. The control of glycogen metabolism in the liver

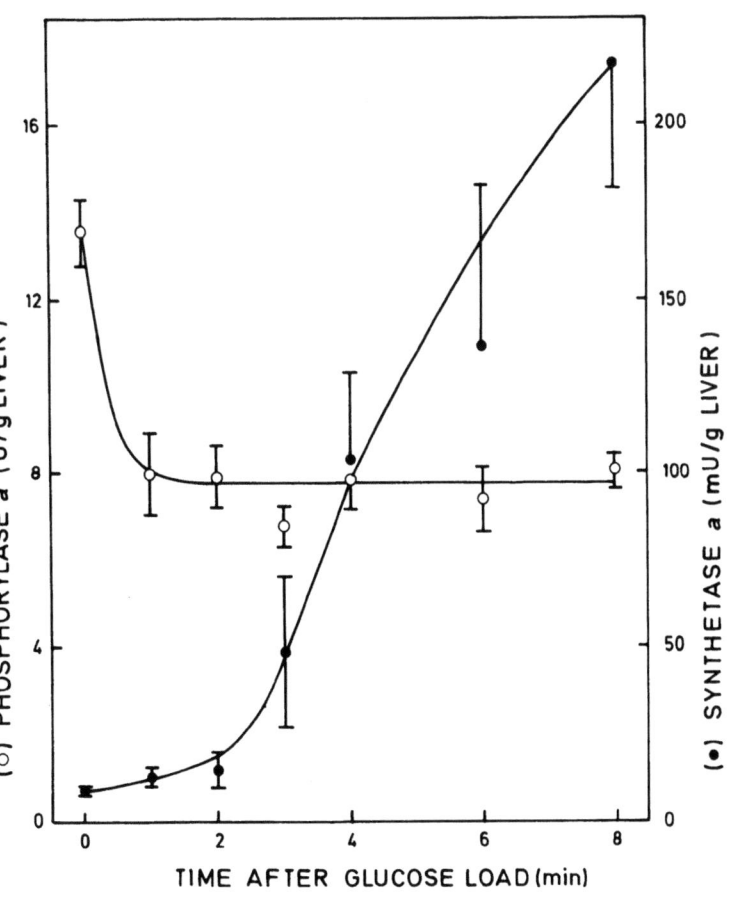

Fig. 9. Sequential changes in the activities of phosphorylase a and synthetase a in the liver after a single glucose load. Groups of 6 to 11 mice were killed at various times after the I.V. administration of glucose (3 mg/g body weight), and the livers were quenched between aluminium blocks, precooled by liquid air (20), within 7 sec after decapitation. Control mice ("0 min") received an equivalent volume of saline 1-2 min before death. The frozen tissue was homogenized with 10 vol. of 0.1 M glycylglycine buffer, pH 7.4, containing 0.1 M NaF, 0.01 M EDTA, and 0.5 % glycogen. Phosphorylase a and synthetase a were assayed as previously described (8), except that 0.5 mM caffeine was present in the phosphorylase assay. Vertical bars represent ± S.E.M

Acknowledgements

W.S., T.d.B. and H.D.W. are fellows of the *Nationaal Fonds voor Wetenschappelijk Onderzoek*. This work was supported by the *Fonds de la Recherche Scientifique Médicale* and by the U.S. Public Health Service (Grant AM-9235).

References

1. HOLMES, P.A. and MANSOUR, T.E., Biochim. Biophys. Acta, 156, 275 (1968).
2. TORRES, H.N. and CHELALA, C.A., Biochim. Biophys. Acta, 198, 495 (1970).
3. CORI, G.T. and CORI, C.F., J. Biol. Chem., 158, 321 (1945).
4. STALMANS, W., DE WULF, H., LEDERER, B. and HERS, H.G., Eur. J. Biochem., 15, 9 (1970).
5. FISCHER, E.H., GRAVES, D.J., CRITTENDEN, E.R.S. and KREBS, E.G., J. Biol. Chem., 234, 1698 (1959).
6. GRAVES, D.J., MANN, S.A.S., PHILIP, G. and OLIVEIRA, R.J., J. Biol. Chem., 243, 6090 (1968).
7. GRAVES, D.J., HUANG, C.Y. and MANN, S.A. In Control of Glycogen Metabolism (Edited by W.J. Whelan), Universitetsforlaget, Oslo and Academic Press, London, 35 (1968).
8. STALMANS, W., DE WULF, H. and HERS, H.G., Eur. J. Biochem., 18, 582 (1971).
9. HASCHKE, R.H., HEILMEYER, L.M.G., MEYER, F. and FISCHER, E.H., J. Biol. Chem., 245, 6657 (1970).
10. CORI, C.F., CORI, G.T. and GREEN, A.A., J. Biol. Chem., 151, 39 (1943).
11. HELMREICH, E., MICHAELIDES, M.C. and CORI, C.F., Biochemistry, 6, 3695 (1967).
12. WANG, J.H., SHONKA, M.L. and GRAVES, D.J., Biochem. Biophys. Res. Commun., 18, 131 (1965).
13. BIRKETT, D.J., RADDA, G.K. and SALMON, A.G., FEBS Letters, 11, 295 (1970).
14. APPLEMAN, M.M., KREBS, E.G. and FISCHER, E.H., Biochemistry, 5, 2101 (1966).
15. WANG, J.H. and BLACK, W.J., J. Biol. Chem., 243, 4641 (1968).
16. KIPNIS, D.M., HELMREICH, E. and CORI, C.F., J. Biol. Chem., 234, 165 (1959).
17. MADDAIAH, V.T. and MADSEN, N.B., J. Biol. Chem., 241, 3873 (1966).
18. SUTHERLAND, E.W. In Methods in Enzymology (Edited by S.P. Colowick and N.O. Kaplan), Academic Press, New York, 215 (1955).
19. HERS, H.G., DE WULF, H. and STALMANS, W., FEBS Letters, 12, 73 (1970).
20. WOLLENBERGER, A., RISTAU, O. and SCHOFFA, G., Arch. Ges. Physiol., 270, 399 (1960).

Properties of Purified Glycogen Synthetase b from Liver[1]

Harold L. Segal, Yukihiro Sanada, and Susan R. Martin

Biology Department, State University of New York, Buffalo, New York/USA

Summary

A procedure has been developed for the acquisition of highly purified rat liver glycogen synthetase b. In density gradient centrifugation experiments 2 major peaks of synthetase b activity were found in the absence of ligands, plus small amounts of higher molecular weight components. The heavier peak appeared to be a dimer of the lighter peak and was of approximately 258,000 to 284,000 molecular weight. In the presence of ligands a single, nearly symmetrical peak of activity was obtained corresponding to the heavier of the 2 peaks. Similar observations were made with disc gel electrophoresis experiments where the presence of G-6-P converted the major component to a species of lower mobility. Preincubation of certain preparations of synthetase b with G-6-P also produced a marked time-dependent increase in activity. In crude extracts of normal livers which had been filtered through Sephadex G-25, Mg^{+2} stimulated the synthetase phosphatase reaction. By contrast Mg^{+2} had no effect on the phosphorylase phosphatase reaction. In addition, phosphorylase phosphatase activity was present in extracts from starved, adrenalectomized animals in which synthetase phosphatase activity was absent. We propose some views of the metabolic significance of interconvertible forms of enzymes.

The glycogen synthetase reaction is a key site of control of glycogen synthesis in the liver (1,2) and is linked to control of glycogen utilization as well, through a reciprocal relationship with phosphorylase (3,4). Regulation of flux through the system depends upon the interconversion of active (a) and inactive (b) forms (5), mediated by a phosphokinase (a to b reaction) and a phosphatase (b to a reaction). The phosphokinase, recently identified as cAMP-dependent[2] protein kinase (6), is responsive

---

[1] Supported by a grant from the National Institutes of Health (AM-08873).

[2] Abbreviations are: cAMP, cyclic adenosine 3',5'-phosphate; UDPG, UDP-glucose; G-6-P, glucose-6-P.

to hormones which affect cAMP levels. Phosphatase activity is absent in starved, adrenalectomized animals, but is restored upon administration of glucocorticoids or glucose (7,8). The phosphatase is also modulated either directly or indirectly by glycogen (3), insulin (for discussion, see (9)), and glucose (3), the latter apparently via its effect on the phosphorylase a to b reaction (10).

In order to progress in the working out of the details of this system and its control, it is obviously necessary that its components be separated and their properties defined. We describe here a procedure for the acquisition of rat liver glycogen synthetase b in a high state of purity and some of its characteristics. Steiner et al. (11), Sevall and Kim (12), and Soderling et al. (6) have recently described purifications of the enzyme from rat liver, tadpole liver, and rabbit muscle, respectively.

Certain comparisons of glycogen synthetase phosphatase and phosphorylase phosphatase are also reported.

Methods and Results

Materials. Male, Sprague-Dawley rats of about 200 g were used. They were provided with 0.9% NaCl in the drinking water. Glycogen (rabbit liver, Type III), UDPG, G-6-P, NADP, glucose-1,6-diphosphate, and all auxiliary and marker enzymes were obtained from Sigma. DEAE-cellulose (DE 52) and cellulose powder (CF 2) were obtained from Whatman. Calcium phosphate tribasic (approx. $Ca_{10}(OH)_2(PO_4)_6$) was obtained from Fisher.

Assays. Unless otherwise noted, glycogen synthetase was assayed by the optical method. Samples were incubated for 10 min at 37° in a volume of 1 ml containing 4 μmoles of UDPG, 5 mg glycogen, 50 μmoles glycylglycine, pH 7.4, with and without 4 μmoles G-6-P for assay of total glycogen synthetase and glycogen synthetase a, respectively. The reaction was terminated by immersing the tubes in boiling water. To measure the amount of UDP formed, there was added 2.2 ml of a solution containing 200 μmoles of glycylglycine, pH 7.4, 10 μmoles of $MgCl_2$, 0.6 μmoles of phosphoenolpyruvate, and 0.4 mg of NADH. After determination of the initial absorbance at 340 nm, 0.05 ml of a mixture of pyruvic kinase and lactic dehydrogenase was added and the decrease in absorbance recorded. Units of glycogen synthetase are expressed as μmoles of UDP formed per min.

Polysaccharide was determined by the anthrone method (13) after precipitation with alcohol. Protein was determined by the Lowry method (14) unless otherwise noted.

Purification. Adrenalectomized rats were fasted for 48 hr before sacrifice. In such preparations the bulk of the total liver glycogen synthetase appears in the soluble extract (as much as 90%) and is virtually all in the _b_ form (7). The livers were homogenized in a blendor in 3 volumes of 0.25 M sucrose containing 5 mM EDTA and 10 mM β-mercaptoethanol, pH 7.0, and the homogenate centrifuged at 30,000 rpm for 90 min. To the supernatant solution was added 20 g of ammonium sulfate per 100 ml. The precipitate was dissolved in the homogenizing solution. Ammonium sulfate precipitation was repeated and the 2nd precipitate dissolved in a solution of 0.1 M glycylglycine, pH 7.4, 1 mM EDTA, and 10 mM β-mercaptoethanol in 25% (v/v) glycerol (buffer I). The enzyme solution was passed through a Sephadex G-25 column equilibrated with buffer I, then applied to a DEAE-cellulose column. The latter was prepared with 2 volumes of DEAE-cellulose and 3 volumes of cellulose powder suspended in the same buffer. The column was washed with buffer I supplemented with 0.1 M NaCl until the 280 nm absorbance fell to a low value, then a gradient of 0.1 M NaCl to 0.5 M NaCl in buffer I was applied to elute the enzyme. The active fractions were concentrated by precipitation with ammonium sulfate and dissolved in and dialyzed against buffer I. The enzyme solution was next applied to a calcium phosphate column prepared with 2 volumes of calcium phosphate tribasic and 3 volumes of cellulose powder suspended in buffer I. The column was washed with buffer as above but with the glycylglycine replaced with 0.2 M potassium phosphate, pH 7.0. The enzyme was eluted by raising the potassium phosphate concentration to 0.5 M. The active fractions were concentrated under vacuum and dialyzed against buffer I in a collodion bag (Brinkmann Instruments).

A typical purification protocol is presented in Table I.

Kinetics. The K_m value for UDPG at 4 mM G-6-P was 0.56 mM, and the K_a value for G-6-P at 4mM UDPG was 0.18 mM (Fig. 1). These values are close to those obtained previously for glycogen synthetase _b_ with a crude preparation using a radioactivity assay procedure (5). Aside from cooperative effects at low ligand concentrations the kinetics of liver glycogen synthetase correspond to a system in which the activator promotes substrate binding (3).

Sedimentation Characteristics. Centrifugations were carried out in a Spinco SW 39 rotor at $0°$. A linear gradient in 4.6 ml was produced with 0% and 12.5% sucrose solutions in 12.5% (v/v) glycerol containing 50 mM glycylglycine, pH 7.4, and 10 mM β-mercaptoethanol, plus additions of ligands as noted. The enzyme was added in 75 μl of the same solution as the gradient in each case except that sucrose was absent and the glycerol concentration was 10%. Yeast alcohol dehydrogenase ($\underline{s}_{20,w}^{0.725}$ = 7.4) and beef liver catalase ($\underline{s}_{20,w}^{0.725}$ = 11.3) (15) were added in all tubes as markers. Thirty-two fractions were collected. The marker enzymes were symmetrically distributed and only the position of the peaks is shown on the figures.

Table I

Purification of Glycogen Synthetase b

Fraction	Activity/ml		Total Units	Specific Activity	Poly-saccharide
	units			units/mg protein	mg/ml
	a	total			
30,000 rpm extract	0.02	0.50	123	0.037	0.60
1st ammonium sulfate precipitate	0.03	0.41	82	0.100	nil[1]
2nd ammonium sulfate precipitate	0.08	2.30	74	0.140	nil
DEAE-cellulose eluate	0.00	6.48	45	1.38	nil
calcium phosphate eluate	0.00	6.80	18	7.56	nil

[1] less than 0.04 mg/ml.

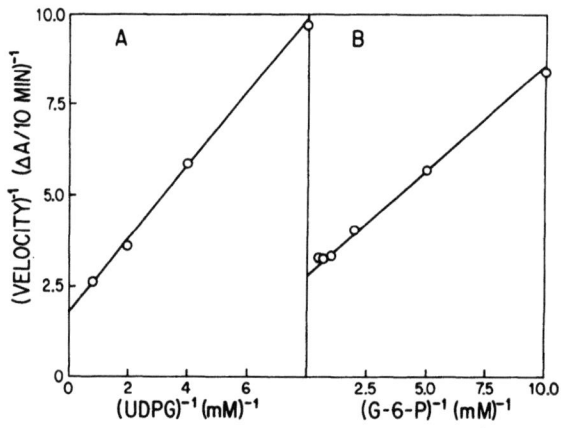

Fig. 1. K_m of UDPG (A) and K_a of G-6-P (B). Assays were performed as described in the text with variable UDPG and 4 mM G-6-P in A, and variable G-6-P and 4 mM UDPG in B. The enzyme was a calcium phosphate gel eluate

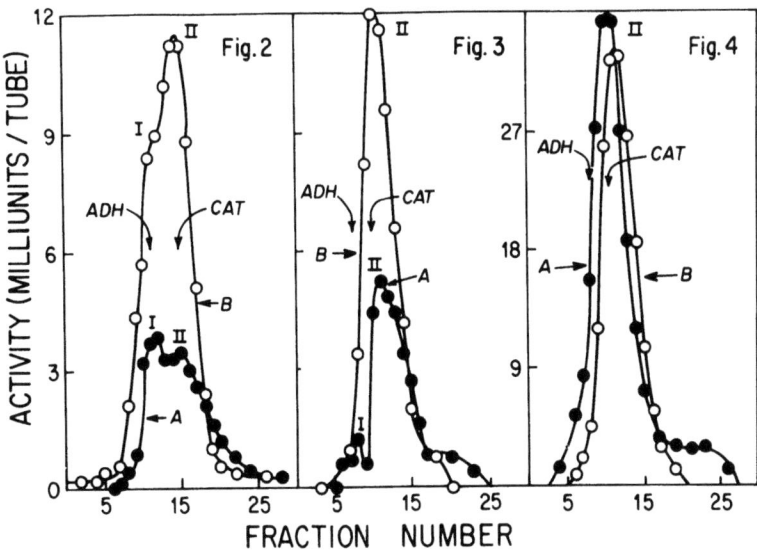

Figs. 2-4. Sedimentation of glycogen synthetase $\underline{b}$. Arrows labeled "ADH" and "CAT" indicate the positions of the alcohol dehydrogenase and catalase peaks, respectively. Roman numerals I and II identify peaks of glycogen synthetase activity. The ordinate scales in Figs. 2 and 3 are the same, but it is 3 times greater in Fig. 4. The enzyme solutions were calcium phosphate gel eluate fractions of different preparations. In Fig. 2, 83 milliunits of enzyme was layered on each tube. Centrifugation was at 35,000 rpm for 13 hrs. In curve A (filled circles) there were no additional components. In curve B (open circles) 5 mM EDTA was present. In Fig. 3, 36 milliunits of enzyme was layered on each tube. Centrifugation was at 32,000 rpm for 12 hrs. In curve A (filled circles) 5 mM $MgCl_2$ and in curve B (open circles) 5 mM ATP were added. In Fig. 4, 73 milliunits of enzyme was layered on each tube. Centrifugation was at 35,000 rpm for 12 hrs. In curve A (filled circles) 5 mM G-6-P was present. In curve B (open circles) 5 mM G-6-P and 5 mM UDPG were present

The effects of various ligands on the sedimentation velocity of glycogen synthetase $\underline{b}$ is shown in Figs. 2-4. In the absence of ligands 2 peaks of activity appeared (I and II), plus a small amount of higher molecular weight components, the total representing only about 40% recovery (Fig. 2, curve A). Addition of 5 mM EDTA resulted in a quantitative recovery of activity with a shift in distribution toward the faster moving peak (II) and a disappearance of the minor high molecular weight components (Fig. 2, curve B). $MgCl_2$ produced a shift from peak I toward the faster moving components with a recovery of about 75% (Fig. 3, curve A). In the presence of ATP (an inhibitor of synthetase competitive with substrate (16)) (Fig. 3, curve B), G-6-P (Fig. 4, curve A), or G-6-P plus UDPG (Fig. 4, curve B), virtually all the activity was obtained in a single symmetrical peak in position II, with a 200% to 300% recovery (see below).

Table II summarizes the sedimentation values obtained. The relative $\underline{s}$ values of peaks I and II correspond to a ratio of molecular weights of 1.8 to 1 (15), suggesting

Table II

$s_{20,w}^{0.725}$ Values of Glycogen Synthetase b Components

Values for peaks I and II were calculated relative to alcohol dehydrogenase and catalase, respectively.

Calculated from	Peak I	Peak II
Fig. 2, curve A	7.7	11.3
Fig. 2, curve B	7.7[1]	10.9
Fig. 3, curve A	7.9	12.4
Fig. 3, curve B	---	11.9
Fig. 4, curve A	---	10.8
Fig. 4, curve B	---	11.8
1 mM EDTA[2]	7.8	11.7
average	7.78	11.54

[1] Shoulder

[2] Not shown in Figures. Recovery was about 30% with most of the recovered activity in peak I.

that peak II is a dimer of peak I. Calculations of the approximate molecular weight (15) of peak II gave values of 284,000 and 258,000 with alcohol dehydrogenase and catalase, respectively, as standards.

Disc Gel Electrophoresis. Enzyme from the calcium phosphate gel step was subjected to disc electrophoresis on acrylamide gel. In the absence of ligands two bands of activity were obtained in the same position as two protein bands (Fig. 5A). In the presence of 5 mM G-6-P, on the other hand, a single major band of activity was detected, in much better yield and migrating to a position intermediate between the activity bands observed in the absence of ligand (Fig. 5B). A protein band also appeared at this location which was only very faintly present in Fig. 5A, while the protein band corresponding to the faster moving peak in Fig. 5A was greatly reduced. The slower moving, smeared-out band was still present and contained some residual activity.

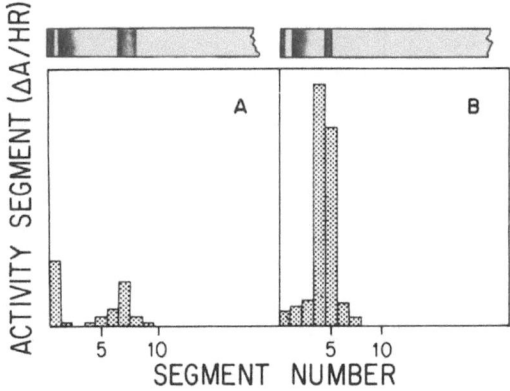

Fig. 5. Disc gel electrophoresis of purified glycogen synthetase b. Gels were prepared from 6% acrylamide in 50 mM glycylglycine, pH 7.4, containing 10 mM β-mercaptoethanol. In B, the enzyme solution and the upper chamber also contained 5 mM G-6-P. No stacking gels were used. The gels were pre-electrophoresed for 1 hr at 4 mA per tube. Twenty-five μl of enzyme solution (calcium phosphate gel eluate containing 21.3 activity units and 2.5 mg of protein per ml) was applied to the gel in A and 10 μl in B. Electrophoresis was at 2 mA per tube for 10 min, then 3 mM for 1.5 hrs. Gels were run in duplicate. One was stained with 0.2% amido black in 5% acetic acid and destained with 2% acetic acid. The other was cut into 3 mm segments which were assayed for activity by adding the usual assay components to the dispersed gel and incubating at 37° with frequent stirring

Reactivation by G-6-P. The changes in sedimentation and mobility distribution of glycogen synthetase b produced by G-6-P had their counterpart in effects on activity. A high speed extract of liver was carried through an ammonium sulfate precipitation and Sephadex G-25 dialysis in the absence of glycerol. As expected under these conditions substantial activity was lost. Upon incubation with G-6-P, however, the major portion of the lost activity was regained (Fig. 6). Stabilization by glycerol is also demonstrated in this figure. No glycogen synthetase a activity was present either before or after reactivation, so what is manifested in this experiment is the interconversion of synthetase b between an active and inactive (or less active) form. While activation by G-6-P was observed in this way on many occasions, there were instances in which it failed to occur. Further experiments are necessary to clarify the conditions under which the reactivatable form is produced.

Comparison of Glycogen Synthetase Phosphatase and Phosphorylase Phosphatase. Hers and his coworkers have introduced the use of Sephadex filtration of liver extracts to reveal more clearly effects of added modulators on the glycogen synthetase and phosphorylase systems (3,4). Using such preparations we have compared some parameters of the synthetase phosphatase and phosphorylase phosphatase reactions.

Fig. 7A demonstrates the effects of Mg^{+2} and glucose on the glycogen synthetase phosphatase reaction. No conversion of synthetase b to a occured in the absence

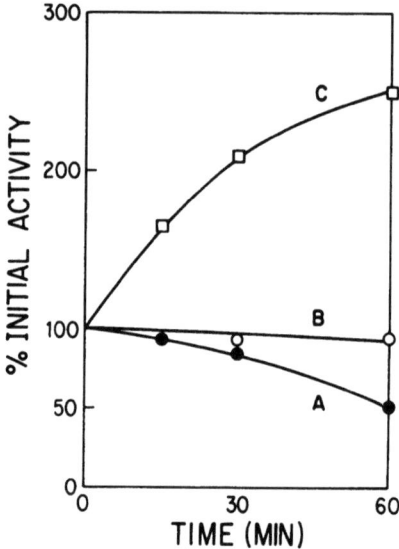

Fig. 6. Reactivation of glycogen synthetase b by G-6-P and stabilization by glycerol. The supernatant solution after centrifugation of the homogenate at 30,000 rpm for 90 min contained 5.4 units of glycogen synthetase b. An ammonium sulfate precipitate (20 g/100 ml) was made and dissolved in 0.1 M glycylglycine, pH 7.4, containing 10 mM β-mercaptoethanol and 1 mM EDTA and passed through Sephadex G-25 equilibrated with the same buffer. Recovery was 1.5 units. To separate portions of the enzyme solution equal volumes of buffer (curve A, filled circles), of 50% (v/v) glycerol (curve B, open circles), or of 10 mM G-6-P in buffer (curve C, squares) were added and the solutions incubated at 37°. Aliquots were removed at the times indicated for assay. No activity was detected in the absence of G-6-P in any of the samples

of added effectors (curve a). Synthetase a formation in the presence of Mg^{+2} followed the typical time course previously observed with unfiltered extracts (7, 18) including the latency period (curve b). The effect of glucose alone on the b to a conversion was minimal (curve c). When glucose was added in the presence of Mg^{+2}, the lag period was reduced, as reported by DeWulf et al. (3), with perhaps an increase in the steady state rate of the phosphatase reaction as well. On the other hand, the effects of Mg^{+2} and glucose on the phosphorylase phosphatase reaction were distinctly different (Fig. 7B). Mg^{+2} was without effect (curve a vs. b and curve c vs. d), whereas glucose stimulated the reaction (curve a vs. c and curve b vs. d) (10). No latency was observed in the phosphorylase phosphatase reaction (10).

A picture very similar to that presented in Fig. 7B prevailed with the phosphorylase system in livers from starved, adrenalectomized rats, except the initial phosphorylase a level was reduced by about half. This is in marked contrast to the glycogen synthetase system in such preparations in which no phosphatase activity whatever could be detected (7,8).

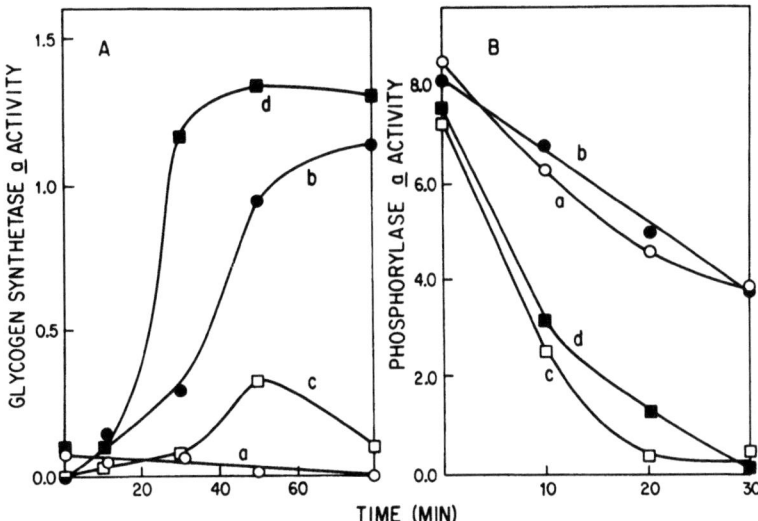

Fig. 7. Effect of Mg^{+2} and glucose on the glycogen synthetase phosphatase and the phosphorylase phosphatase reactions. Livers were homogenized in 2 volumes (w/v) of buffer composed of 0.1 M glycylglycine, pH 7.4, 10 mM glutathione, and 1 mM EDTA. 3 ml of the extract obtained after centrifugation for 10 min at 8,500 x g was passed through a 30 ml column of Sephadex G-25 previously equilibrated with the homogenizing buffer. Fractions containing protein were pooled and the dilution of the extract determined by measurement of protein (17) before and after chromatography. Effectors were added to 0.8 ml of the filtered extract to a final volume of 1.0 ml. Aliquots were removed at the times shown for assay of glycogen synthetase a (A) or phosphorylase a (B). Additions to the preincubation mixtures were as follows (as final concentrations): a, no additions; b, 10 mM $MgCl_2$; c, 10 mM glucose; d, 10 mM $MgCl_2$ and 10 mM glucose. Assays for glycogen synthetase a (no added G-6-P) were by the radioactivity method previously described (7). For phosphorylase a assays 0.1 ml of the enzyme preparation was added to 0.9 ml of a solution containing 50 μmoles potassium phosphate, pH 7.5, 5 mg glycogen, 10 μmoles NaF, and 1 μmole EDTA, and incubated at 37° for 5 min. The reaction was stopped by the addition of 0.1 ml of 5 M $HClO_4$, and the solution neutralized with 0.2 ml of 2.5 M $KHCO_3$. After centrifugation an 0.1 ml aliquot was added to 0.9 ml of a solution containing 50 μmoles of imidazole, pH 7.5, 1.2 μmoles of $MgCl_2$, 0.4 μmoles of NADP, and 0.01 μmoles of glucose-1,6-diphosphate. After an initial absorbancy measurement at 340 nm, about 0.1 unit each of phosphoglucomutase and glucose-6-phosphate dehydrogenase was added in 10 μl and the amount of NADPH formation determined as a measure of glucose-1-phosphate (plus G-6-P). Units of glycogen synthetase (A) and phosphorylase (B) are in μmoles of product formed per min per g of liver. Total glycogen synthetase activity (assayed with 4 mM G-6-P added) was 1.2 units per g liver initially and was essentially unchanged at 80 min when Mg^{+2} was present (curves b and c) but declined markedly in its absence (curves a and c)

Discussion

The procedures described here have provided a glycogen synthetase preparation over 200-fold purified relative to the soluble extract of rat liver, and entirely in the b form and free of polysaccharide. Steiner et al. (11) have reported a somewhat higher final specific activity in a purification of glycogen synthetase from the same source. Their preparation appears to be predominantly in the a form and contains small amounts of polysaccharide. The preparation of Sevall and Kim (12) from tadpole liver also contains glycogen. Rabbit muscle glycogen synthetase, recently purified to near homogeneity (predominantly in the a form) by Soderling et al. (6), is of approximately the same specific activity as the preparation described here, when allowance is made for the lower temperature and less favorable buffer composition in their assay procedure.

From the results presented here it can be concluded that liver glycogen synthetase b can exist in multiple states of association of different activity. We tentatively identify the lighter peak obtained in density gradient centrifugations in the absence of ligands (Fig. 2, curve A, peak I) with the more rapidly migrating band of activity observed on disc gel electrophoresis under these conditions (Fig. 5A), and the heavier peak II (presumably the dimer of peak I), which is the predominant form in the presence of ligands (Fig. 4), with the more slowly migrating single band obtained on electrophoresis in the presence of G-6-P (Fig. 5B). The activity near the origin in both gels would then correspond to the minor high molecular weight components observed in the density gradient centrifugations. Concomitant with the physical changes produced by G-6-P, as well as other ligands, are the kinetic changes observed upon preincubation with G-6-P (Fig. 6).

Steiner et al. (11) have reported a similar co-occurence of activation and increased sedimentation rate after exposure to G-6-P of their preparation, which as noted above appears to be predominantly glycogen synthetase a. An elevation of liver glycogen synthetase a activity upon preincubation with G-6-P has been observed in this laboratory as well[3]. In density gradient experiments with a partially purified preparation of rat muscle synthetase containing substantial amounts of both a and b forms, Staneloni and Piras (19) have also observed shifts toward higher molecular weight forms in the presence of G-6-P and UDPG. ATP reversed these effects. As with the present preparation, muscle glycogen synthetase a purified by Soderling et al. (6) appeared as a single band on disc gel electrophoresis in the presence of G-6-P, and dissociated into 2 bands in its absence, with greatly reduced activity. We have also

---

[3] D.C. Lin and H.L. Segal, unpublished experiments.

observed multiple peaks of activity in sucrose density gradient centrifugation experiments with the trout liver enzyme[3].

Thus it appears that both the phosphorylated (b) and nonphosphorylated (a) forms of glycogen synthetase are capable of undergoing reversible conformational changes between active and inactive (or less active) states. What, if any, significance in respect to the physiological control of glycogen synthetase activity may attach to these properties remains to be discovered.

A possible relationship between the glycogen synthetase and the phosphorylase systems has emerged from the findings of Hers and his coworkers (4,10). In particular they have proposed that phosphorylase a is an inhibitor of the synthetase phosphatase. An alternative possibility is that phosphorylase phosphatase and glycogen synthetase phosphatase are the same enzyme, but that phosphorylase a is a superior substrate. The stimulating effect of glucose uniquely on the dephosphorylation of phosphorylase a could reflect an interaction of the ligand with the substrate rather than with the phosphatase, similarly for the effect of Mg^{+2} uniquely on synthetase dephosphorylation as reported here.

The loss of glycogen synthetase phosphatase activity, in contrast to phosphorylase phosphatase activity, in starved, adrenalectomized rat livers (7), remains to be fitted in. We earlier postulated that synthetase phosphatase (or an activating enzyme thereof) was absent in these preparations and induced by glucocorticoids, since the glucocorticoid effect was blocked by actinomycin and cycloheximide (8). The restoration of activity by glucose, however, was inhibited only by cycloheximide and not actinomycin. A different interpretation of these findings can be proposed. Namely, that the loss of synthetase phosphatase activity is the result of the depletion of a factor (perhaps G-6-P) essential for the conversion reaction, which can arise from glucose, or by gluconeogenesis in response to glucocorticoids. It would then be the latter which depended upon enzyme induction by the glucocorticoid. It is clear that a definitive selection among these possible interpretations will require purification of the phosphatase or phosphatases involved.

We wish finally to propose some views of the metabolic significance of interconvertible enzymes. We suggest that in several instances one of the forms - the less active, or b form - is responsive to restraints which reflect conditions internal to the cell, that is, it is a unit integrated into the cell's overall economy and adapted to function in what might be termed a resting state; while the more active, or a form, is released from intracellular restraints, permitting maximum flow through the reaction. The conversion to the a form is in response to a signal external to the cell, which has the effect of overriding the internal signals. A simple example is the ability of glucagon to promote glucose release from the liver by conversion of phosphorylase $\underline{b}$ to $\underline{a}$ (20). The consequent increased rate of glycogenolysis serves no metabolic need of the liver per se, but rather subordinates those needs to respond to a situation elsewhere in the organism. A similar argument applies to the promotion by high glucose levels of glycogen deposition through a conversion of glycogen synthetase $\underline{b}$ to $\underline{a}$ (21), and the promotion by glucagon and epinephrine of lipolysis through a conversion of lipase $\underline{b}$ to $\underline{a}$ (22). In these instances the $\underline{b}$ form would appear to be the more primitive, with the capability of conversion to $\underline{a}$ forms arising as an adaptive mechanism in evolutionary and embryonic development.

In other instances the interconversions appear simply to be in response to alternative nutritional conditions. For example, when carbohydrate is the main available energy source, pyruvic dehydrogenase is in the $\underline{a}$ form, while the presence of high levels of free fatty acids promotes conversion to the $\underline{b}$ form, thereby sparing the utilization of carbohydrate (23). Similarly with glutamine synthetase of E. coli where the $\underline{a}$ form prevails when N is limiting and C supplies are in excess, but is converted to the $\underline{b}$ form when N is in excess and the product of the reaction accumulates (24).

Acknowledgement

Some of the experiments reported here were carried out in the laboratory of Prof. Tag E. Mansour of the Pharmacology Department, Stanford University, whom we wish to thank for his help and generosity.

References

1. HORNBROOK, K.R., BURCH, H.B., and LOWRY, O.H., Biochem. Biophys. Res. Commun., 18, 206 (1965).
2. HORNBROOK, K.R., BURCH, H.B., and LOWRY, O.H., Mol. Pharmacol., 2, 106 (1966).
3. DE WULF, H., STALMANS, W., and HERS, H.G., Europ. J. Biochem., 15, 1 (1970).
4. STALMANS, W., DE WULF, H., and HERS, H.G., Europ. J. Biochem., 18, 582 (1971).
5. MERSMANN, H.J., and SEGAL, H.L., Proc. Nat. Acad. Sci., 58, 1688 (1967).
6. SODERLING, T.R., HICKENBOTTOM, J.P., REIMANN, E.M., HUNKELER, F.L., WALSH, D.A., and KREBS, E.G., J. Biol. Chem., 245, 6317 (1970).
7. MERSMANN, H.J., and SEGAL, H.L., J. Biol. Chem., 244, 1701 (1969).
8. GRUHNER, K., and SEGAL, H.L., Biochim. Biophys. Acta, 222, 508 (1970).
9. HERS, H.G., DE WULF, H., and STALMANS, W., FEBS Letters, 12, 73 (1970).

10. STALMANS, W., DE WULF, H., LEDERER, B., and HERS, H.G., Europ. J. Biochem., 15, 9 (1970).

11. STEINER, D.F., YOUNGER, L., and KING, J., Biochemistry, 4, 740 (1965).

12. SEVALL, J.S., and KIM, K.H., Biochim. Biophys. Acta, 206, 359 (1970).

13. HASSID, W.Z., and ABRAHAM, S., Methods in Enzym., 3, 34 (1957).

14. LOWRY, O.H., ROSEBROUGH, N.J., FARR, A.L., and RANDALL, R.J., J. Biol. Chem., 193, 265 (1951).

15. MARTIN, R.G., and AMES, B.N., J. Biol. Chem., 236, 1372 (1961).

16. GOLD, A.H., Biochemistry, 9, 946 (1970).

17. GORNALL, A.G., BARDAWILL, C.J., and DAVID, M.M., J. Biol. Chem., 177, 751 (1949).

18. GOLD, A.H., and SEGAL, H.L., Arch. Biochem. Biophys., 120, 359 (1967).

19. STANELONI, R.J., and PIRAS, R., Biochem. Biophys. Res. Commun., 42, 237 (1971).

20. RALL, T.W., SUTHERLAND, E.W., and WOSILIAT, W.D., J. Biol. Chem., 218, 483 (1956).

21. DE WULF, H., and HERS, H.G., Europ. J. Biochem., 2, 50 (1967).

22. HUTTENEN, J.K., STEINBERG, D., and MAYER, S.E., Proc. Nat. Acad. Sci., 67, 290 (1970).

23. WIELAND, O., SIESS, E., SCHULZE-WETHMAR, F.H., v. FUNCKE, H.G., and WINTON, B., Arch. Biochem. Biophys., 143, 593 (1971).

24. HOLZER, H., SCHUTT, H., and MASEK, Z., Proc. Nat. Acad. Sci., 60, 721 (1968).

Discussion:

Sols

Does glucose inhibit phosphorylase $\underline{a}$ iso-or allosterically?

Helmreich

I would like to comment on Dr. Sols question:
Glucose competes with glucose-1-phosphate for the same site, but the release from inhibition by AMP is an allosteric effect.

Sols

But glucose, glucose-1-phosphate and AMP are not chemically related.

Helmreich

True, but it seems very unlikely that AMP would compete with glucose for the same site to which glucose-1-phosphate binds. However, Dr. Stalmans has shown that you can counteract the glucose effect by AMP. One might assume then that 5'-AMP induces an allosteric transition to a form of the enzyme which binds glucose-1-phosphate better than glucose and thus relieves inhibition.

Fischer

This is a question to either of the two speakers, Dr. Stalmans and Dr. Segal. Both have indicated the possible interaction of the synthetase phosphatase with phosphorylase $\underline{a}$. Do you see any interaction using other methods. For instance, any change in the behaviour of one of these enzymes on Sephadex chromatography? Do you have any information on the molecular weight of this phosphatase? Although I understand it has not been purified I am asking this question, because rabbit muscle phosphorylase phosphatase has been purified approximately a 1,000-fold or a little more. It has in its purest form a very low molecular weight - 33,000 at the maximum. Would you care to comment on that?

Cori

I would like to ask, what is the ratio of activities in the liver when you have both phosphorylase and the synthetase in the $\underline{a}$ form?

Stalmans

The ratio would be about 10 : 0.5. In fact the total amount of phosphorylase $\underline{a}$ one measures is much greater than the amount needed to account for the highest rates of glycogenolysis that one can observe. I suppose, there are two possibilities: either phosphorylase $\underline{a}$ is never completely active or phosphorylase $\underline{a}$ in the liver is inhibited quite extensively by metabolites, such as glucose and glucose-6-phosphate.

Segal

As shown in Fig. 7 of my talk the phosphorylase activity in liver expressed as μmoles/min/g of liver is 8, and the maximum synthetase activity after full conversion is a little over 1, so that the ratio is about 8 : 1 even considering that not all of the phosphorylase has been converted to the active form $\underline{a}$. Madsen has calculated that there's enough phosphorylase activity in resting muscle to deplete all the glycogen in a third of an hour whereas approximately 12 hours are required for the depletion of the liver glycogen in a starved rat. Hence phosphorylase was about 35-36 times too active, so there must be something restraining the enzyme and preventing the expression of 97 - 98 % of its potential activity.

The Mechanism of Activation of the Lac Operon

Geoffrey Zubay

Department of Biological Sciences, Columbia University,
New York, New York/USA

The lac operon consists of a cluster of genes found on the single chromosome of the bacterium Escherichia coli. Starting from one side the cluster contains a promoter, an operator, the structural genes for β-galactosidase, permease and thiogalactoside transacetylase.[1] Such a grouping of regulating and structural genes is termed an operon. In 1961 Jacob and Monod proposed a mechanism for the regulation of the activity of this operon. At this time no promoter region had been defined although no reasonable person would have denied that there must be preferred regions for the initiation of RNA synthesis. The Jacob-Monod mechanism proposed that a repressor molecule produced from a nearby i gene forms a strong complex with the operator which prevents gene expression, i.e. RNA and protein synthesis. The complex can be broken by the low spontaneous rate of dissociation of the complex or more effectively by interceding of a small molecule inducer which is a derivative of lactose. It so happens that lactose is the natural substrate of the enzymes encoded by the operon indicating the immediate value of such a system of regulation. Thus if substrate is available for a certain class of enzymes, one produces the enzymes necessary for action on this substrate but not otherwise. Since the Jacob-Monod model was proposed, this negative control mechanism functioning between repressor and operator has been verified by direct isolation of the repressor and demonstration of its conditional affinity for operator both on purified DNA[2] and in in vitro systems for transcription and translation of lac messenger RNA.[3]

In spite of its elegance and correctness the Jacob-Monod model was only half the story about how the lac operon becomes activated. Events which have taken place in the past seven years have shown that the promoter locus is essential for gene expression[4] and is a site where an elaborate positive control system functions to make gene expression possible. This positive control system functioning at the promoter locus consists of the σ-core RNA polymerase, 3'5' cyclic AMP and CAP, the catabolite gene activator protein.[5] Guanosine tetraphosphate, ppGpp, provides additional stimulus to the positive control system.[6] In this paper I shall describe the present state of our understanding on how this positive control system works. We shall ignore further mention of the negative control exerted at the operator site since it is clear that the repressor-operator complex prevents initiation or propagation of RNA synthesis in the way described. Little more can be said about the repressor-operator complex until we know the detailed molecular structure.

The first indication that a promoter locus existed arose out of genetic studies in which it was shown that deletions or alterations in this region of the chromosome could seriously impair the expression of the operon even though it was clear that the structural genes were still intact.[4] Later more subtle mutations showed that the promoter locus could be modified so that gene expression is no longer inhibited by the general phenomenon known as catabolite repression.[7] Catabolite repression involves the inhibition of synthesis of a wide variety of catabolic enzymes (including β-galactosidase) when cells are grown on glucose or a variety of other related carbon sources.[3] The extent of inhibition may be great or small depending upon the exact growth conditions. The genetic observations showing that catabolite repression could be eliminated for the lac operon (and only the lac operon) by modification of the lac promoter, pinpointed the ultimate site at which the catabolite repression mechanism operates. It seemed likely that the elucidation of the catabolite repression mechanism for the lac operon would have general implications for how this mechanism operates for other catabolite sensitive genes.

The next clue as to how the catabolite repression mechanism works came from the work of Makman and Sutherland in 1965.[8] These workers showed that in the presence of sufficient glucose to cause catabolite repression in E. coli, the cyclic AMP concentration falls drastically from about 10^{-4}M to 10^{-7}M. This observation was followed up by others[9] who showed that cyclic AMP (cAMP) added with glucose inhibits the onset of catabolite repression. Thus a reasonable level of cAMP seemed to be required for expression of catabolite sensitive genes. The level of cAMP could be controlled either by its rate of synthesis from ATP or its rate of breakdown to 5'-AMP or both.[10]

Further progress in understanding how AMP functions was made primarily through the use of a cell-free DNA-directed system for protein synthesis which had been developed in our laboratory.[11] This type of cell-free system is proving to be of value in the study of many gene systems both at the transcriptional and translational level. The DNA-directed cell-free system for β-galactosidase (β-gal) synthesis contains DNA with the lac operon, a cell-free extract of E. coli and the cofactors and substrates essential for RNA and protein synthesis. In 1968 we were able to show that normal initiation of lac operon messenger synthesis had a strict requirement for cAMP.[12] Actually cAMP provided a twenty-fold stimulus to the amount of β-gal enzyme synthesized in the cell-free system but the small amount of residual synthesis occurring in the absence of cAMP was attributed to <u>false initiation</u> since it was insensitive to the state of the lac promoter. The cAMP stimulated β-gal synthesis is completely dependent upon the presence of the normal lac promoter. The sensitivity of the cell-free coupled system to cAMP suggested its usefulness as an assay tool for isolating the cAMP receptor. Whereas we knew that the promoter was the ultimate site of action of cAMP it seemed likely that this action must be mediated by a protein. The immediate task was to attempt to isolate E. coli mutants with defective cAMP receptors which could be studied in the cell-free system. This proved to be an easy task provided the proper selection techniques were applied. Mutagenized E. coli cells which could not metabolize arabinose or maltose were selected. Such cells did not have defects in the arabinose or maltose operons but rather in the general machinery for activating catabolite sensitive genes. About half of the mutants isolated could be phenotypically corrected by growing on cAMP. Apparently these mutants had defects in their ability to maintain a useful level of cAMP. The other half of the mutants could not be phenotypically corrected with cAMP. It seemed likely that these latter mutants had their defects in the cAMP receptor protein. Extracts of these mutants were studied in the coupled cell-free system. When used in conjunction with normal lac operon DNA such extracts produced only five percent of the normal level of β-gal synthesis and were unaffected by the presence of cAMP. Thus it appeared that the cell-free behavior was mimicking the whole cell behavior. Next a small amount of cell-free extract from a normal E. coli strain was found to greatly augment the amount of β-gal synthesis. This augmentation was completely dependent upon the presence of cAMP. Ultimately the cAMP receptor protein was purified from the normal extract using the cell-free stimulation of β-gal synthesis to trace its isolation.[13]

The cAMP receptor protein or more correctly the catabolite gene activation protein (since 80 per cent of the proteins which binds cAMP in E. coli is other protein) has been found to be a **dimer** consisting of identical 22,000 M.W. subunits.[14] The dimer binds cAMP with a formation constant of 0.6×10^5 liters moles^{-1}.[13] Stimulation of β-gal synthesis by the catabolite gene activation protein (CAP) in the defective cell-free system described above requires the presence of cAMP and the quantitative dependence is consistent with the cAMP-CAP formation constant. Thus about half maximum stimulation of β-gal synthesis occurs at a concentration of cAMP at which it may be calculated from the formation constant that the CAP binding sites for cAMP are half occupied.

In order to further clarify the mechanism of action of CAP, interaction studies were carried out with RNA polymerase and DNA. CAP does not have any measurable affinity for RNA polymerase. On the other hand CAP strongly binds to DNA but only in the presence of cAMP with a concentration dependence on cAMP consistent with the binding curve for cAMP to CAP. The dependence of CAP on cAMP for binding to DNA, strongly implicates the interaction between CAP and DNA as the significant factor in activation. Thus CAP appears to be a new type of regulatory protein -- a DNA binding activator. An unexpected observation in the CAP-DNA binding studies is that the CAP appears to bind equally well to all parts of the DNA with no special affinity for the promoter region. Studies directed towards finding conditions under which CAP might bind preferentially to the lac promoter region are continuing.

Another approach to demonstrating the importance of CAP and cAMP to lac messenger transcription has been to construct a purified system for RNA synthesis. This system contains λlacDNA, σ-core polymerase, cAMP, CAP, Mg^{+2} and the four ribotriphosphates, ATP, UTP, CTP and GTP.[5] The RNA synthesized in this system is characterized by hybridization of radioactive RNA to DNA template. As in the coupled system it is found that both CAP and cAMP are required for lac messenger synthesis. Use of the purified system for RNA synthesis demonstrates most clearly that the control operates at the transcriptional level but it does not rule out the possibility that there is some control at the translational level. It also shows that both σ factor and CAP are required. For some phage genes at least it has been clearly demonstrated that σ factor alone is sufficient for initiation but for the lac messenger the two protein factors, σ and CAP, are both required.

It was mentioned earlier that ppGpp stimulates the lac operon. This factor also stimulates the arabinose operon but has no effect on the tryptophan operon or the tRNA genome.[15] We do not know how this factor operates in the system.

Except for ppGpp we now appear to understand what factors are required for activating the lac operon for transcription. Still to be determined is the temporal sequence of events in initiation. The next major phase of research in understanding this system will require that we determine the detailed molecular structure of the interacting molecules.

ACKNOWLEDGMENT

I have had many collaborators and helpers throughout the course of this research. I would particularly like to mention J.R. Beckwith, D.A. Chambers, J.K. DeVries, M. Lederman and A.D. Riggs. This work was supported by grants from the National Institutes of Health, the National Science Foundation, the American Cancer Society and the Damon Runyon Memorial Fund.

References

1. ZUBAY, G. and CHAMBERS, D., Regulation of the Lac Operon in Metabolic Regulation, Academic Press (1971); The Lac Operon, ed. D. Zipser and J.R. Beckwith, Cold Spring Harbor, N.Y., Laboratory of Quantitative Biology (1970).

2. GILBERT, W. and MULLER-HILL, B., Proc. Natl. Acad. Sci., 58, 2415 (1967).

3. ZUBAY, G., CHAMBERS, D.A. and CHEONG, L.C., in The Lac Operon, ed. D. Zipser and J.R. Beckwith, Cold Spring Harbor, N.Y., Laboratory of Quantitative Biology (1970).

4. JACOB, F., ULLMAN, A. and MONOD, J., Compt. Rend., 258, 3125 (1964); IPPEN, K., MILLER, J.H., SCAIFE, J. and BECKWITH, J., Nature, 217, 825 (1968).

5. ARDITTI, R., ERON, L, ZUBAY, G., TOCCHINI-VALENTINI, G., CONNAWAY and BECKWITH, J., Cold Spring Harbor Symp. Quant. Biol., 35, 437 (1970); DECROMBRUGGHE, B., CHEN, R., ANDERSON, M., NISSLEY, B., GOTTESMAN, M., PASTAN, I. and PERLMAN, R., Nature N.B., 231 139 (1971); ERON, L. and BLOCH, K., Proc. Natl. Acad. Sci., 68, 1828 (1971).

6. TRAVERS, A., KAMEN, A. and SCHLEIF, R., Nature, 228, 748 (1970).

7. SILVERSTONE, A.E., MAGASANIK, B., REZNIKOFF, W.S., MILLER, J.H. and BECKWITH, J.R., Nature, 221, 1012 (1969).

8. MAKMAN, R.S. and SUTHERLAND, E.W., J. Biol. Chem, 240, 1309 (1965).

9. PERLMAN, R.L. and PASTAN, I., J. Biol. Chem., 243, 5420 (1968); ULLMAN, A. and MONOD, J., FEBS Letters, 2, 57 (1968).

10. JOST, J.P. and RICKENBERG, N.V., An. Rev. Biochem., 40, 741 (1971).

11. DEVRIES, J.K. and ZUBAY, G., Proc. Natl. Acad. Sci., 57, 1010 (1967).

12. CHAMBERS, D.A. and ZUBAY, G., Proc. Natl. Acad. Sci., 63, 118 (1969); ZUBAY, G. and CHAMBERS, D.A., Cold Spring Harbor Symp. Quant. Biol., 34, 753 (1969).

13. ZUBAY, G., SCHWARTZ, D. and BECKWITH, J., Proc. Natl. Acad. Sci., 66, 104 (1971).

14. ZUBAY, G., SCHWARTZ, D. and BECKWITH, J., Cold Spring Harbor Symp. Quant. Biol., 35, 433 (1970); RIGGS, A.D., REINESS, G. and ZUBAY, G., Proc. Natl. Acad. Sci., 68, 1222 (1971).

15. ZUBAY, G., GIELOW, L. and ENGLESBERG, E., Nature N.B., 233, 164 (1971); ZUBAY, G., CHEONG, L. and GEFTER, M., Proc. Natl. Acad. Sci., 68, 2195 (1971); ZUBAY, G.; unpublished results.

Discussion:

Hartmann

You have evidence for the role of CAP in the start of transcription process - that is, genetic evidence that mutated promoters have no sensitivity to CAP anymore. Have you evidence of any other mutations, not in the promoter region, which obviate the requirement for CAP?

Zubay

No

N.N.

Do you know if anyone has tried to look for transcription with non-specific DNA - say, poly-dAT - to look whether transcription of such DNA is also stimulated by CAP and cyclic AMP?

Zubay

We haven't done anything quite that simple, but we have looked at other operons including the tryptophan operon. There is no effect of c-AMP or CAP in this case.

N.N.

I ask because sigma, which was supposed to be specific, also enhances transcription of poly-dAT.

Zubay

Yes, but we now know that sigma isn't as specific as it was supposed to be.

Dixon

Has anyone done the experiment of taking the CAP - cAMP - DNA complex and treating it with phosphodiesterase to see if the cAMP could be broken down and thus the binding reversed?

Zubay

No.

Regulatory Mechanism of Enzyme Catabolism

N. Katunuma, E. Kominami, S. Kominami, K. Kito, and T. Matsuzawa

Department of Enzyme Chemistry, Institute for Enzyme Research, School of Medicine, Tokushima University, Tokushima City/Japan

Abstract: Three kinds of new enzymes were discovered, those specifically inactivate pyridoxal enzymes or NAD-dependent enzymes or FAD-dependent enzymes in their apoforms in small intestine, liver and skeletal muscle of rats, respectively. These inactivating enzymes were purified until almost homogeneous protein and their some enzyme-chemical properties were studied. Each purified inactivating enzyme does not act on the other types of coenzyme-dependent enzymes or any other proteins tested. These inactivating enzymes split each apo-enzyme to a homogeneous smaller protein and an oligopeptide in alkaline condition. The inactivation by these enzymes is protected by the addition of each respective coenzyme. The activities of these inactivating enzymes in small intestine increase only under the condition of the respective vitamin deficiency. The large part of these inactivating enzymes exist as latent enzyme of inactive forms by binding with specific inhibitor in normal dietary condition. The amount of the inhibitor decreases in the vitamin-deficient condition.

These inactivating enzymes might play very important roles in the control of degradation of coenzyme-dependent enzymes. It may be considered that these inactivating enzymes can serve as an initiator of these enzyme degradation, and the products formed by our inactivating enzymes are subsequently degradated into amino acids by lysosomal or other non-specific proteinases.

INTRODUCTION

In general, it is well known that intracellular level of enzymes is determined by the velocity balance between the synthesis and the degradation of apo-enzyme protein. Understanding of protein synthesis and its regulation is now fairly extensive and developed, but the information on the mechanism and control of the degradation at the molecular level is rather scarce. The possibility of controlling enzyme levels by changing the degradation rate has been brought out by the work of Schimke et al.(1). Segal et al.(2), and de Duve et al.(3) also emphasized the importance of degradation rate for the control of the enzyme levels in the cell. But our knowledge on the mechanism of enzyme protein degradation have been not enough to explain the individual fate of various enzyme proteins.

We discovered three kinds of new enzymes which specifically inactivate pyridoxal enzymes or NAD-dependent enzymes or FAD-dependent enzymes in their apo-forms in small intestine, liver and skeletal muscle of rats. These inactivating enzymes were purified until almost homogeneous protein. Each purified inactivating enzyme does not act on the other types of coenzyme-dependent enzymes or any other proteins tested. These inactivating enzymes split each apo-enzyme to a homogeneous smaller protein than the apo-enzyme as substrate and an oligopeptide in alkaline condition. The activities of these inactivating enzymes increase only under the condition of the respective vitamin deficiency. These inactivating enzymes might play very important roles in the control of degradation of coenzyme dependent enzymes.

Part I. Inactivating enzyme for pyridoxal enzymes

Purification and assay method. The small intestine of the vitamin B_6 deficient rats was homogenized with 0.05M potassium phosphate buffer(pH 7.5) and the homogenate was sonicated. Centrifugation was carried out at 10,000xg for 10 min. The supernatant was passed over a column of Sephadex G-25. The protein fraction(crude extract) were used for the further purification. The crude extract was fractionated with 40-70% saturation of ammonium sulfate, then fractionated with 40-70% saturation of acetone. The fraction dissolved in 0.05M potassium phosphate buffer(pH 7.5) was passed through Sephadex G-25 column chromatography, and then applied to DEAE-cellulose column cgromatography which had been equilibrated with 0.05M potassium phosphate buffer(pH 7.5). The eluate with 0.2M of the buffer was collected and concentrated with addition of 70% ammonium sulfate. The precipitate dissolved in a small volume of 0.05M phosphate buffer(pH 7.5) was applied onto a column of Sephadex G-100 (2.0x80 cm), equilibrated with the same buffer. A summary of purification of the inactivating enzyme for pyridoxal enzymes is shown in Table 1. The purity increased 300-400 fold with an average yield of 10%. Inactivating enzyme from control rats was also purified using the same procedures. At this time the enzyme was approximately 2,000-3,000 times pure than crude extract. Both purified preparations showed homo-

Table I. Purification of the inactivating enzyme for pyridoxal enzymes from small intestine of B_6 deficient rats

Fraction	Volume (ml)	Protein concentration (mg/ml)	Total activity (units)	Specific activity (units/mg)	Recovery (%)
Crude extract	420	4,410	76,000	17	
Ammonium sulfate	75	1,125	61,000	54	80.5
Acetone	25	250	37,000	148	48.7
DEAE-cellulose	2.5	37	27,500	680	35.2
Sephadex G-100	0.9	1.9	9,540	5,050	12.5

geneous band on cellulose acetate membrane electrophoresis at pH 8.0. Molecular weight of the inactivating enzyme is determined to be around 30,000 by gel filtration using ovalbumin and chymotrypsin as the markers.

Usual assay of the inactivating enzyme was performed as follows; reaction mixture contained in a final 0.5 ml, 0.05M Tris-HCl buffer(pH 9.0), 40-60 units of ornithine transaminase(OTA) and various amounts of the inactivating enzyme. After incubation at 37°C, the reaction was terminated by 10 fold dilution with cold buffer. Subsequently, the remaining activity of OTA was assayed in the presence of pyridoxal phosphate($20\mu M$). One unit of the inactivating enzyme was defined as the amount inactivating 50% of substrate enzymes after 30 min. reaction(6).

Mode of action. The gel filtration were performed to elucidate the nature of the products of the inactivating reaction. Crystalline apo-enzyme was inactivated in about 75% by incubation with the purified inactivating enzyme, and applied onto a column of Sephadex G-100. The elution profile is presented in Fig. 1(b).

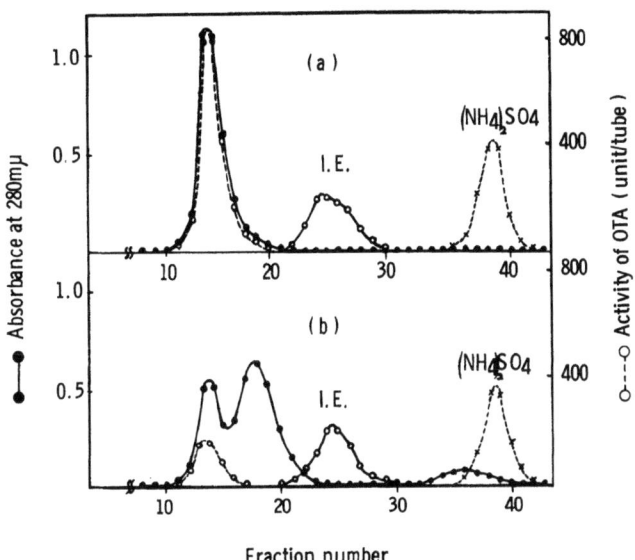

Fig. 1. Chromatography of the reaction products containing OTA and the inactivating enzyme on Sephadex G-100. Sephadex G-100(1.8x50 cm) was equilibrated with 0.05M potassium phosphate buffer(pH 7.5) containing $10^{-3}M$ mercaptoethanol. The flow rate was adjusted to 8.0 ml/hr. 12 mg of crystalline holo-OTA(a) or apo-OTA(b) was incubated with 2.0 mg of the purified inactivating enzyme for 12 min. After the reaction was stopped by standing at 0°C, the reaction mixture was applied onto the chromatography. $(NH_4)_2SO_4$ was added as a marker

The first peak in solid line indicates remaining apo-OTA and the second peak in solid line shows the product protein derived from OTA by the enzyme reaction. No OTA activity was observed in the second peak. The activity of the inactivating enzyme appears in the third peak indicated in solid line, and the forth peak seems to be an oligopeptide. The elution pattern showed in Fig. 1(a) is the case that holo-

OTA was used as substrate. All the OTA activity added was recovered in the first peak, and no other proteins or peptide were detected. These results indicate that the inactivating enzyme splits OTA into two products, a homogeneous smaller protein and oligopeptide. Thus the inactivation reaction is one kind of specific endopeptidase reaction. Ultracentrifugal analyses of the first and second peaks which are separated by Sephadex G-100, were performed. Both proteins were homogeneous in ultracentrifugal analysis. The sedimentation constants in 0.1M potassium phosphate buffer of pH 7.5 were 9.9 S and 5.1 S for first and second peaks, respectively. Electrophoretic pattern of both proteins on cellulose acetate membrane also proved to be homogeneous, and mobilities are quite similar each other. This second peak has no activity of OTA, but it is still precipitable by the addition of antibody for OTA. These observations suggest that the product is major remainder after liberating oligopeptide from OTA monomer unit. The other product, oligopeptide, was also homogeneous and the following amino acids composition was obtained from duplicate preliminary analyses; Lys. Asp. Thr. Ser. Glu. Gly. Ala. Leu..

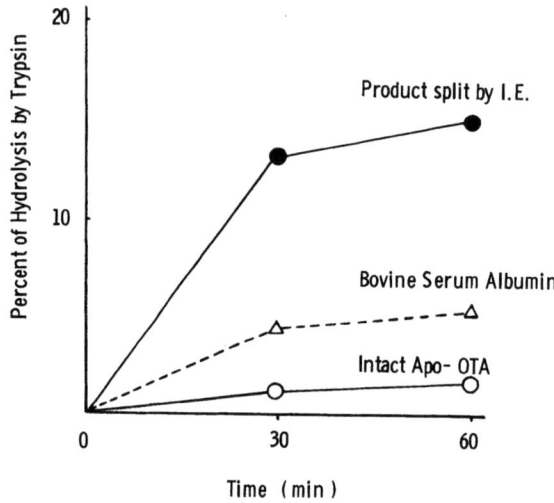

Fig. 2. Tryptic digestion of apo-OTA and its product split by the inactivating enzyme for pyridoxal enzymes

Apo-OTA and its product split by the inactivating enzyme was digested by the trypsin as shown in Fig. 2. Apo-OTA was unable to be degradated by trypsin treatment, while its product was markedly degradated by trypsin. These data suggest that the inactivating enzymes can serve as an initiator of these enzyme degradation, and the products formed by our inactivating enzymes are subsequently degradated into amino acids by other non-specific proteinases.

Properties. We studied the substrate specificity on the purified inactivating enzyme. The inactivating enzyme specifically inactivates ornithine transaminase, serine dehydratase and tyrosine transaminase in their apo-forms and PAMP forms at various degree, but all the non-pyridoxal enzymes tested were not affected by the inactivating enzyme as shown in Table II. Michaelis constant of the inactivating enzyme for apo-OTA as substrate was calculated from the double reciplocal plots. Since the molecular weight of OTA was 132,000(4), Km for OTA was estimated to be 1.28×10^{-10}M. The inactivating enzyme has quite small Km value for OTA as substrate. Optimal pH of the inactivating enzyme was studied on both crude and purified preparations and both case gave around pH 9.0.

Table II. Substrate specificity of the inactivating enzyme for pyridoxal enzymes. Values indicate the activities against substrate enzymes when that of ornithine transaminase was 100%

Pyridoxal enzymes	%	Non Pyridoxal enzymes	%
Ornithine transaminase	100	Glutamic dehydrogenase	0
Tyrosine transaminase	21	Lactic dehydrogenase	0
Serine dehydratase	140	Urease	0
		Glutaminase (phosphate independent)	0

Effects of vitamin B_6 derivatives and other related compounds to the activity of inactivating enzyme were examined and the results were shown in Table III and Fig.3.

The inactivation was suppressed markedly by the addition of pyridoxal phosphate and weakly by pyridoxal. No protecting effect was observed in addition of pyridoxamine phosphate, pyridoxine phosphate and other compounds tested. The concentration of PALP or PAL required to 50% prevention of the action of inactivating enzyme was calculated to be 6.0×10^{-6}M or 1.6×10^{-4}M. Peraino(4) reported that 58 lysine residues were contained in one molecule of OTA, and only two of them form the shiff base with pyridoxal phosphate. The special 2 lysine residues must be kept free for the present inactivating reaction. It is reasonable from above data that 30 times more PAL than PALP was required to pretect against the inactivating reaction. Tissue distribution of the inactivating enzyme in B_6 deficient rat was studied. The activity was high in small intestine and weak in skeletal muscle, but no activity was detected in liver, kidney, brain, heart muscle and spleen. The activities of the inactivating enzymes increase under the condition of B_6 deficiency(6), but the activity of the inactivating enzyme does not increase under the niacin deficiency or B_2 deficiency as shown in Table IV. On the contrary the activity of inactivating enzyme for NAD-dependent enzyme is not induced under B_6 deficiency.

Conclusions. It might involve very important control mechanisms that PALP form enzyme is unable to split by the inactivating enzyme but PAMP form enzyme is easily

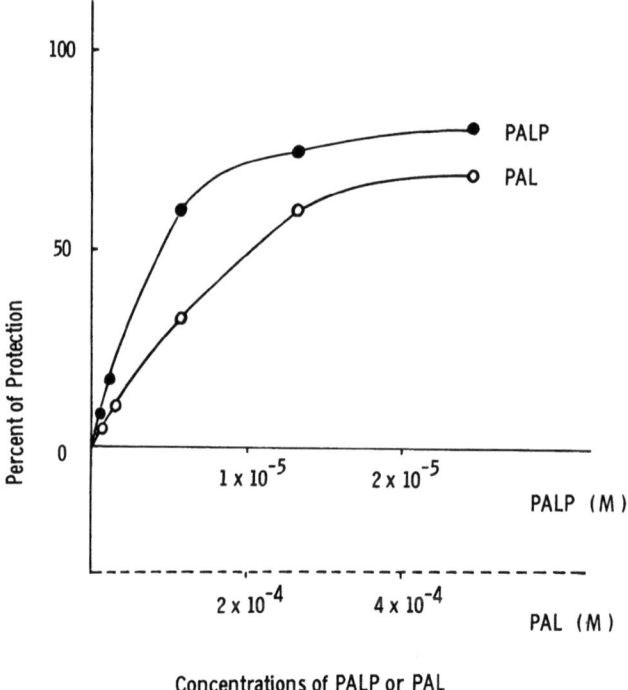

Fig. 3. Protective effect of PALP or PAL on the activity of inactivating enzyme for pyridoxal enzymes at the diverse concentrations

split by the inactivating enzyme in spite of the both forms are freely interconvertible during the transaminase reaction. When the concentration of amino acids in a given tissue increases by feeding high protein diet, relative amount of PAMP form enzyme may increase. The extents of decrease in various pyridoxal enzymes activities in B_6 deficient rats were compared in feedings with high protein diet and with low protein diet by Okada et al.(5). Marked decreases of rat liver pyridoxal enzymes were observed in the case of high protein diet. We would like to offer the following explanation for the regulatory mechanism of pyridoxal enzymes catabolism as shown in Fig. 4.

Part II. Inactivating enzyme for NAD-dependent enzymes

Purification. Almost the same procedure was applied to the purification of inactivating enzyme for NAD-dependent enzymes as that for pyridoxal enzymes, as shown in Table V. A few characteristics of the inactivating enzyme are pointed out as follows: the inactivating enzymes for pyridoxal enzymes and for NAD-dependent enzymes are able to separate in different ammonium sulfate fractions and their behaviors on DEAE-cellulose column chromatography are also different. The major part of the inactivating enzyme activity for apo-NAD dependent enzymes is precipitated between 35-

Table III. Protective effect of vitamin B_6 derivatives and other compounds on the activity of inactivating enzyme for pyridoxal enzymes

Addition	Final concentration(M)	Enzyme activity	% of protection
None	-	4.00	0
PAMP	2.5×10^{-4} 2.5×10^{-3}	3.81 3.91	5.0 2.0
PINP	2.5×10^{-4} 2.5×10^{-3}	3.74 3.82	6.4 4.4
NAD	2.5×10^{-4} 2.5×10^{-3}	4.00 4.01	0 0
FAD	2.5×10^{-4} 2.5×10^{-3}	3.96 4.01	1.1 0
ATP	2.5×10^{-4} 2.5×10^{-3}	4.00 3.95	0 1.3
PALP	2.5×10^{-5}	0.63	83.7
PAL	2.5×10^{-4}	1.35	66.2

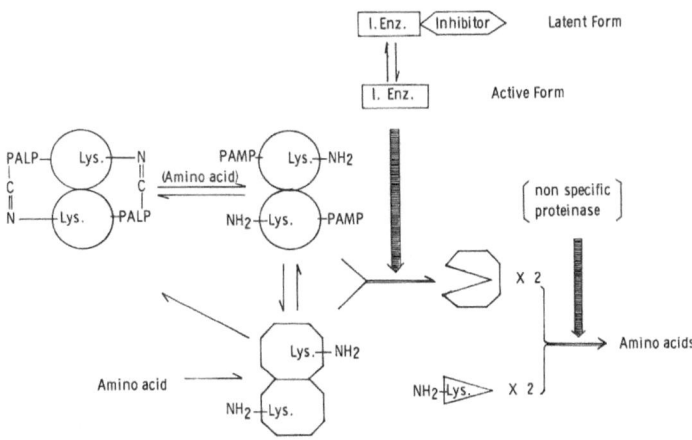

Fig. 4. Regulatory mechanism on catabolism of pyridoxal-dependent enzymes

Table IV. The levels of the inactivating enzyme activity in small intestine under the various dietary conditions.

Rats were fed on laboratory chow ad libitum before the experiment. Values are expressed as mean values ± standard errors

Dietary condition	No. of rats	Activity of the inactivating enzyme(units/mg protein)
Chow	18	0.78±0.21
5% protein diet(5 days)	5	0.12±0.01
70% protein diet (5 days)	5	0.66±0.05
Starvation (2 days)	3	0.63±0.03
Niacin-deficient diet (2 weeks)	33	1.11
Vitamin B_2-deficient diet (4 weeks)	16	1.50
Vitamin B_6-deficient diet (4 weeks)	12	15.5±2.3

Table V. Purification of the inactivating enzyme for NAD-dependent enzymes from small intestine of niacin deficient rats

	Volume (ml)	Protein concentration (mg/ml)	Total activity (units)	Specific activity (unit/mg)	Recovery (%)
Crude extract	370	3,700	238	0.0645	100
1st ammonium sulfate	105	714	735	1.05	308
IInd ammonium sulfate	26	161	565	3.50	237
Sephadex G-100	0.7	2.1	133.5	6,410	56

55% ammonium sulfate saturation, while the inactivating enzyme for pyridoxal enzymes is fractionated between the 50-70% saturation. Marked increase of total activity is observed after the ammonium sulfate fractionation. The phenomenon may suggest the existence of inhibitor in 55-70% fraction which can be separated from the inactivating enzyme.

The purified inactivating enzyme for NAD-dependent enzymes was proved almost homogeneous by cellulose acetate membrane electrophoresis in acid or alkaline pH. Assay method and definition of enzyme unit was almost same as that of inactivating enzyme for pyridoxal enzymes(7). Remaining activity of NAD-dependent enzymes was measured by following the rate of the decrease in absorbance of $NADH_2$ at 340 mµ.

Mode of action. The gel filtration experiments were carried out as same as that in the case of the inactivating enzyme for pyridoxal enzymes in order to elucidate the products after the inactivation reaction. After incubation of apo-glutamic dehydrogenase(GDH) with the inactivating enzyme, the reaction mixture was applied onto Sephadex G-100 and the elution pattern was as shown in Fig. 5(b). The first peak of solid line indicates remaining GDH and the second peak of solid line shows one product protein from GDH by the enzyme reaction. No GDH activity was observed in the second peak and the molecular weight of it is obviously smaller than the intact GDH but it is still acid insoluble. The inactivating enzyme itself appears in the third peak indicated by solid line and the forth peak seems to be oligopeptide, released from the degradation reaction. The pattern of column chromatography showed in Fig. 5(a) is that in the case of zero time incubation. All the GDH activity added was recovered in the first peak and no other protein or peptides were detected.

Properties. Km for GDH as substrate is calculated to be 1.63×10^{-6}M as shown in Fig. 6. We have studied the substrate specificity on the purified inactivating enzyme and the inactivating enzyme specifically inactivates malic dehydrogenase(MDH),

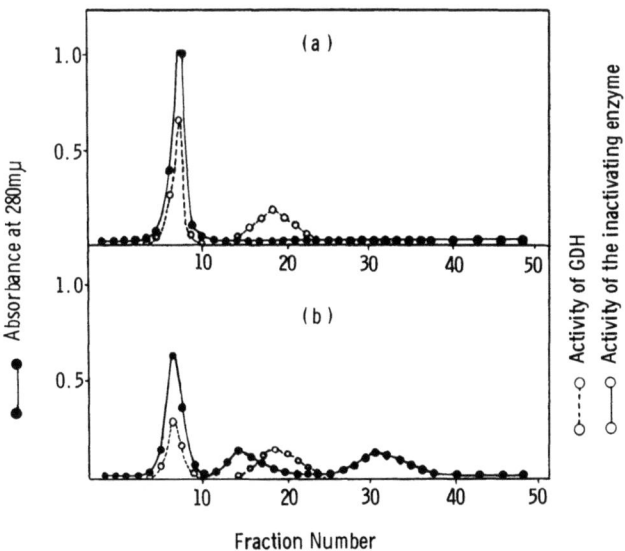

Fig. 5. Chromatography of the reaction products containing glutamic dehydrogenase and the inactivating enzyme on Sephadex G-100. The same reaction condition and method were used as in the experiment described in Fig.1

GDH and lactic dehydrogenase(LDH) in their apo-forms at various degree, but all the non-NAD dependent enzymes tested were unaffected by the inactivating enzyme as shown in Table VI.

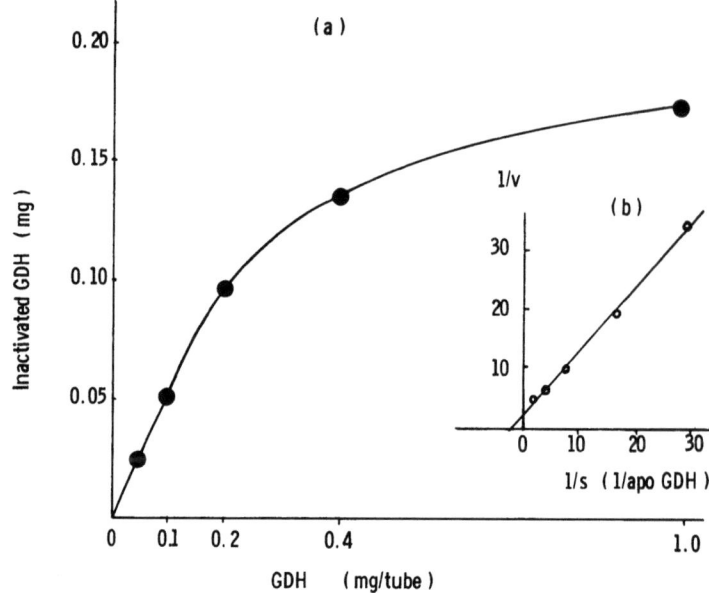

Fig. 6. Kinetic studies of the inactivating enzyme for NAD-dependent enzymes

Table VI. Sustrate specificity of the inactivating enzyme for NAD-dependent enzymes. Values indicate the activities against substrate enzymes when that of MDH is 100%. MDH, GDH and LDH were obtained commercially and were preparations from pig heart muscle, bovine liver and rabbit muscle, respectively

NAD-dependent enzymes	%	Non-NAD dependent enzymes	%
Malic dehydrogenase	100	Urease	0
Glutamic dehydrogenase	53	Glutaminase(phosphate independent)	0
Lactic dehydrogenase	36	Ornithine transaminase	0
		D-amino acid oxidase	0

The inactivation of apo-NAD dependent enzymes by the inactivating enzyme is inhibited by the addition of NAD or $NADH_2$. Rather high concentration of NAD or $NADH_2$ is required to prevent the inactivation reaction, namely 50% protection was observed in the presence of 1 mM of NAD or $NADH_2$. The parallel relationship beween the appearance of the activity of inactivating enzyme and the lapse of niacin deficiency are observed. The inactivating enzyme activity for pyridoxal enzymes does not increase in the niacin deficiency(7). On the contrary, the enzyme activity for NAD-dependent enzymes is not elevated in the B_6 deficiency.

Inhibitor studies. We found the specific inhibitor which is fractionated in 55-70% ammonium sulfate, while the inactivating enzyme for NAD-dependent enzymes precipitated in 35-55% ammonium sulfate fraction. The inhibitor is able to purify by passing through Sephadex G-100, and this is detected just after the inactivating enzyme on the column chromatography. The mode of inhibition of this inhibitor on the inactivating enzyme is shown with crude inhibitor preparation in Fig. 7. After preincubation for 30 min. in co-existence of the diverse amounts of inhibitor indicated and constant amount of the inactivating enzyme, and then constant amount of GDH as substrate was added.

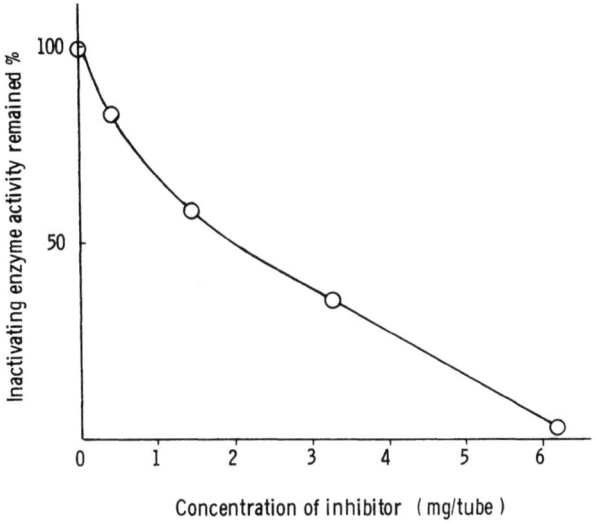

Fig. 7. Inhibition of the inactivating enzyme for NAD-dependent enzyme by the specific inhibitor. The reaction mixture contained in final 0.5 ml: 0.05M potassium phosphate buffer(pH 7.5), 0.2 mg of glutamic dehydrogenase, 3 mg of inactivating enzyme and inhibitor as indicated

No inhibitor was found in the case of niacin deficient rats, however various amount of the inhibitor was detected in normal rats as shown in Fig. 8(a). On the content of the inactivating enzyme in small intestine, on the other hand, a reciplocal movement could be recognized. The appearance of the inactivating enzyme activity were observed in small intestine and liver in niacin deficiency. The most parts of these inactivating enzymes exist as latent enzyme in inactive forms by binding with the specific inhibitor protein in normal rats. In order to demonstrate this mechanism more directly, following experiment was performed as shown in Fig. 9. Considerable amounts of the inactivating enzyme activity from small intestine of normal rats appeared by following treatments: 5 times repeated freezing-thawing in the presence of 50% ammonium sulfate, or ammonium sulfate precipitation in the presence of

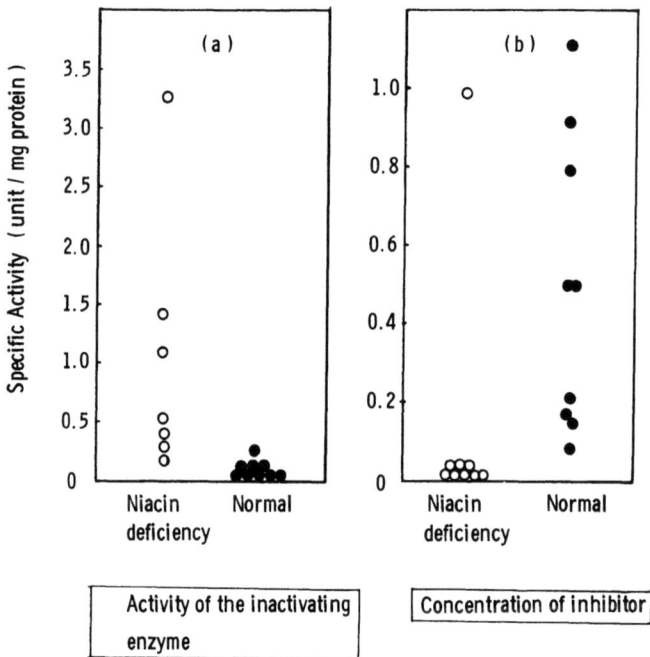

Fig. 8. Reciprocal relationship between the activities of inactivating enzyme and inhibitor. The levels of the inactivating enzyme activity for NAD-dependent enzymes in niacin-deficient and normal rats(left side). The levels of inhibitor in niacin-deficient and normal rats(right side)

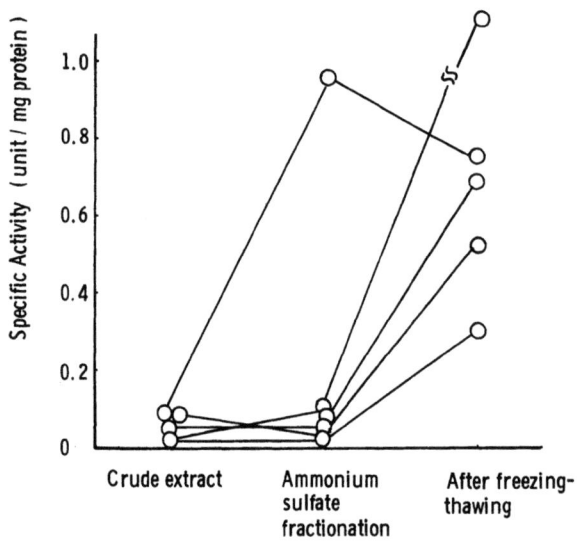

Fig. 9. Appearance of the inactivating enzyme activity for NAD-dependent enzymes from the latent form. Homogenates of small intestine of the normal rats were centrifuged at 105,000xg for 60 min. The levels of inactivating enzyme activity of the supernatant were described in the left column. The supernatant was fractionated with ammonium sulfate and precipitate of the 50-70% saturation was dissolved in 0.05M potassium phosphate buffer(pH 7.5) and then passed through a column of Sephadex G-25 equilibrated with same buffer. The activities of the protein fractions were shown in the middle column. In the right column, the activities appeared after repeated freezing-thawing in the presence of ammonium sulfate were shown

0.3% deoxycholate or 0.5% Triton X-100. From these experimental results, the activity of the inactivating enzyme is considered to be regulated by the amount of inhibitor which is controlled by the content of coenzyme level. Regulatory mechanism of NAD-dependent enzymes catabolism is offered in Fig. 10.

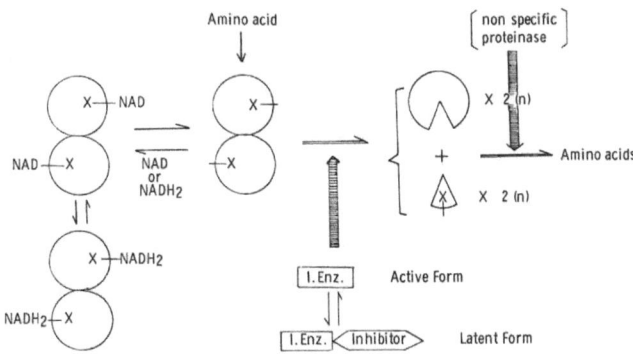

Fig. 10. Regulatory mechanism on catabolism of NAD-dependent enzymes

Part III. Inactivating enzyme for FAD-dependent enzymes

The crude preparation was precipitated with 35 to 50% saturation of ammonium sulfate. The fraction was applied onto Sephadex G-100 as the same procedure as the other inactivating enzymes. The elution pattern is shown in Fig. 11.

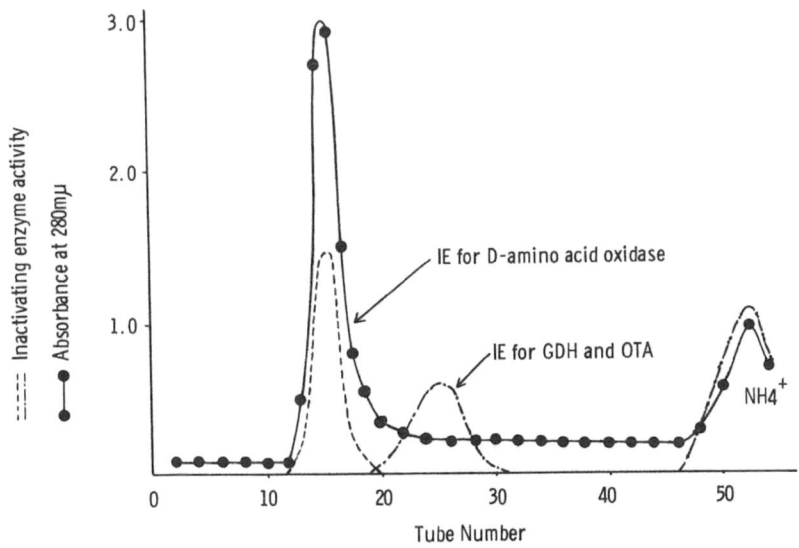

Fig. 11. Separation of respective inactivating enzymes on Sephadex G-100 column chromatography

The inactivating enzyme for FAD-dependent enzymes is able to be separated from the inactivating enzymes for NAD-dependent enzymes and for pyridoxal enzymes on the column chromatography. From the elution pattern of Sephadex G-100, the specific inactivating enzyme for FAD-dependent enzyme was recognized to be the highest molecular weight among these inactivating enzymes discovered. On the substrate specificity of the inactivating enzyme, the inactivating enzyme for FAD-dependent enzyme was able to inactivate D-amino acid oxidase, but the enzyme was unable to inactivate ornithine transaminase and glutamic dehydrogenase. Effect of coenzymes on the inactivating reaction was also studied. Protection against the inactivation was observed by addition of 1.0×10^{-4} to 1.3×10^{-5} M of FAD, but FMN or riboflavin had no any protecting effect for this inactivating reaction. Increases of the inactivating enzyme activity in small intestine were observed in riboflavin-deficiency, at the same time the liver D-amino acid oxidase determined as an indicator of riboflavin-deficiency was lowered extremely as shown in Fig. 12.

Finally, several characteristics of these three kinds of inactivating enzymes for pyridoxal enzymes, NAD-dependent enzymes and FAD-dependent enzymes are summarized in Table VII. Their molecular weights, substrate specificities, heat stabilities, ammonium sulfate fractionations are quite different and these inactivating enzymes increase only under respective vitamin deficiencies, but their optimal pH are about 9 in common. The inhibitor of the inactivating enzyme for NAD-dependent enzymes acts only on the inactivating enzyme for NAD-dependent enzymes among the inactivating enzymes tested. These results support three kinds of inactivating enzymes are separate enzymes each other.

<u>Table VII.</u> Summary of the enzyme-chemical natures of the inactivating enzymes for pyridoxal-dependent, NAD-dependent and FAD-dependent enzymes

	PALP	NAD	FAD
Molecular weight	≒30,000	≒30,000	100,000-150,000
$(NH_4)_2SO_4$ fraction	50-70%s	35-55%s	25-50%s
Optimal pH	9.0	9-10	9.0
Km for substrates	10^{-10}M (for OTA)	10^{-6}M (for GDH)	---
Protection by coenzymes	10^{-6}-10^{-7}M	10^{-3}M	10^{-4}-10^{-5}M
Stability (storage at frozen state, heat treatment)	unstable	stable	unstable
Organ distribution	small intestine skeletal muscle	small intestine liver	small intestine ---
Activity increased in	B_6 deficiency	niacin deficiency	B_2 deficiency
Inhibitor of the inactivating enzyme for NAD-dependent enzymes	no inhibition	inhibition	no inhibition

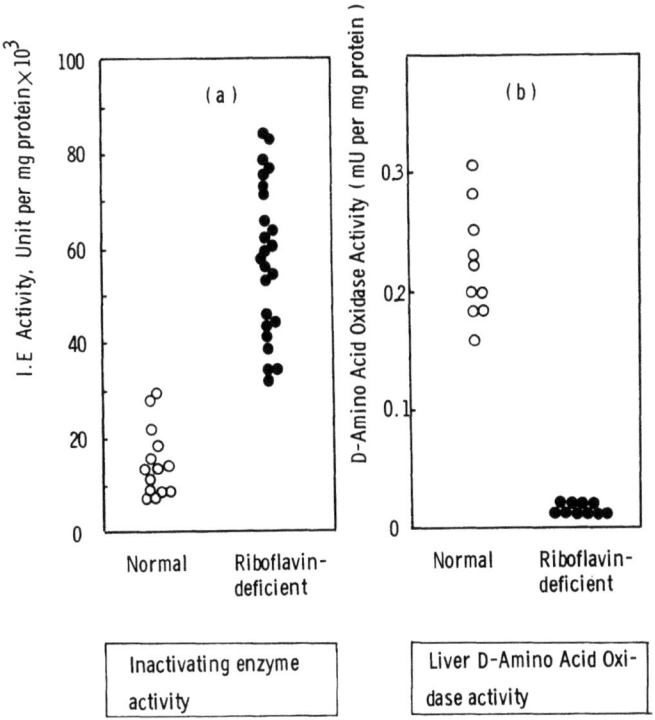

Fig. 12. Inactivating enzyme activity in small intestine of riboflavin-deficient rats (right side). Liver D-amino acid oxidase activity as a marker of riboflavin-deficiency (left side).

The assay method for the inactivating enzyme specific for FAD-dependent enzyme was carried out as follows; the reaction mixture contained 1.5-2.0 units of D-amino acid oxidase, appropriate amount of the inactivating enzyme and 100 µmoles of Tris-HCl(pH 8.5) in final 0.5 ml. The incubation was carried out for 15-60 min. at 37°C. An aliquot(0.1 ml) was taken in another tube containing 2 volumes of buffer and 1.0×10^{-4} M of FAD, this brings the reaction to terminate. For the reconstitution of remaining apo-enzyme into holo-enzyme, further incubation was continued for 10 min. The D-amino acid oxidase activity was determined spectrophotometrically with $NADH_2$ and coupling enzymes, catalase and lactic dehydrogenase. The enzyme unit of the inactivating enzyme was defined as follows; unit= $1/T(1/2) \times 10^3$, $T(1/2)$= the time required for a half inactivation expressed in min. The substrate enzyme, homogeneous D-amino acid oxidase was purified from pig kidney

It is possible that these group specific inactivating enzymes which we have discovered can serve as an initiator of these respective enzyme group degradation, and the products formed by our inactivating enzymes are subsequently degradated into amino acids by various non-specific proteinases. We would like to offer the following hypothesis for regulatory mechanisms on catabolism of enzyme protein as shown in Fig. 13.

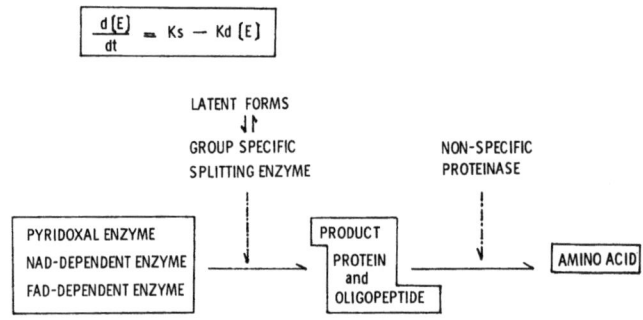

Fig. 13. Regulatory mechanism of enzyme catabolism

References

1. SCHIMKE, R.T., SWEENEY, E.W. and BERLIN, C.M., Ann.N.Y.Acad.Sci., 102, 587(1964)
2. SEGAL, H.L. and KIM, Y.S., J.Cell and Comp.Physiol.,Sup.1 to V,66, No.2, October,11(1965).
3. COFFEY, J.W. and de DUVE, C., J.Biol.Chem., 243,3255(1968)
4. PERAINO, C. BUNRILLE, L.G. and TAKMISIAN, T.M., J.Biol.Chem., 244, 2241(1969)
5. OKADA, M. and OCHI, A., J. Biochem.(Tokyo), 70, 581(1971).
6. KATUNUMA, N., KOMINAMI, E. and KOMINAMI, S., Biochem.Biophys.Res.Commun., in press.
7. KATUNUMA, N., KITO, K. and KOMINAMI, E., Biochem. Biophys. Res. Commun., in press.

Discussion:

Segal

I think that these are two very remarkable enzymes that Prof. Katunuma has uncovered, and it will be very interesting to see what physiological role they play in protein turnover. We have approached the problem from a slightly different point of view - asking what is the rate limiting step in protein degradation - and have focused on lysosomes, because in liver, at least, virtually all the activity we could find in inactivation of native enzymes was in the lysosomal fraction. Lysosomal extracts produced a first order inactivation of crystalline alanine aminotransferase at pH 5.0, under conditions where the enzyme was completely stable in the absence of the lysosomal fraction (Segal et al.: Biochem. Biophys. Res. Commun. 36, 764 (1969)). A similar result was obtained with partially purified arginase (Haider and Segal: Arch. Biochem. Biophys. 148, 228 (1972)). Added pyridoxal phosphate did not inhibit the inactivation of alanine aminotransferase, but bovine serum albumin did that of this enzyme and to an even greater extent that of arginase. In addition inactivation of arginase was inhibited by Mn^{2+}, also known to inhibit arginase turnover in vivo, and by amino acids, which may have its physiological counterpart in the diminuation of arginase turnover in starvation.

Katunuma

I think that the degradation depends on the tertiary structure of the apo- and holo-enzyme. In the case of aspartate transaminase, there is no great difference between the apo- and holo-forms: both are very resistant to degradation. On the other hand there are great differences between the apo- and holo-forms of ornithine transaminase and serine dehydratase. In such cases it is the apo-enzyme that is very easily degraded.

Krebs

Did you use an apoenzyme from the one type as a control protein for the other type, in other words, did you use an apo-enzyme from a B_6 enzyme as a control for a NAD^+-enzyme?

Katunuma

Yes, but there was not inactivation at all.

Pette

What is the tissue distribution of the FAD-enzyme inactivating enzyme?

Katunuma

I have not tested the tissue distribution of the FAD splitting enzyme. In the case of the splitting enzyme for NAD^+ dependent enzymes, however, the activity in the liver and in small intestine is very strong, but it does not exist in the heart and brain. I think that this is one kind of economizing the vitamins available.

Holzer

Prof. Katunuma spent several weeks in our laboratory in Freiburg and he was working with Dr. Afting. We have found enzymes in yeast that are very similar to the splitting enzyme described by Prof. Katunuma in rat intestine. Fig. 1 shows the effect of the rat splitting enzyme

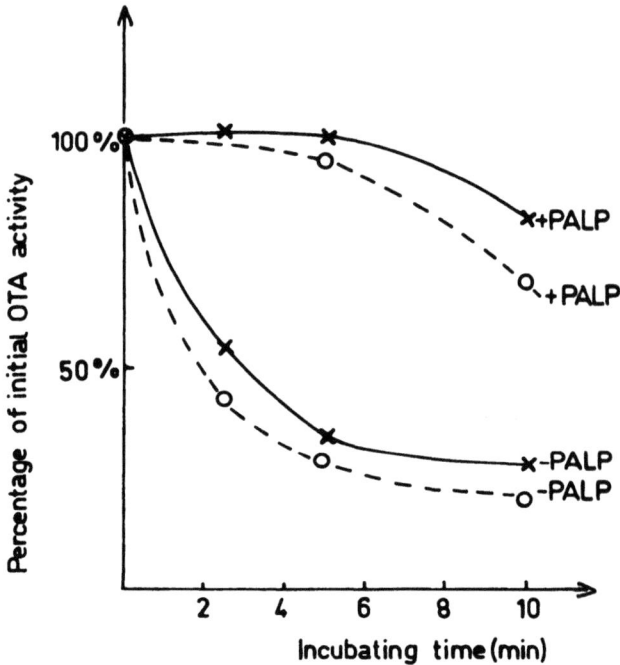

Fig. 1. Inactivation of ornithine transaminase (OTA) from yeast and rat by rat splitting enzyme. o--o = rat OTA. x--x = yeast OTA. pH 8.o. T = 37° C. GSA = glutamic semialdehyde

on crystallized rat ornithine transaminase and partially-purified yeast ornithine transaminase at pH 8.o. The apo-enzymes from both organisms are inactivated by the rat-splitting enzyme; inactivation is prevented by pyridoxal phosphate. In the absence of inactivating enzyme both substrate enzymes are stable. The apo-enzyme of threonine dehydratase and aspartate aminotransferase from yeast are not inactivated. It seems therefore that there is some specificity in the coenzyme binding site of substrate enzymes, but no specificity in the organism from which the enzyme comes.

The yeast inactivating enzyme inactivates the apo-ornithine transaminase from both yeast and rat as shown in Fig. 2. Pyridoxal phos-

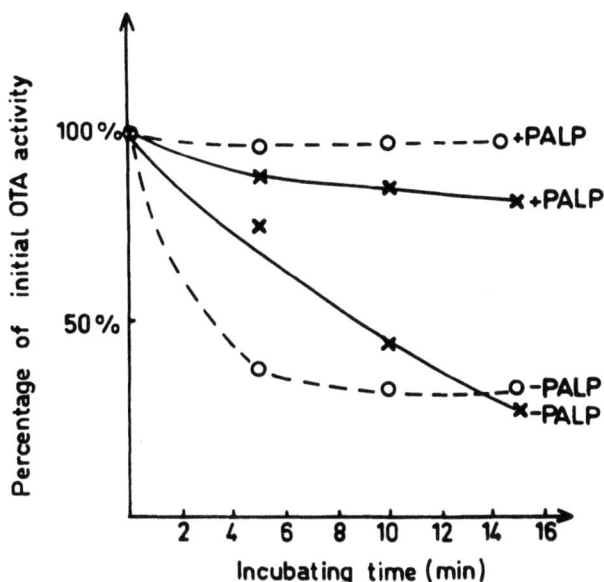

Fig. 2. Inactivation of ornithine transaminase (OTA) from yeast and rat by yeast inactivating enzyme. o--o = rat OTA. x--x = yeast OTA. pH = 7.o. T = 37° C. GSA = glutamic semialdehyde

phate protects in both cases. However, the inactivating enzymes do differ in their pH optima. The yeast inactivating enzymes has a pH optimum of about 7 irrespective of the source of the substrate enzymes. The rat splitting enzyme has a pH optimum of about 8. I think this is useful because the intracellular pH of yeast is about pH 6.o.

Fischer

Prof. Holzer, are proteins which are not pyridoxal phosphate dependent inactivated by these enzymes?

Holzer

The problem of specificity has to be studied much more extensively with a purified inactivating enzyme. We have studied so far only pyridoxal phosphate dependent enzymes.

Interconversion of Two Forms of Leucyl-tRNA Synthetase

Pierre Rouget and Francois Chapeville

Institut de Biologie Moléculaire, CNRS, Université Paris VII, Paris/France

Abstract : Two forms of leucyl-tRNA synthetase were isolated from E. Coli. Both forms, named E_I and E_{II}, catalyse the formation of leucyladenylate and the ATP-pyrophosphate exchange reaction with identical Km for leucine and ATP. Each form binds specifically tRNAs acceptor of leucine but only E_{II} is able to transfer the activated aminoacid onto tRNA. Moreover if E_{II} is highly specific for leucine, E_I catalyses pyrophosphate exchange also in the presence of valine, isoleucine and methionine. The titration of sulfhydryl groups shows that two groups are masked in E_I. One SH group in E_{II} is rapidly reacting with pMB and is essential to the transfer of leucine to tRNA. E_{II} with one SH blocked resembles E_I by its catalytic properties but remains specific for leucine and cannot be separated from native E_{II}. The two forms E_I and E_{II} are interconvertible in the presence of cell extracts. It was found that two factors are involved in this process ; F_1 which is a protease and releases a peptide of 3 000 MW from E_{II}, thereby converting E_{II} into E_I. This peptide which is the factor F_2 can reassociate with E_I leading to an enzyme with catalytic and chromatographic properties of E_{II}. The results indicate that E_{II} is constituted from two subunit-like fragments which are in E_I associated by low-energy bonds and in the E_{II} also by two peptide bonds at the ends of peptide F_2. The association of F_2 with the two "subunits" is necessary for full E_{II} activity. Low-energy bonds are sufficient to keep this association (E_I-F_1) and reconstituted enzyme from E_I and F_1 maintains a conformation and properties close to those of native E_{II}.

I. INTRODUCTION

Two forms of leucyl-tRNA synthetase, differing in their catalytic and chromatographic properties have been isolated from E. Coli (1). Both forms of the enzyme, named E_I and E_{II} catalyse the formation of leucyladenylate and the ATP-pyrophosphate exchange reaction in the presence of leucine, with the same Km values for ATP and also similar Km's for leucine. However, whereas E_{II} is very specific for leucine, E_I catalyses the ATP-pyrophosphate exchange also with isoleucine, valine and to a lesser extent with methionine. Moreover each enzyme can bind specifically tRNA's acceptor of leucine, even though only E_{II} is able to transfer the activated aminoacid onto tRNA (1).

As shown in Table 1 (a, b), the affinities of the two enzyme forms for $tRNA^{Leu}$ are identical. No difference was observed between $tRNA_I^{Leu}$ and $tRNA_{II}^{Leu}$. The rate constants of the tRNA-enzyme binding reaction were the same with E_I as with E_{II}. However in the case of E_{II}, these rate constants were increased fivefold by the presence of ATP and leucine, whereas in the case of E_I they remained unaffected by the presence of these substrates. This observation indicates that some interactions between the site of the enzyme for leucyladenylate and its site for tRNA have disappeared in E_I (2).

II. SULFHYDRYL GROUPS OF E_I AND E_{II}

The differences between the catalytic properties of E_I and those of E_{II}, as reported in Table 1 (a, b), presented some parallelism with the differences observed between the catalytic properties of N-ethylmaleimide-treated isoleucyl-tRNA synthetase and those of untreated isoleucyl-tRNA synthetase - Iaccarino and Berg (3) -. The question therefore arose as to whether the two forms of leucyl-tRNA synthetase differed in the reactivity of their sulfhydryl groups.

The inactivation of leucyl-tRNA synthetase (E_{II}) by para-chloro-mercuri benzoate (pMB) was examined. As shown in Fig. 1b, the binding of one pMB per enzyme sufficed to completely inactivate the enzyme for the formation of leucyl-tRNA without modifying its activity for the ATP-pyrophosphate exchange and its ability to bind ^{14}C-leucyl-tRNA. Moreover, as reported in Table 1 (b, d), when one pMB was bound per molecule of enzyme, the affinities of the enzyme for ATP and $tRNA^{Leu}$, its Km for leucine and the rate constants of the tRNA-enzyme binding reaction remained unaffected. However, as in the case of untreated E_I (Table 1 a), these rate constants were no longer enhanced by the presence of ATP plus leucine, as if interactions between the enzyme sites for leucyladenylate and for tRNA were abolished. Results reported in Fig. 1a and in Table 1 (b, c) showed that the incubation of leucyl-tRNA synthetase (E_{II}) with N-ethylmaleimide (NEM) led to results somewhat similar to those obtained with the pMB-inactivation.

From these results it appears that one highly reactive sulfhydryl group of leucyl-tRNA synthetase (E_{II}) is essential for interactions between the enzyme sites for leucyladenylate and for tRNA, and is necessary for the transfer of leucine onto tRNA. This sulfhydryl group seems not to be involved in the other enzymatic properties of leucyl-tRNA synthetase. Moreover, strong similarity is observed between the enzymatic properties of E_I and those of pMB-E_{II}.

TABLE 1
COMPARATIVE PROPERTIES OF E_I, E_{II} AND TREATED E_{II}

	(a) E_I	(b) E_{II}	(c) (NEM) E_{II}	(d) (pMB) E_{II}	(e) trypsin-treated E_{II}
Km (Leu) (M)	1.2×10^{-4}	8.7×10^{-5}	1.0×10^{-4}	9.4×10^{-5}	1.0×10^{-4}
Km (ATP) (M)	7.2×10^{-4}	7.1×10^{-4}	6.8×10^{-4}	7.3×10^{-4}	7.5×10^{-4}
K_A (ATP) (M^{-1})	4.9×10^6	4.5×10^6	4.2×10^6	4.8×10^6	5.1×10^6
tRNA$_I^{Leu}$	– ATP – Leu :+ ATP + Leu 0.4 mM	– ATP – Leu :+ ATP + Leu 0.4 mM	– ATP – Leu :+ ATP + Leu 0.4 mM	– ATP – Leu :+ ATP + Leu 0.4 mM	– ATP – Leu :+ ATP + Leu 0.4 mM
K_A (M^{-1})	1.2×10^8 : 1.1×10^8	1.1×10^8 : 1.0×10^8	9.4×10^7 : 9.8×10^7	1.3×10^8 : 1.2×10^8	9.7×10^7 : 9.1×10^7
k_d (sec^{-1})	8.8×10^{-3} : 8.6×10^{-3}	8.4×10^{-3} : 4.7×10^{-2}	8.7×10^{-3} : 9.1×10^{-3}	9.0×10^{-3} : 8.8×10^{-3}	8.9×10^{-3} : 8.5×10^{-3}
k_a (sec^{-1})	1.1×10^6 : 9.5×10^5	4.7×10^6 : 9.2×10^5	8.6×10^5 : 8.9×10^5	1.1×10^6 : 1.2×10^6	8.6×10^5 : 7.8×10^5

Purified E_I and E_{II} enzymes were prepared as described (2). The (NEM) E_{II} was E_{II} enzyme treated with NEM for 30 min in the conditions of Fig. 1a ; the (pMB) E_{II} was obtained from the binding of one mole of pMB per mole of enzyme as described under Fig. 1b ; the trypsin-treated E_{II} was the first DEAE-cellulose peak eluted as reported in Fig. 5 after an incubation of E_{II} with trypsin for 30 min. The Km values were determined by the leucine-dependent ATP-pyrophosphate exchange. Affinity constants for ATP were measured on the basis of an equilibrium dialysis, using Sephadex G-75 filtration according to the method described by Pfleiderer and Auricchio (6), the results being plotted according to Scatchard (7). The binding of acylated or unacylated tRNALeu to the enzymes was followed by filtration through nitrocellulose filters, using the method developped by Yarus and Berg (8) ; the affinity constants were calculated from a Scatchard plot (7) and the rate constants were measured in the absence or in the presence of 4×10^{-4} M ATP and 4×10^{-4} M Leu, by following the exchange between enzyme-bound acylated and free unacylated tRNALeu or reciprocally (results to be published indicate that the rate-limiting step in this exchange is the dissociation of the first complex).

The number of sulfhydryl groups in each form of leucyl-tRNA synthetase was determined according to Boyer (4). The results indicated that E_I presented 9 pMB-reactive sulfhydryl groups per enzyme molecule, whereas with E_{II} 11 were titrated. However, after 8M urea (or 6M guanidinium chloride) denaturation, 16 sulfhydryl groups per molecule were titrated with each enzyme form. This suggests that the content in sulfhydryl groups was identical for both enzyme forms, but that two of these groups were masked in E_I. It should be noted that no molecular weight variations were observed under the pMB treatment of each enzyme(2).

The difference between the content in sulfhydryl groups of E_I and E_{II}, added to the analogy between the enzymatic properties of E_I and those of pMB-E_{II} suggest that E_I differs from E_{II} by the absence of reactivity of two sulfhydryl groups, one of which would be necessary

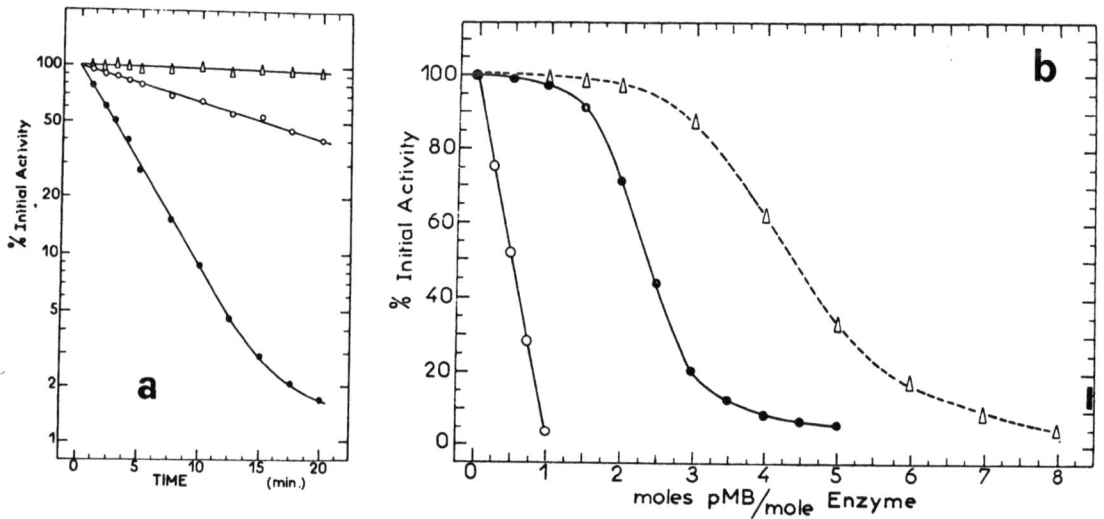

Fig. 1. Inactivation of leucyl-tRNA synthetase (E_{II}) by sulfhydryl blocking reagents :

The enzyme was reduced by incubation for 30 min at 20° C with 0.01M 2-mercaptoethanol, the excess reagent being then removed by Sephadex G-75 filtration.
  a) Kinetics by NEM-inactivation : The enzyme (0.1 mg/ml) was incubated at 20° C with 1mM NEM. At various times, aliquots were removed, one volume of 0.05M 2-mercaptoethanol was added and their activities were assayed : ●——● leucyl-tRNA formation, o——o ATP-pyrophosphate exchange, ▵——▵ binding of ^{14}C-leucyl-tRNA.
  b) pMB-inactivation : Increasing amounts of pMB were bound to leucyl-tRNA synthetase in the conditions described by Boyer (4). The formation of mercaptide was measured by following the absorption at 250 nm. Enzyme activities were assayed (2) after fiftyfold dilution in 0.1M potassium phosphate pH 7 : o——o leucyl-tRNA formation, ●——● ATP-pyrophosphate exchange, ▵---▵ binding of ^{14}C-leucyl-tRNA

for the interactions between the sites of the enzyme for leucyladenylate and for t-RNA and required for the transfer of leucine to t-RNA.

The specificity of leucyl-tRNA synthetase (E_{II}) for leucine and its chromatographic properties were not modified by pMB or NEM inactivation, whereas it was already reported (1) that the chromatographic properties of E_I were different from those of E_{II} and that E_I could catalyze the ATP-pyrophosphate exchange not only with leucine as E_{II} does but also with isoleucine, valine and, to a lesser extent with methionine. This suggests that the variations in the reactivity of the sulfhydryl groups does not explain all the differences existing between E_I and E_{II}.

III. CONVERSION OF E_{II} INTO E_I AND OF E_I INTO AN E_{II}-LIKE ENZYME STIMULATED BY FACTORS ISOLATED FROM E. COLI

When treated with E. Coli 105 000 g supernatant (previously heated to inactivate endogenous leucyl-tRNA synthetase), a proportion of the purified E_I acquired the catalytic and chromatographic properties of E_{II} and, in the same conditions, the purified E_{II} appeared to be partly converted into E_I (1). Accordingly, we have tried to isolate from the E. Coli supernatant a factor responsible for these conversions.

It was observed that in fact (Fig. 2), two different factors were involved in the enzyme transformations : factor F_1 which catalyzed the conversion of E_{II} into E_I, eluted from Sephadex G-100 at a position corresponding to a molecular weight of 20 000 and the factor F_2, stimulating an apparent transformation of E_I into E_{II}, eluted as a 3 000 molecular weight component (Fig. 3).

When E_{II} was first incubated with the factor F_1 it lost rapidly more than 90 % of its transfer activity. After subsequent addition of F_2 to the incubation mixture in 3 hours 50 % of the initial transfer activity were recovered (Fig. 4). The heat denaturation studies of these factors showed that at 55°C, F_1 was slowly inactivated but F_2 remained unchanged. At 70°C both lost about 90 % of activity, F_1 in 5 min and F_2 in 10 minutes. This fact and the absence of nucleic acids in partially purified F_1 suggested that this factor is a protein. As to the nature of factor F_2 its lost of activity after incubation in the presence of pronase suggested that it is a peptide.

The rate of the conversion of E_{II} into E_I was proportional to the amount of factor F_1 present, but the final extent of conversion did not depend on the F_1 concentration, suggesting that the action of factor F_1 on the E_{II} form of leucyl-tRNA synthetase might be enzymatic.

As regards to the activation of factor F_2, it was observed that both the rate and the final extent of the transformation of E_I into an E_{II}-like enzyme increased with F_2 concentration.

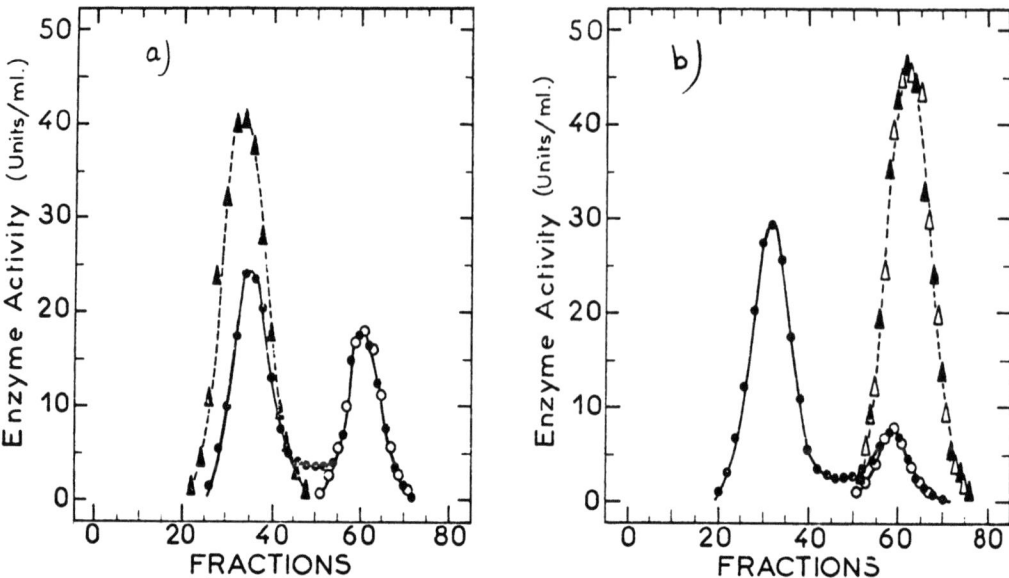

Fig. 2. DEAE-cellulose chromatography of E_I and E_{II} following their incubation with F_2 or F_1

a) Enzyme E_I (1 mg/ml) was incubated for 2 hours at 37°C with 0.1 mg of partially purified factor F_2 and chromatographed at 4°C on a 1 x 20 cm DEAE-cellulose column. A linear gradient (100:100 ml) from 0.05 to 0.3 M potassium phosphate pH 6.8 was applied to the column ; 2 ml fractions were collected at a flow rate of 20 ml/h. As control, 1 mg of the enzyme was incubated for the same time in the absence of F_2 and chromatographed as above.

b) Enzyme E_{II} (1 mg/ml) was incubated for 2 hours at 37°C with 1 mg of partially purified factor F_1 and chromatographed as above. As control 1 mg of enzyme was incubated without factor F_1 and chromatographed.

Enzyme plus factor : ●——● ATP-PPi exchange
○——○ Leucyl-tRNA formation

Enzyme alone : ▲---▲ ATP-PPi exchange
△---△ Leucyl-tRNA formation

IV. MILD TRYPTIC PROTEOLYSIS OF LEUCYL-tRNA SYNTHETASE (E_{II})

When leucyl-tRNA synthetase (E_{II}) was treated by trypsin under mild conditions, its activity for the leucine-dependent ATP-pyrophos-

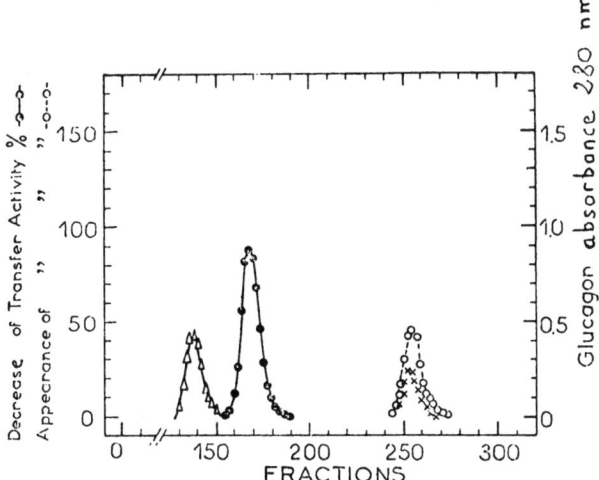

Fig. 3. Sephadex G-100 filtration of factors F_1 and F_2

The 105 000 g supernatant was kept for 2 hours at 55°C to inactivate leucyl-tRNA synthetase, then chromatographed on DEAE-cellulose column. An aliquot of the fraction eluted at 0.16 M phosphate and containing F_1 and F_2 activities was filtered on Sephadex G-100 column (1.8 x 100 cm). 1 ml fractions were collected, concentrated and their conversion capacity of E_I to E_{II} and of E_{II} to E_I was determined.

o——o E_I to E_{II} (factor F_2) ; •——• E_{II} to E_I (factor F_1)
xxxxx Glucagon ; △——△ Horseradish peroxydase

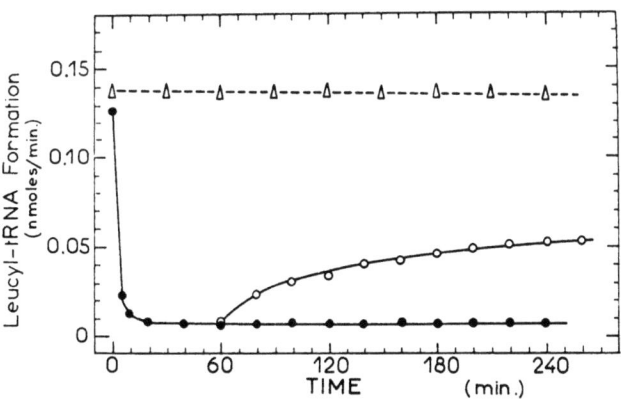

Fig. 4. Effect of F_2 on F_1 treated E_{II} enzyme

After incubation of E_{II} (4µg/ml) with F_1 at 37°C for one hour the factor F_2 (40 µg/ml) was added. Aliquots were taken at various times and assayed for leucyl-tRNA formation and ATP-PPi exchange.

•——• $E_{II} + F_1$; leucyl-tRNA formation
o——o $E_{II} + F_1 + F_2$ (at 60 min) ; leucyl-tRNA formation
△——△ $E_{II} + F_1 + F_2$ (at 60 min) ; APT-PPi exchange

phate exchange and its tRNA binding capacity were unaffected during at least one hour whereas, after only 10 minutes, about 90 % of its activity for leucyl-tRNA formation was abolished.

Results reported in Table 1 (b, e) indicate that the affinity constant of the enzyme for tRNA was not modified by this tryptic treatment, nor were the rate constants of the tRNA-enzyme binding reaction. As observed for E_I (Table 1, a) and for NEM or pMB treated E_{II} (Table 1, c, d), these rates were no longer increased by the presence of ATP plus leucine.

The results reported in Fig. 5 show that incubation with trypsin of E_{II} led to an enzyme with chromatographic properties similar to those of E_I.

The conversion in the presence of trypsin increased with time. The enzyme found in the first peak (Fig. 5) which has the same position as E_I, was unable to charge the tRNA with leucine and its amino acid specificity decreased ; it catalysed the ATP-pyrophosphate ex-

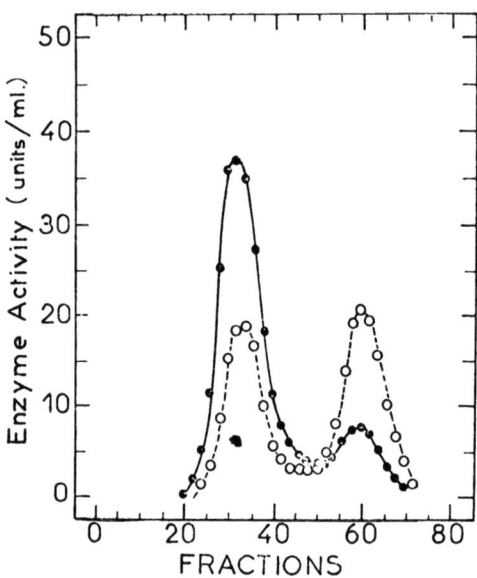

Fig. 5. DEAE-chromatography of trypsin-treated E_{II}

The enzyme (1 mg/ml) was incubated at 37°C with 2 µg trypsin for 3 and 10 min., at which times 5 µg of soja-bean trypsin inhibitor were added. The mixtures were then chromatographed on DEAE-cellulose columns under conditions described in Fig. 2. Eluted fractions were assayed for ATP-PPi exchange and leucyl-tRNA formation.

o---o elution profil resulting from the 3 min. incubation
•——• elution profil resulting from the 10 min. incubation

The position of E_I and E_{II} when analysed in the same conditions are indicated in Figure 2

change also with isoleucine, valine and methionine with Km values identical to those of E_I (Table 2).

TABLE 2

AMINO-ACID SPECIFICITY OF TRYPSIN-TREATED E_{II}
AS COMPARED TO THAT OF E_I

Amino-acids	Trypsin-treated enzyme Km (M).10^4	E_I Km (M).10^4
Leucine	1.0	1.2
Isoleucine	1.4	1.3
Valine	1.9	1.5
Methionine	7.6	8.1

The E_I enzyme was prepared as described (2). Trypsin-treated enzyme was the first DEAE-cellulose peak eluted as reported under Fig. 5 after an incubation of E_{II} with trypsin for 30 min in conditions of Fig. 5. The Km values were determined by following the ATP-pyrophosphate exchange in the presence of various amino-acids. No enzyme activity was observed with amino acids other than those indicated here.

Sephadex G-200 filtrations showed that its molecular weight was similar to that of E_I. It was checked that in the conditions described above trypsin did not modify the catalytic and chromatographic properties of E_I. The addition of factor F_2 partially restored the activity of the trypsin-treated E_{II} for leucyl-tRNA formation.

Titration of the sulfhydryl groups of the trypsin-treated E_{II} indicated that this enzyme presented, as did the E_I enzyme, 9 sulfhydryl groups, whereas, as noted above, 11 sulfhydryl groups per enzyme molecule were titrated with untreated E_{II}. After urea denaturation, 16 sulfhydryl groups were titrated as well with the trypsin-treated E_{II} as with E_I or E_{II}.

These results show a strong similarity between the properties of the enzymes obtained from treatment of E_{II} by factor F_1 or by trypsin, and suggest that F_1 is a protease. The fact that there was no appreciable variation of the molecular weight of E_{II} submitted to factor F_1 or to trypsin treatment, suggests that F_1 and trypsin act on only very few peptide bonds, their breaking being responsible for the loss of interaction between the enzyme sites for leucyladenylate and for tRNA, possibly due to the removal of a small peptide. Therefore, the question arose as to whether this hypothetic peptide could be factor F_2.

Accordingly, the E_{II} enzyme was trypsin-treated for 30 min. in the conditions described under Fig. 5 and then heated at 55°C for 30 min. to destroy E_I and residual E_{II} activities. After centrifugation to remove denatured material, increasing amounts of the supernatant were added to the E_I purified from crude cell extracts or obtained from E_{II}. In each case after addition of the supernatant, tRNA charging activity was observed and was proportional to the amount of the supernatant. The results reported in Fig. 6 show that after addition of the supernatant as in the case of F_2 a part of E_I acquires E_{II} chromatographic properties.

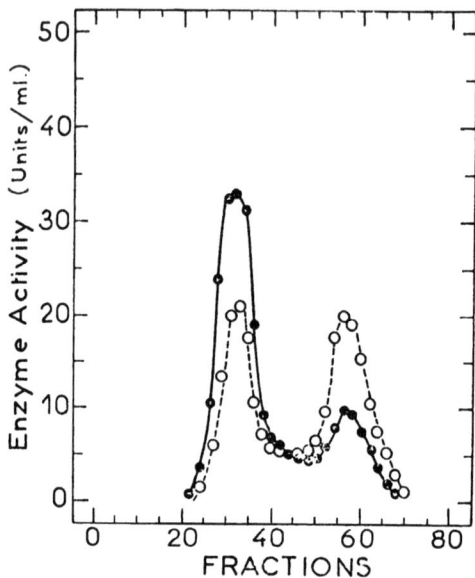

Fig. 6. Effect of E_{II} supernatant on chromatographic properties of E_I

Leucyl-tRNA synthetase (E_{II}) was treated with trypsin for 30 min. Then heated at 55°C and centrifuged. The supernatant (40 or 80 μg) was added to E_I, the mixture incubated at 37°C for 2 hours and chromatographed on DEAE-cellulose as indicated under Fig. 2.

●——● E_I + 40 μg of supernatant from trypsin-treated E_{II}
o——o E_I + 80 μg of supernatant from trypsin-treated E_{II}

The separated E_{II} like enzyme has also recovered its specificity for leucine. When the supernatant was filtered on Sephadex G-100 in the conditions similar to those described under Fig. 3 the molecular weight of the component active in the E_{II} reconstitution appeared to be of about 3 000 and similar to that of F_2.

Consequently all these results and those presented in the preceding pages strongly indicate that E_I is a proteolysed form of E_{II} which is the native enzyme and from which a peptide has been removed. Association of this peptide with E_I appears to take place by low-energy bonds sufficient to maintain a conformation and properties close to those of the native E_{II} molecule.

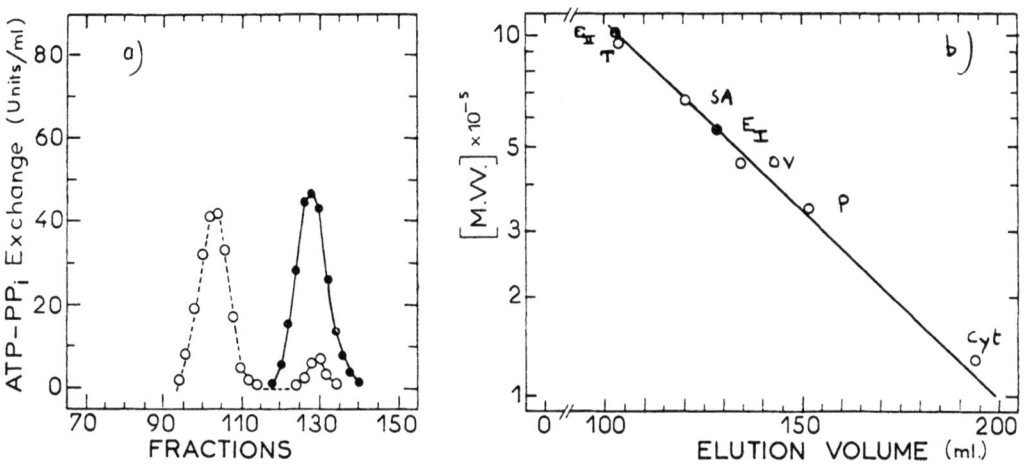

Fig. 7. Sephadex G-100 filtration of E_I and E_{II} in the presence of urea.

Each form of the enzyme (E_I or E_{II}) was incubated for one hour at 20°C with 8 M urea and then dialysed against 0.02 M potassium phosphate pH 6.8 buffer containing 0.05 M KCl, 0.01 M 2-mercaptoethanol and 6 M urea. Each form was then filtrated on a Sephadex G-100 column previously equilibrated with the above solution and calibrated with proteins of known molecular weights. The fractions collected were dialysed against 0.02 M potassium phosphate pH 6.8, 0.01 M 2-mercaptoethanol and their activity was assayed for leucine-dependent ATP-pyrophosphate exchange.

a) Elution profiles of E_I (●——●) and of native E_{II} (o---o)
b) Molecular weight determination : cytochrome C (Cyt), Pepsine (P), Ovalbumine (OV), Bovine serum albumine (SA) and Tyrosyl-tRNA synthetase (T) were used for calibration

It has been previously shown (1) that the molecular weights of E_I and E_{II} as determined by gel filtration and other methods is similar and approximately 105 000. In order to determine from what part of E_{II} the peptide F_2 is removed during the conversion of E_{II} into E_I both these forms of enzyme were treated with 8 M urea or 6 M guanidinium chloride and filtered in the presence of these reagents on Sephadex G-100 (Fig. 7a). It was found that in these conditions, E_I prepared by treatment with F_1 or with trypsin dissociated into two parts both of about 55 000 molecular weight (Fig. 7b). About 90 % of native E_{II} did not dissociate but E_{II} constituted from E_I and F_2 furnished two 55 000 fragments. The small amount of dissociation (less than 10 %) observed with native E_{II} was probably due to its partial proteolysis during storage and handling.

V. CONCLUSIONS

The results presented in this paper suggest that the leucyl-tRNA synthetase of E. Coli is composed of two compact protein fragments which are associated by low-energy bonds and linked covalently each by a peptide bond to the peptide F_2. These subunit-like fragments remain strongly associated after removal of the peptide, but the form E_I which results does not retain all of the original catalytic activity. The reconstitution experiments showed that the peptide association with E_I through low-energy bonds is sufficient to maintain a conformation and properties close to those of native enzyme E_{II} and which are characterised by : 2 sulfhydryl group reactivity, catalytic site interactions, specificity for leucine and different chromatographic behaviour from E_I. The study of association equilibrium between E_I and peptide should provide information on the nature of this association.

A schematic representation of the enzyme structure is proposed in Fig. 8.

For simplification, the two subunit-like fragments are suggested as being identical. Experiments which will be published elsewhere showed that leucyl-tRNA synthetase presents only one site for the binding of each substrate (ATP, leucine, $tRNA^{Leu}$) and only one sulfhydryl group essential for the transfer of leucine to tRNA. This strongly suggests that the molecule is asymmetric. However preliminary experiments to separate the two 55 000 molecular weight fragments were unsuccessful. This might mean that there is some similarity in their structure and that observed asymmetry is induced by the peptide F_2.

We do not know what is the role of this peptide in the conformation of the enzyme as related to its catalytic properties.

When all possible care was taken to avoid proteolysis during the separation of the enzyme, it was found that the extract from cells harvested in exponential phase growth contained about 60 % of the native enzyme E_{II} and 40 % of E_I. It is difficult to prove that this is the situation in vivo but if it is, the process of proteolysis and dissociation-association between E_I and F_2 must have an important physiological role in the regulation of leucyl-tRNA synthetase transfer function.

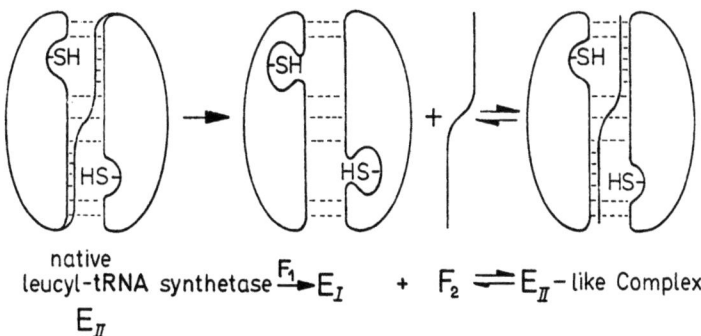

Fig. 8. Schematic representation of the 2 forms of tRNA synthetase and their apparent interconversion

We are happy to dedicate this paper to Prof. Alexandre E. Braunstein, Head of the Laboratory on Chemical Principles of Biocatalysis of the Institute of Molecular Biology, Academy of Sciences of the USSR, for his seventieth anniversary.

References

1. ROUGET, P., and CHAPEVILLE, F., Europ. J. Biochem., 14, 498 (1970).
2. ROUGET, P., and CHAPEVILLE, F., Europ. J. Biochem., in press.
3. IACCARINO, M., and BERG, P., J. Mol. Biol., 42, 151 (1969).
4. BOYER, P.D., J. Amer. Chem. Soc., 76, 4331 (1954).
5. Worthington Biochemical Corporation, Manual n° 11, Freehold, New Jersey 1961.
6. PFLEIDERER, G., and AURICCHIO, F., Biochem. Biophys. Res. Commun. 16, 53 (1964).
7. SCATCHARD, G., Ann. N. Y. Acad. Sci., 51, 660 (1949).
8. YARUS, M., BERG, P., J. Mol. Biol., 28, 479 (1967).

Discussion:

NN

Does a shift between the two forms of the enzyme E_I and E_{II} have a regulatory function?

Chapeville

We have no idea as to a possible physiological role. When *E. coli* cells were harvested in the exponential growth phase and all precautions were taken to avoid proteolysis, it was found that about 40 % of the enzyme activity was in the E_I-form and about 60 % in the E_{II}-form.

Holzer

In the light of what Dr. Katunuma told us yesterday, I propose the following biological function for your system. The proteolytic enzyme F_1 catalyzes the first step of degradation of the native, i.e. active, enzyme E_{II}. The E_I formed in this reaction is then further degraded to amino acids. The E_{II}-like complex of E_I with F_2 demonstrated *in vitro* would then be only an artefact without biological significance. A study on the content of intact cells on E_I, E_{II}, F_1 and F_2 under different growth conditions could perhaps substantiate this speculation.

Fischer

Is the E_{II} to E_I reaction inhibited by inhibitors of proteolytic enzymes like diisopropylfluorophosphate or phenylmethylsulfonyl fluoride? Is there appearance of an N-terminal group during the reaction?

Chapeville

These points were not checked.

NN

Can you characterize the peptidase reaction more closely?

Chapeville

Methionine tRNA synthetase looses about 35 % of its amino acids by incubation with *E.coli* extracts or with proteolytic enzymes such as trypsin, chymotrypsin, subtilisin, etc. in contrast to this the parts of leucine tRNA synthetase which are cleaved off by F_1 are not further degraded by trypsin.

Hasilik

You have mentioned that sooner or later E_{II} to E_I conversion is completed independently of the amount of F_1. The E_{II}-like activity should, however, not completely disappear if E_I can combine with the oligopeptide produced during the conversion. According to your model, the degree of the conversion of the activity should increase with the dilution.

Diphtheria Toxin-catalyzed Adenosine Diphosphoribosylation of Aminoacyl Transferase II from Rat Liver

Tasuku Honjo*, Kunihiro Ueda, Tadashi Tanabe, and Osamu Hayaishi

Department of Medical Chemistry, Kyoto University, Faculty of Medicine, Kyoto/Japan

Diphtheria toxin is a simple protein consisting of a single polypeptide chain of 62,000 daltons. It is lethal for certain species of laboratory animals such as guinea pigs and rabbits in doses as low as 0.1 µg per kg body weight. The biochemical basis for this extremely high toxicity has been the subject of intensive investigation in a number of laboratories but has not been clearly understood for many years.

In 1959, Strauss and Hendee (1) reported that low concentrations of the toxin completely blocked the incorporation of ^{35}S labeled methionine into the protein fraction of cultured HeLa cells. This observation was soon confirmed and extended by Pappenheimer and his coworkers (2, 3) at Harvard who found that in cell-free systems NAD was required as an essential factor for the toxin-dependent inhibition of polypeptide synthesis. Subsequently Collier (4) and also Goor and Pappenheimer (5) showed that toxin specifically inactivated aminoacyl transferase II which participated in the translocation of peptidyl tRNA on ribosomes. However, the role of NAD and the molecular mechanisms by which the toxin exerts its inhibitory effect still remained completely unexplained. Being generally interested in NAD metabolism, Drs. Honjo, Nishizuka and myself attacked this problem about 1967 and, in collaboration with Professor Iwao Kato of Tokyo University, we were able to show that diphtheria toxin catalyzes the transfer of the ADP-ribose portion of NAD to aminoacyl transferase II and ADP-ribosylation results in a concomitant inactivation of this particular enzyme. These findings are schematically presented in Scheme I.

Available evidence indicates that diphtheria toxin is a kind of transribosylase of unique specificity and catalyzes the reversible transfer of the adenosine diphosphate ribose portion of NAD to trans-

---

* Recipient of Postgraduate Fellowship of Sigma Chemical Company.

ferase II through the covalently bound ribosyl linkage thereby inactivating transferase II. These results were recently confirmed and extended by Collier and Cole (6) and also by Goor and Maxwell (7). In this lecture I shall try to present briefly some of our experimental results supporting this equation and to discuss more recent preliminary data on the chemical nature of the adenosine diphosphate ribosyl transferase II linkage.

$$TII + \underset{RPPR}{\overset{+}{N}} \overset{CONH_2}{\underset{}{A}} \xrightleftharpoons[]{\text{Diphtheria Toxin}} TII-RPPRA + \underset{N}{\bigcirc}CONH_2 + H^+$$

Active   Inactive

Our initial studies were essentially made with protein synthesizing systems from rat liver according to the method described by Moldave and his co-workers (8). In confirmation of the finding of Collier and Pappenheimer, poly U-directed polyphenylalanine synthesis was reduced to less than 10% in the presence of both diphtheria toxin and NAD (Table I). The individual addition of either NAD or toxin did not inhibit polyphenylalanine synthesis. Diphtheria toxoid was ineffective and diphtheria antitoxin counteracted the effect of toxin completely. Nicotinamide protected polyphenylalanine synthesis from the inhibition by toxin and NAD.

Table I. NAD-dependent inhibition of protein synthesis by diphtheria toxin

Addition	Poly phenylalanine synthesis (%)
none	100
NAD (5×10^{-6} M)	111
Toxin (20 µg/ml)	92
NAD + Toxin	6

It soon became apparent that when NAD-(adenine)-^{14}C was incubated with diphtheria toxin and transferase II, radioactivity was quickly incorporated into an acid-insoluble fraction. The time course of this reaction is shown in Fig. 1. This incorporation was dependent on the presence of both diphtheria toxin and transferase II. When either one of these proteins was omitted from the reaction mixture, incorporation of radioactivity into acid-insoluble material was almost negligible. Transferase I was incapable of replacing transferase II. The amount of NAD-(adenine)-^{14}C incorporated was not affected significantly by the addition of poly U, GTP or ribosomes. Diphtheria toxoid was inactive. Diphtheria antitoxin inhibited the reaction completely at a concentration equivalent to that of employed toxin. Now the question arises; what are the functions of diphtheria toxin and transferase II in this reaction?

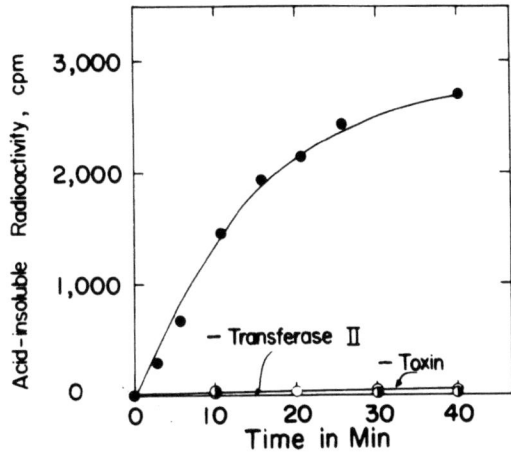

Fig. 1. Incorporation of radioactivity into acid-insoluble fraction from NAD-(adenine)-C^{14}

As shown on the left of Fig. 2, when the concentration of transferase II was kept constant and the amount of diphtheria toxin in the reaction mixture was varied from 0.4 to 2 μg, the initial rate of the incorporation increased but the maximal amount of ADP-ribose incorporated remained essentially constant. On the other hand, when the toxin concentration was kept constant and the amount of transferase II in the reaction mixture was varied, as shown on the right of Fig. 2, both the rate and the extent of the incorporation of radioactivity appeared to be dependent on the amounts of transferase II. These results indicate that toxin plays a catalytic function, the NAD molecule or a portion thereof being bound to transferase II.

We then synthesized NADs labeled at different positions in order to determine which portion of NAD molecule is incorporated into the acid-insoluble product. It can be seen from Table II that when NAD is

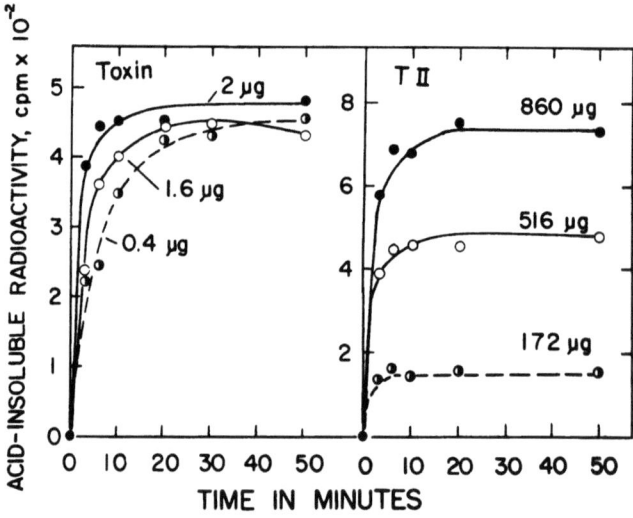

Fig. 2. Effect of toxin and T II

labeled with ^{14}C in adenine, with tritium in adenosine, with ^{32}P, or with ^{14}C in ribose of nicotinamide mononucleotide, used as substrate, essentially the same amounts of radioactivity were incorporated into the acid-insoluble fraction. However, when NAD labeled with ^{14}C in nicotinamide was used as substrate no radioactivity was observed in the acid-insoluble fraction; rather a stoichiometric amount of nicotinamide was found released in the acid-soluble fraction. These results clearly indicate that the ADP-ribose portion of NAD was incorporated into an acid-insoluble product in the presence of toxin and transferase II and nicotinamide is released into the medium. Neither toxin nor transferase II alone showed any hydrolytic activity toward NAD.

The ADP-ribosylated product was then prepared in a large scale incubation mixture containing transferase II, diphtheria toxin-^{125}I, and NAD-(both phosphates)^{32}P. The reaction product was chromatographed on a hydroxylapatite column and the radioactivities of ^{125}I and ^{32}P in

Table II. Incorporation of ADP-ribose unit into acid-insoluble fraction

NAD employed	μμmoles reacted
NAD-(adenine)-^{14}C	55.3
NAD-(adenosine)-^{3}H	54.9
NAD-(both phosphates)-^{32}P	57.0
NAD-(ribose in NMN)-^{14}C	56.5
NAD-(nicotinamide)-^{14}C	0.0

each fraction were measured. As shown in Fig. 3, ^{125}I-labeled diphtheria toxin was eluted in Fractions 7 to 19, and ^{32}P-ADP-ribosylated protein in Fractions 20 to 25. The above results strongly suggest that ADP-ribose is transferred to transferase II, and not to toxin.

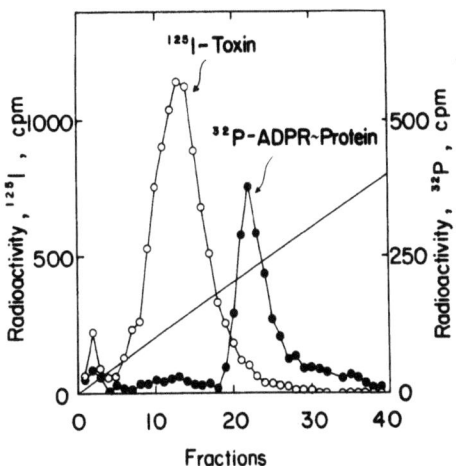

Fig. 3. Hydroxyapatite column chromatography of the ADPR-protein complex

Upon hydroxylapatite column chromatography, the ADP-ribosylated product was eluted as a single peak as shown in Fig. 4, while the transferase II activity was found almost negligible in all fractions. However, when each fraction was treated with toxin and nicotinamide,

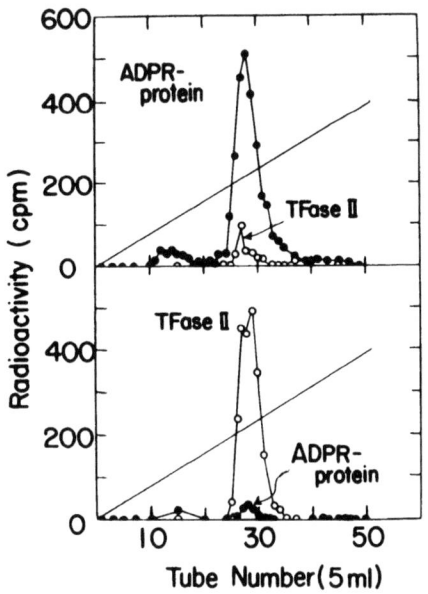

Fig. 4. Binding of the ADPR moiety to transferase II

the transferase II activity appeared with a concomitant disappearance of the acid-insoluble radioactivity in the corresponding fraction as shown in the lower part of Fig. 4. The results indicate that ADP-ribose is transferred to transferase II, resulting in a concurrent inactivation of this enzyme and furthermore that the treatment of ADP-ribosyl transferase II with both toxin and nicotinamide restored the transferase II activity with a simultaneous disappearance of the acid-insoluble radioactivity.

The requirements of the reverse reaction are shown in Table III. Both toxin and nicotinamide were required to release the radioactivity from the ADP-ribosylated transferase II and to restore the transferase II activity. However, the same treatment with nicotinamide or toxin alone caused neither the disappearance of acid-insoluble ADP-ribose nor the appearance of transferase II activity.

Table III. Requirement of the reverse reaction

Addition	ADP-ribosyl transferase	Transferase activity
	c.p.m.	c.p.m.
none	1,028	29
Nicotinamide, 0.05M	1,027	26
Toxin, 4 μg	1,021	7
Nicotinamide + Toxin	76	328

The product of the reverse reaction was NAD as shown in Fig. 5. When ^{14}C-ADP-ribosylated transferase II was incubated with toxin and nicotinamide, and the acid-soluble product was isolated by Dowex 1 column chromatography, almost all radioactivity was eluted in coincidence with NAD added as a marker. The product thus isolated was further identified as NAD by paper chromatography in three different solvent systems and also by high voltage paper electrophoresis.

The time course of incorporation of the ADP-ribose moiety of NAD and inactivation of transferase II is shown in Fig. 6. Again a reaction mixture containing transferase II, toxin and NAD-adenine-^{14}C was incubated. Aliquots were removed and assayed for the transferase activity and for radioactivity incorporated into acid-insoluble material. The incorporation of radioactive ADP-ribose is exactly proportional to the degree of inactivation of the transferase II activity. Essentially similar results were obtained when the reaction was run in reverse, indicating that the ADP-ribosylation reaction is quantitatively proportional to and is closely associated with the inactivation of the transferase activity.

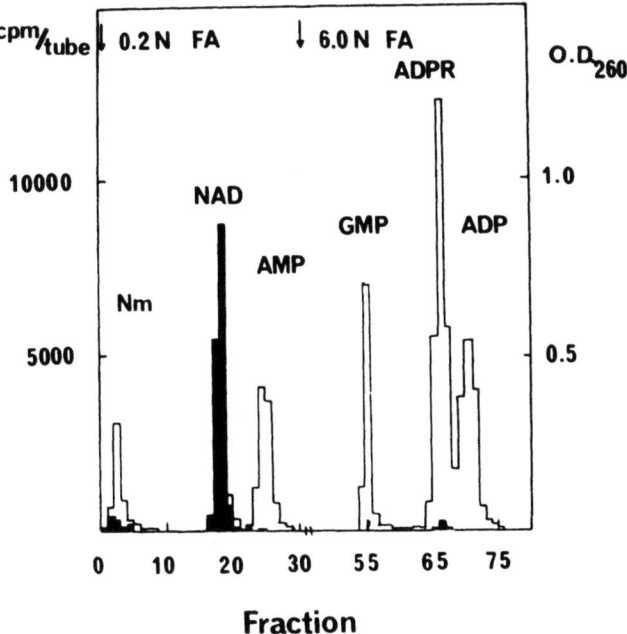

Fig. 5. Isolation of the released product

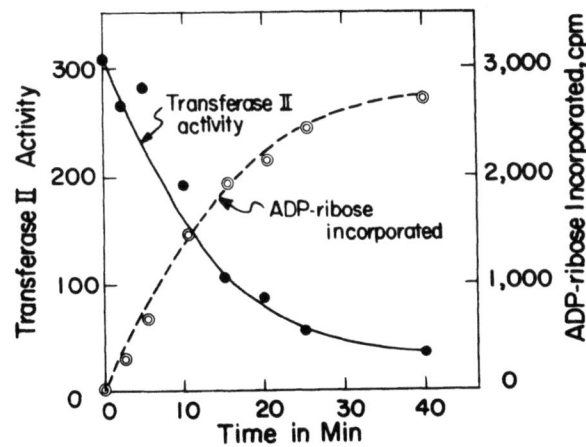

Fig. 6. Incorporation of ADP-ribose and inactivation of transferase II

It is interesting to note that no protein other than the transferase II of eucaryotic cells has been found capable of accepting ADP-ribose from NAD in the presence of diphtheria toxin. Thus when a crude NAD-free extract from eucaryotic cells is incubated with ^{32}P or ^{14}C adenine labeled NAD together with excess toxin, the acid precipitable radioactivity will provide a measure of its transferase II content. In fact this method provides a useful quick assay procedure of transferase II instead of the troublesome protein synthesis system.

We were then able to purify transferase II from rat liver to a homogeneous protein by ammonium sulfate, DEAE-Sephadex, hydroxylapatite and electrofocusing as shown in Table IV. The ratios of specific activity of transferase II to those of the ADP-ribose acceptor in these fractions were essentially identical throughout the purification steps indicating that transferase II is the sole protein species to be ADP-ribosylated in the high speed supernatant fraction from rat liver.

Table IV. Purification of transferase II from rat liver

Fraction	Total Protein (mg)	Total Act. (units·10^{-3})	Specific Act. (units/mg)
Crude Extract	91,200	—	—
$(NH_4)_2SO_4$ Precipitate (0.4 - 0.6)	11,900	900	75
DEAE-Sephadex Eluate	605	575	950
Hydroxylapatite Eluate	152	460	3,250
Electrofocusing Fraction	3.86	225	58,000

The final preparation thus obtained was essentially homogeneous upon polyacrylamide gel electrophoresis. One microgram of the most highly purified preparation of transferase II was capable of accepting 5.2 and 16.5 pmoles of ADP-ribose based on the colorimetric and spectrophotometric determinations of protein concentration, respectively. The molecular weight of transferase II was estimated to be approximately 60,000 to 70,000 by its elution profile from a Sephadex G-100 column, which coincided well with the value (64,600) reported by Galasinski and Moldave (8). It follows, therefore, that 0.34 and 1.1 moles of ADP-ribose bind to 1 mole of transferase II, depending on the method of protein determination employed. The equilibrium constant was therefore, calculated on the assumption that 1 mole of transferase II accepted 1 mole of ADP-ribose.

In a series of experiments the concentration of the nicotinamide varied from zero to 36.8 m$\underline{M}$ and the incorporation of ^{14}C-ADP-ribose into acid insoluble material was followed. As shown in Fig. 7 the equilibrium was reached within 30 min. From this type of experiment both in forward and reverse directions, the equilibrium constant of this reaction has been calculated as follows.

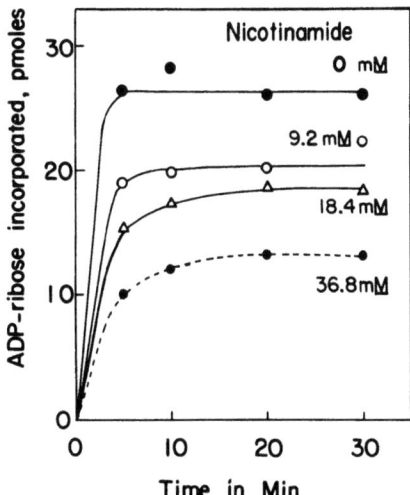

Fig. 7. Equilibrium of the reaction

At pH 7 and 25° the equilibrium constant of the ADP-ribosylation reaction K' was determined to be 6.3 X 10³ and from this equilibrium constant, the standard free energy change, $\Delta G^{o'}$, accompanying the ADP-ribosylation of transferase II was determined to be -5,200 cal per mole under these conditions. The nicotinamide ribose linkage of NAD is a high energy bond which contains about 9,200 cal per mole at pH 7 at 25°. The free energy of hydrolysis of the ADP-ribosyl transferase II linkage is therefore approximately 4,000 cal per mole. This value is close to the free energies of hydrolysis reported for the ribosidic linkage of purine ribosides and that of the α (1-4) glycosidic linkage in the glycogen molecule. It is reasonable to assume that in concentrations of NAD and transferase II comparable to those found in living cells, the reaction goes almost to completion. However, the ADP-ribo-

$$NAD^+ + T_{II} \rightleftharpoons ADPR-T_{II} + Nm + H^+$$

$$K' = \frac{(ADPR-T_{II})(Nm)}{(T_{II})(NAD)} = 6.3 \times 10^3$$

(pH 7, 25°)

$$\Delta G^{o'} = -RT \ln K' = -5160 \text{ cal/mole}$$

$$NAD^+ + H_2O \rightleftharpoons ADPR + Nm + H^+$$

$$\Delta G^{o'} = -9200 \text{ cal/mole} \quad (pH 7, 25°)$$

(Zatman et al. 1953)

$$ADPR-T_{II} + H_2O \rightleftharpoons ADPR + T_{II}$$

$$\Delta G^{o'} = -9200 - (-5160) = -4040 \text{ cal/mole}$$

sylation reaction is accompanied by the release of a proton and therefore the $\Delta G^{o'}$ value of the reaction increases as the pH of the reaction mixture is lowered. Indeed at pH 5 the reaction is readily reversible since $\Delta G^{o'}$ accompanying the forward reaction is about -2,500 cal per mole at pH 5 and 25°. The chemical stability of the ADP-ribosidic linkage was then examined.

^{14}C-NAD labeled in the ribose portion in nicotinamide mononucleotide was used to ADP-ribosylate transferase II and then the ADP-ribosylated protein was isolated and subjected to various pHs under the conditions shown in Table V. It can be seen that this linkage is re-

Table V. Stability of ADP-ribosyl transferase II

pH	Acid-insoluble Radioactivity Remaining		
	37°, 15 hrs	50°, 15 hrs	82°, 6.5 hrs
	%	%	%
0.1N HCl	100	100	80
3	100	100	95
5	100	100	100
7	100	100	93
9	100	100	—
0.1N NaOH	100	95	40

latively stable and with 0.1 molar sodium hydroxide at 50° for 15 hours essentially no ^{14}C became acid-soluble. It has been reported in literature that one of the typical linkages of glycoproteins is the O-glycosyl linkage attached to the hydroxyl group of serine or threonine. But this type of linkage is easily hydrolyzed by dilute alkali and therefore our results seem to exclude the possibility that the ADP-ribose is attached to the hydroxyl group of serine or threonine in the protein. Fig. 8 shows the stability of the ADP-ribosyl linkage at 100° under acidic or alkaline conditions. It can be seen that this linkage is more stable in acid and relatively unstable in alkaline conditions. The fact that this linkage is easily hydrolyzed in one normal sodium hydroxide at 100° within 10 min seems to exclude the possibility that the ribosyl linkage is on the hydroxyl group of hydroxylysine because the latter linkage has been known to be extremely stable under these conditions. If one calculates from these data pseudo-first order rate constants of the hydrolysis under these conditions, the following numbers were obtained (Fig. 9).

The rate constants of the hydrolysis in 1 normal HCl or 0.2 normal NaOH are both about 5×10^{-2} per min and almost 10 times as much as

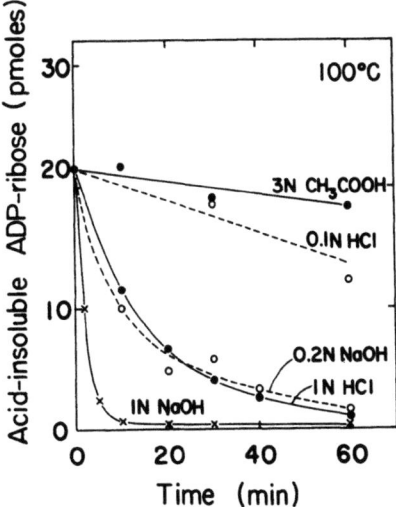

Fig. 8. Hydrolysis of ADP-ribosyl transferase II

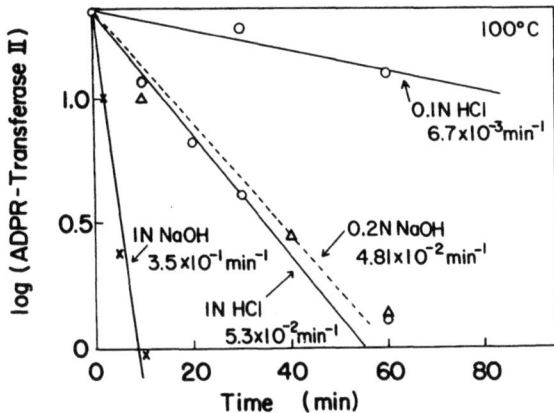

Fig. 9. Hydrolysis of ADP-ribosyl transferase II

those found for the N-glycosidic linkage in asparagine N-glucosides. The ADP-ribosidic linkage is also stable against hydroxylamine treatment which seems to exclude the possibility that it is attached to the carboxyl group. These results seem to indicate that the stability of the ADP-ribosidic linkage is different from the previously known linkages of sugars with serine, threonine, hydroxylysine, asparagine or carboxyl groups. In an attempt to investigate other possibilities we have used various reagents to chemically modify the amino acid residues of transferase II in order to see if these chemical modifications would affect the ability of transferase II to accept ADP-ribosyl moiety of NAD.

Fig. 10 shows the result of an experiment with TNBS (2,4,6-trinitrobenzene sulfonate) that has been established to be a specific modifier of the amino group. It can be seen that when the enzyme is incubated with 5 mM TNBS, ADP-ribose incorporation diminished with time and after about 3 hours inactivation was almost completely. The

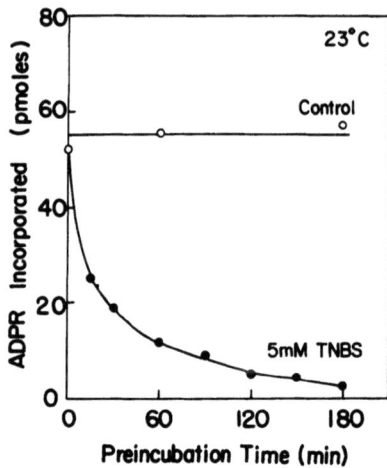

Fig. 10. Effect of amino-group modification of aminoacyl transferase II on ADPR acceptor activity

results indicate that the amino group is probably involved in this linkage. The treatment with acetic anhydride or succinic anhydride also supports this interpretation. Encouraged by this finding we synthesized NAD which is labeled with ^{14}C in the ribose of the nicotinamide mononucleotide portion and ADP-ribosylated a large amount of transferase II using an excess amount of NAD synthesized. The ADP-ribosylated transferase II was then subjected to the treatment of chymotrypsin and trypsin. The fragment thus obtained, was purified on an AG-50 column. Only two major peaks were seen together with several minor fractions (Fig. 11). The second peak was isolated and subjected to thermolysin treatment. A single radioactive peak was obtained upon chromatography on AG-50, and this fraction was subjected to high voltage paper electrophoresis (Fig. 12). It can be seen that the label moved towards the cathode indicating the basic nature of this component. When this component was further subjected to the action of phosphodiesterase and alkaline phosphatase in order to remove the ADP portion leaving only the ribose moiety on the protein, then the remaining fragment migrated to the cathode at the rate essentially similar to that of lysine. We are currently synthesizing ribosyl lysine, ribosyl arginine and ribosyl histidine and also are trying to determine whether this fragment is composed of a single amino acid or still contains a variety of amino acids. Further investigation is necessary to determine if the ribosyl linkage is on some particular basic amino acid or some basic amino acids are in the vicinity of the ADP-ribosylated amino acid.

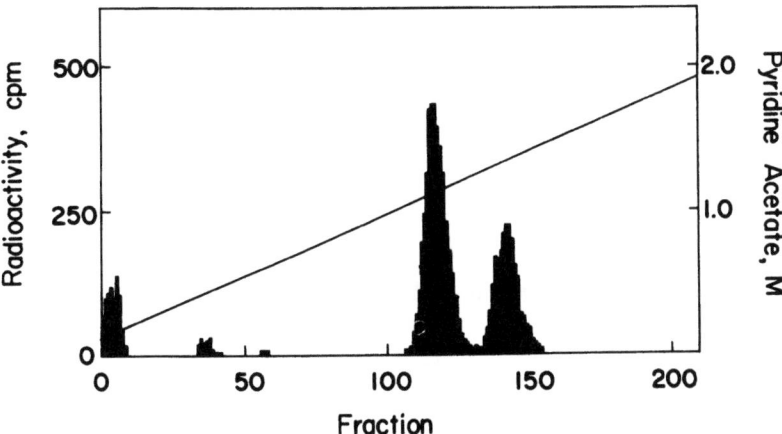

Fig. 11. Chromatography of ADPR transferase II digested by trypsin and α-chymotrypsin on AG-50 column

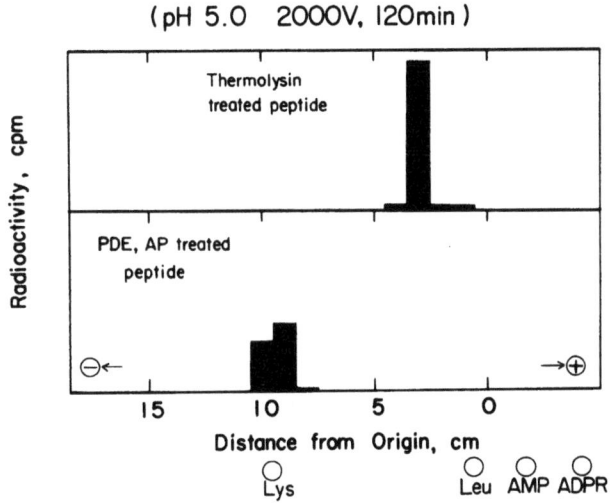

Fig. 12. Paper electrophoresis of peptides

Lastly a word about the mechanism of this unique trans ADP-ribosylation reaction. Since diphtheria toxin is a kind of transglycosidase, the reaction may proceed through the formation of ADPR-toxin as shown in the first equation, followed by the second reaction in which transferase II accepts ADPR from this hypothetical intermediate. However, our results show that there is no appreciable exchange reaction between NAD and nicotinamide-^{14}C in the absence of transferase II. Therefore, the presence of ADP-ribosyl toxin as an intermediate in the reaction appears unlikely. This interpretation is consistent with the recent observations by Goor and Maxwell (7) as well as by Everse and his co-workers (10) that the ADP-ribosylation of transferase II follows a concerted mechanism, forming a ternary complex consisting of diphtheria toxin, transferase II, and NAD.

A Possible Mechanism

(1) NAD + Toxin $\rightleftharpoons$ Toxin–ADPR + Nicotinamide + H$^+$

(2) Toxin–ADPR + T II $\rightleftharpoons$ ADPR – T II + Toxin

sum

NAD + T II $\underset{\text{Toxin}}{\rightleftharpoons}$ ADRR – T II + Nicotinamide + H$^+$

In summary, the ADP-ribosylation reaction proceeds through the replacement of the nicotinamide moiety of NAD with transferase II. The ADP-ribosylated transferase II is catalytically inactive. Although the reaction is reversible, the equilibrium lies far over on the side of transferase II inactivation under physiological conditions. There is no evidence that this or a similar reaction plays any role in the regulation of protein synthesis in vivo. But it is likely that the primary action of diphtheria toxin in the living animal is to effect the inactivation of aminoacyl transferase II, and the resulting inhibition of protein synthesis leads ultimately to death of the animal.

This investigation was supported in part by United States Public Health Service Research Grants CA-04222, from the National Cancer Institute, and AM-10333, from the National Institute of Arthritis and Metabolic Diseases, and by grants from the Jane Coffin Childs Memorial Fund for Medical Research, the Squibb Institute for Medical Research, the Scientific Research Fund of the Ministry of Education of Japan and the Waksman Foundation of Japan, Inc.

References

1. STRAUSS, N., and HENDEE, E. D., *J. Exp. Med.*, **109**, 145 (1959).
2. KATO, I., and PAPPENHEIMER, A. M. Jr., *J. Exp. Med.*, **112**, 329 (1960).
3. COLLIER, R. J., and PAPPENHEIMER, A. M. Jr., *J. Exp. Med.*, **120**, 1019 (1964).
4. COLLIER, R. J., *J. Mol. Biol.*, **25**, 83 (1967).
5. GOOR, R. S., and PAPPENHEIMER, A. M. Jr., *J. Exp. Med.*, **126**, 899 (1967).
6. COLLIER, R. J., and COLE, H. A., *Science*, **164**, 1179 (1969).
7. GOOR, R. S., and MAXWELL, E. S., *J. Biol. Chem.*, **245**, 616 (1970).
8. GALASINSKI, W., and MOLDAVE, K., *J. Biol. Chem.*, **244**, 6527 (1969).
9. ZATMAN, L. J., KAPLAN, N. O., and COLOWICK, S. P., *J. Biol. Chem.*, **200**, 197 (1953).
10. EVERSE, J., GARDNER, D. A., KAPLAN, N. O., GALASINSKI, W., and MOLDAVE, K., *J. Biol. Chem.*, **245**, 899 (1970).

Discussion:

NN

How does the toxin get into the cell?

Hayaishi

I don't think that anyone knows the mechanism.

Sols

I was amused by a rapid mental calculation which indicates that the lethal dosage mentioned by Dr. Hayaishi corresponds to hardly one molecule of enzyme per animal cell. So the efficiency of interconversion is well founded in this case.

Hayaishi

Yes.

Helmreich

I was just wondering whether there is perhaps another possibility to explain this reaction. Wouldn't it be possible to say - if it is the toxin that actually undergoes the protein-protein interaction - that the transferase acts as a kind of specifier protein which induces N-glucosidase activity in the peptidyltransferase.

Hayaishi

I think that this is a very important and pertinent point. We once postulated that ADP-ribosylated toxin is an intermediate in this reaction, but we were not able to show any exchange reaction in the absence of transferase II. This result is in full agreement with the data of Maxwell et al. suggesting that the components form a quarternary complex. I think your suggestion is probably right.

Segal

Is there any evidence for a reversibility of the inactivation of the transferase II back to an active form so that one might consider the possibility that this is a control mechanism in the organism.

Hayaishi

This is an interesting possibility but we haven't investigated it as yet.

NN

Is the E.coli protein synthesizing system sensitive to diphtheria toxin?

Hayaishi

There is a report by Pappenheimer that *E.coli* protein synthesis is inhibited, but that is due to an entirely different mechanism and much higher concentrations of diphtheria toxin are required.

NN

Is the mitochondrial protein synthesizing system also inhibited by diphtheria toxin?

Hayaishi

I don't believe so, but it has not been tested by us.

Haschke

Pappenheimer published about 12 years ago that diphtheria toxin reacts with cytochrome b2 of heart muscle.

Chapeville

Pappenheimer has shown that the toxin is enzymically inactive, but has to be split into two pieces. This happens probably *in vivo.* One of the pieces is probably required for the transport of the enzyme into the cell. The remaining piece carries the enzymically active part. Is this right?

Hayaishi

Yes, I am aware of that experiment of Pappenheimer, but the splitting occurs at a very different position. For instance the MW doesn't change very much, and still we have to face the question of how the molecule enters the animal.

NN

Coming back to the previous question concerning the sensitivity of the mitochondrial system, it should be mentioned that the mitochondrial system is not inhibited.

Dixon

What is the species-specificity of this reaction?

Hayaishi

I personally haven't done experiments along this line but Pappenheimer has made a very thorough study of this and he believes that all protein synthesizing systems from eucaryotic cells are inhibited, so they must have the same amino acid sequence in the vicinity of the ATP-ribosylated active centre.

Sols

I would like to ask if you have obtained reasonably pure transferase II from liver. Could you tell us what the molecular activity of the diphtheria toxin as an enzyme is?

Hayaishi

I think we have calculated it but I don't remember the value.

Chapeville

Could you please say how the transferase II activity was measured?

Hayaishi

We used polyU as messenger, purified washed ribosomes, partially purified transferase I and transferase II, started with phenylalanyl-tRNA labeled in the aminoacid moiety and then measured the incorporation of radioactivity into acid-insoluble material.

NN

And these were rat-liver ribosomes?

Hayaishi:

Yes, this was a rat liver system.

Krebs

Can the reaction be reversed *in vivo* by using sufficientyl high concentrations of nicotinamide?

Hayaishi

Yes, I believe so – this might be an interesting way to cure diphtheria. But we couldn't do the whole animal experiments as this would require a lot of animals. So we felt we might do it in another way. If we inject a small amount of toxin we get reddening and inflamation. When we injected nicotinamide together with toxin into the skin of the rabbit then we observed very variable results – and contrary to our expectations sometimes the reddening and inflamation increased tremendously – presumably because when you inject a rabbit with nicotinamide you produce a lot of NAD and increase toxicity – so we stopped this.

Haschke

Looking at your graphs demonstrating the chromatographic behaviour on hydroxyl apatite, I noticed a little peak in the toxin area. Does it mean that the toxin itself is an acceptor for ADPR?

Hayaishi

That's what we hoped for but we couldn't get any evidence for ADP-ribosylated toxin.

NN

It is my impression that moderately high concentrations of the toxin produce very rapid death within a matter of minutes. Is this consistent with the inhibition of protein synthesis?

Hayaishi

I don't think I am competent to answer that question. I would refer you to the recent review by Dr. Pappenheimer which I think appeared in Experimental Medicine.

Fischer

I realize that you have not identified yet the specific group that is modified on the transferase II but some of your evidence suggests that it might be a lysine residue. Lysine residues can be modified by many types of reagents such as fluorodinitrobenzene and pyridoxal phosphate. Is it possible to inhibit transferase II by chemical reagents which are known to react with lysine residues?

Hayaishi

We have tested a number of modifying reagents including pyridoxal phosphate, but somehow it didn't work. We also tried PNBS.

O. Wieland

Is there anything known about the specific groups of diphtheria toxin which might be essential for its action? Are there SH groups which are important? Has anybody tried to modify the toxin?

Hayaishi

I think that this work is going on in many places, including Dr. Pappenheimer's laboratory, but we are not doing anything along these lines.

NN

What is known about the temperature dependence of the different toxin reactions and what is known about the thermostability of the toxin?

Hayaishi

The toxin itself is thermolabile, but as I said, the enzyme activity appears to be thermostable or activated by thermal treatment under certain conditions.

Regulation of Transcription by T4 Phage Induced Chemical Alteration and Modification of Transcriptase (EC2.7.7.6)

Dietmar Rabussay, Reinhard Mailhammer, and Wolfram Zillig

Max-Planck-Institut für Biochemie, München/Germany

Transcription, that is the synthesis of a RNA-chain complementary to a codogenic DNA strand, is one of the fundamental processes in the expression of genetic information. Since a growing and developing organism or phage needs different gene products at certain times in defined economical amounts transcription has to be regulated.

In fact, several positive and negative regulatory principles of transcription have been elucidated. In most cases the point of action of the regulatory processes is the initiation of transcription. The control of this process occurs at various levels which are organized in an only incompletely understood hierarchical system (compare ref. 1).

Certain elements of transcriptional regulation, some of which are simultaneously also components of the transcription machinery, are:

1) Repressors (λ, lac) which block initiation by direct binding to the DNA.

2) Promotor activating elements (ara C gene product).

3) Modification of the DNA (phage T4).

4) Transcription factors (CAP, ψ(?)), necessary for initiation at a certain class of initiation sites.

5) Modification of RNA polymerase (T4) or synthesis of a completely new one (T3, T7).

6) Modification of initiation factor σ or replacement by analogous factors (T4).

Moreover, antitermination factors (T4, λ) could be involved in transcriptional control (2).

In this paper, a special case of regulation will be discussed, namely, the switch from the transcription of the E.coli genome to the transcription of the T4 genome which occurs after infection of the bacterium by the phage.

Pattern of RNA products in vivo

Within a relatively short time after infection one T4 phage particle causes the production of 100 - 200 identical offsprings. This is feasible by a sophisticated time schedule for the preparation and assembly of the phage components: after the shut off of host transcription at first those T4 genes are transcribed whose products are necessary to start phage DNA replication. Shortly after the beginning of DNA replication the synthesis of another class of gene products, mostly phage structural proteins, becomes dominant (3).

In agreement with these findings two main classes of T4-RNA are found in T4 infected cells: early and late RNA (4,5). Early RNA is mainly synthesized prior to the beginning of DNA replication while late RNA is efficiently transcribed only from replicating DNA.

Early RNA can be divided into two subclasses: immediate early (IE) and delayed early (DE) RNA (6,7). Immediate early RNA starts about 1 min. after infection at 25° (8) and is, moreover, also synthesized in the presence of chloramphenicol (6,7). The transcription of the delayed early RNA starts about 2 - 2,5 min. after infection at 25° (8) (2 min. after infection at 30° (6,7)) and requires prior phage-specific protein synthesis (4,7).

Late RNA is also known to consist of at least two subclasses, namely true late RNA which is synthesized effectively only during T4 DNA replication and quasi-late RNA whose synthesis starts, in fact, early after infection (7).

The switch from early to late transcription involves also a switch from one codogenic strand to another: while early RNA is mainly transcribed from the l strand, late RNA is mainly transcribed from the r strand of T4-DNA (9,10). Fig. 1 shows a simplified graph of this situation. There is still the question of how IE and DE genes are arranged within the early region. Some observations favour an arrangement where IE and DE genes alternate (2,11,12,13). Nevertheless, the other model where all IE genes and all DE genes are clustered cannot be excluded.

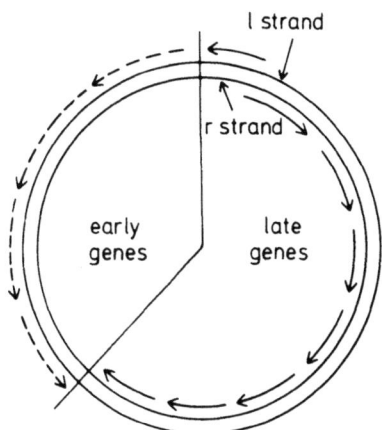

Fig. 1. Schematic diagram demonstrating orientation and codogenic strands in the transcription of the different genes of the T4 genome. The orientation of DNA replication is identical with that of the transcription of true late genes (= clockwise)

Changes in the transcription machinery during phage development

How is the synthesis of the different RNA species turned on and off? It is not yet possible to answer this question in detail. However, three important findings were made in an attempt to answer the question.

a) Walter, Seifert and Zillig showed a structural change in the host RNA polymerase after infection of E.coli with phage T4 (14). This effect has recently been investigated in more detail (15,16).

b) Geiduschek and co-workers found that some requirements regarding the structure of DNA and the function of genes 55 and 33 must be fulfilled to allow efficient transcription of the late genes (17,18,19).

c) Travers (20) and Hager, Hall and Fields (21) have detected proteins which stimulate E.coli core polymerase or T4 modified core polymerase in transcribing certain parts of the T4 genome.

Our interest has been focused on the fate of the host RNA polymerase after phage infection and the correlation of the observed structural

changes with functional changes. Two successive structural and
functional changes of the host RNA polymerase could be differentiated: a first one called alteration and a second one termed modification (22,15).

The first change of host RNA polymerase (alteration)

RNA polymerase isolated from normal E.coli cells consists of four
(or five [+]) different subunits, the relative amounts of which are
expressed by the formula $\beta\beta'\alpha_2\sigma_{(0,2-1)}$ (23,24). The particle
$\beta\beta'\alpha_2$ is called core enzyme. σ is termed initiation factor because it is necessary for efficient and specific, i.e. non random,
initiation on native DNA templates (23,24). The amount of σ varies
from enzyme preparation to enzyme preparation and seems to depend on
the physiological state of the cells used.

When E.coli is infected with phage T4 (multiplicity of infection =
5-10) and the cell functions are blocked immediately after infection
by the addition of NaN_3 and ice, the enzyme isolated from these cells
is structurally and functionally altered. Alteration occurs in the
presence of chloramphenicol as well and seems, moreover, also to be
triggered by phage ghosts (22). The alteration reaction is complete
30 sec. after infection at 25° as can be seen by the rapid decrease
of the enzymatic activity (22,8) and, in a more direct way, by the
appearance of the altered α subunit described below.

T4 altered enzyme (E_A) can be prepared (25,26) either from cells
killed immediately after infection or from cells infected with phage
ghosts or from cells infected in the presence of chloramphenicol
(100 µg/ml) and harvested 3 min. after infection. Normally, we make
use of the last possibility.

The structural change in the host enzyme becomes visible after electrophoresis on cellulose acetate sheets (cellogel) (Fig. 2b): Compared with the pattern for normal E.coli enzyme an additional new
band is observed migrating between α and σ (15). The material
migrating in this band turned out to be an altered α subunit (α_A).
A quantitative estimation of the protein bands separated by cellogel

---

[+] It is still not clear whether another subunit, ω, belongs to the
enzyme or not. No function is known for ω.

electrophoresis and stained with Fast Green shows equal amounts of α and $α_A$.

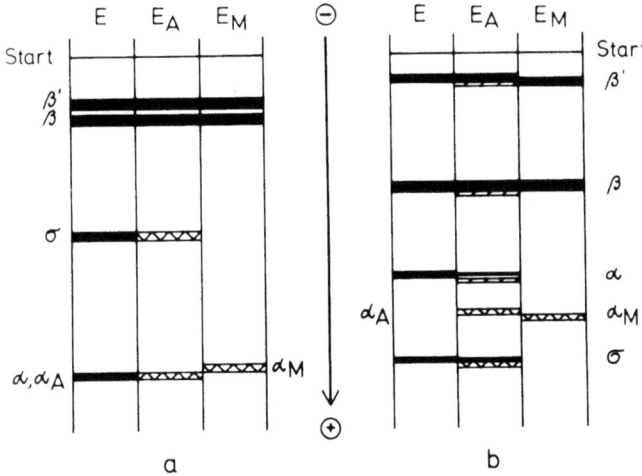

Fig. 2. Schematic representation of (a) polyacrylamide gel electrophoresis of E.coli RNA polymerase holoenzyme (E), T4 altered enzyme (E_A) and T4 modified enzyme (E_M) in the presence of 0,1 % Sodium Dodecylsulfate (SDS) and 6M urea at pH 9 (27). (b) Electrophoresis of E, E_A and E_M on cellogel (cellulose acetate sheets) in 0,6 M ammonium borate (pH 9), 0,01 M EDTA, 0,02 M ß-mercaptoethanol and 6M urea (28). All protein bands were stained with Amido Black or Fast Green. The hatched bands are those which show a significant label of ^{32}P when cells from which the RNA polymerase has been purified were grown on ^{32}P containing medium for several generations prior to phage infection

What is the chemical difference between α and $α_A$? As already described in a previous publication (15) we have found two ^{32}P atoms per $α_A$ when E.coli was grown on ^{32}P containing medium for a sufficient time prior to infection with T4. Normal E.coli RNA polymerase prepared under the same conditions contains only traces of ^{32}P (much less than 1 ^{32}P per subunit). In ^{32}P labeled E_A another strong radioactive band can be observed which migrates in the σ position on SDS polyacrylamide gels and slightly ahead of σ in cellogel electrophoresis (Fig. 2) (15). Neither the amount of P residues was estimated nor tryptic fingerprint maps were made of this material so that we cannot say definitely whether this material is a phosphorylated σ (perhaps comparable with that described by Orlando et al.(29)) or not.

Only in the case of T4 altered enzyme have weak protein bands carrying a considerable ^{32}P label and migrating slightly in front of ß', ß and α been observed in cellogel electrophoresis (Fig. 2). No attempts were made to clarify the meaning of these bands.

The difference spectrum of α and $α_A$ (Fig. 3) and its quantitative evaluation gives strong indication for the presence of one adenine residue per $α_A$ (15).

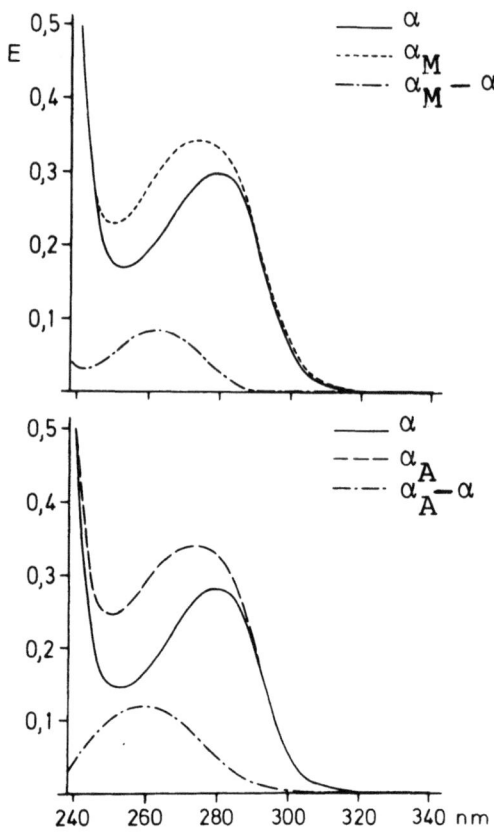

Fig. 3. Spectra of α, $α_A$ and $α_M$ and difference spectra of $α_A$ and $α_M$ versus α. Difference spectra of T4 altered and T4 modified versus normal E.coli α were directly obtained using a solution of normal α as a blank (Beckman DB spectrophotometer). RNA polymerase subunits were obtained in pure state by preparative electrophoresis of highly purified polymerases on cellogel blocks under the conditions described in the legend of Fig. 2. They were removed from the blocks as previously described (30), precipitated by ammonium sulfate (60 % saturation at 0°), redissolved in 0,01 M tris-HCl(pH8) + 0,02 M β-mercaptoethanol, dialysed, reprecipitated by ammonium sulfate, dissolved in and thoroughly dialysed against 0,01 M tris-HCl, pH 8. For the estimation of the nucleotide content in $α_A$ and $α_M$ the concentration of α was determined using the micromethod previously described (30) and that of $α_A$ and $α_M$ assumed to be equal when the extinction values at 295 nm were identical with that of α. Specific extinctions ($ε_{280}$/mg/ml): α = 0,92; $α_A$ = 1,53; $α_M$ = 1,31

The question arises whether and how the P and adenine residues are linked to one another and to the α protein. We can only say that about 80 % of the ^{32}P label occurs in a defined weakly ninhydrin positive spot in the tryptic fingerprint (15) and that the ^{32}P label is not removed by SDS, 6M urea or 10 % acetic acid[+].

T4 altered enzyme is eluted from a phosphocellulose column as a homogeneous peak at about 0,25 M KCl. This means that T4 altered enzyme has the formula $\beta\beta' \alpha \alpha_A \sigma_{(A)}$ and is not a mixture of $\beta\beta' \alpha_2 \sigma_{(A)}$ and $\beta\beta' (\alpha_A)_2 \sigma_{(A)}$. E_A is also not split into an α_A containing and an α_A lacking complex during DEAE cellulose chromatography.

The second change of host RNA polymerase (modification)

Modification of RNA polymerase requires T4 DNA dependent protein synthesis (22). It does not occur with phage ghosts or in the presence of chloramphenicol. The reaction is half complete about 2 min. after infection at 37°C (22). We normally prepare T4 modified enzyme (E_M) from cells which have been infected for 5 - 10 min. with T4 am N 82. T4 modified enzyme has the formula $\beta_M \beta'_M (\alpha_M)_2$. T4 modified α (α_M) contains, just like α_A, 2 P atoms and 1 adenine residue (15) (Fig. 3) but it behaves differently from α_A and α in respect to its electrophoretic mobility (Fig. 2.a). In a tryptic fingerprint of ^{32}P labeled α_M the radioactivity appears in one defined ninhydrin positive spot which is significantly more acidic and hydrophobic than the radioactive spot in the fingerprint map of α_A (15). It remains unclear whether E_A is the substrate for the modification reaction or whether normal E.coli enzyme can be modified directly. α_M migrates more slowly in SDS polyacrylamide gel electrophoresis than α and α_A (22). This can be due to either a special arrangement of the P and adenine residues different from that in α_A or an additional compound, not present in α and α_A, which is linked to α_M. α_M was also investigated by Goff and Weber (31). They first showed that ^{32}P label is incorporated into α after phage infection and they were able to split off a 5'-AMP residue by digestion with high amounts of commercial phosphodiesterase. Since we estimated 2 P atoms and 1 adenine residue per subunit the question arises how this second P atom is linked in α_M.

[+] room temperature, 10 hours.

Fate of the host σ factor

What happens to the E.coli σ factor after T4 infection? T4 altered enzyme still contains σ factor as can be seen in SDS polyacrylamide gel electrophoresis and cellogel electrophoresis. As already stated it is not clear, however, to what extent σ becomes altered. In T4 modified enzyme, σ is completely absent or only observed in traces (22). This fact can be due either to a degradation of σ or to a lowered affinity of the (altered (?)) σ to the modified core enzyme. To decide between these two possibilities we homogenized T4 infected cells (harvested 10 min. after infection at 37°C), which in a control experiment, were shown to contain only T4 modified enzyme. To the homogenate we added normal E.coli core enzyme (0,3 mg/g infected cells), stirred gently for 20 min. and purified RNA polymerase from this homogenate in the usual way (26,25). The obtained enzyme examined in SDS polyacrylamide gel electrophoresis and cellogel electrophoresis showed clearly the presence of a σ band. Moreover, the isolated enzyme was 2,5 times more active with T4 DNA as template than with calf thymus DNA while the input E.coli core enzyme had a 5 times greater activity on calf thymus DNA than on T4 DNA. (Assay conditions as described in ref. 32). This indicates that the E.coli core enzyme has picked up a σ factor from the homogenate of T4 infected cells which behaves like normal host σ in respect to its electrophoretic mobility, molecular weight and stimulating activity. We conclude that σ is not degraded after T4 infection but is less tightly bound to the T4 modified core enzyme than to E.coli core enzyme. The same experiment but performed with ^{32}P labeled cells will show whether the σ present 10 min. after infection in the infected cell is phosphorylated or not.

The function of alteration

Since alteration is extremely rapid, not affected by the presence of chloramphenicol and occurring also with phage ghosts it seems to be a normally dormant capacity of the host cell. Alteration (and modification) can also be triggered by infection with bacteriophage T6 (33). Colicin K (33), however, as well as an extract which contained tail fibers but not intact tails (prepared from E.coli B cells infected with T4 am B 17 am 120 (gene 23 and 27) (34)), did not cause alteration. These findings suggest that perhaps drastic mechanical damage such as perforation of the cell wall by phage is necessary to trigger the alteration reaction. Three other findings suggest the opposite: Fuchs has already observed a decrease in enzymatic activity of RNA polymerase when cells become stationary (35). Goff and Weber (36) found ^{32}P label in the α band when RNA polymerase, isolated from bacteria which were grown to extremely advanced stationary phase on ^{32}P containing medium, was applied to SDS polyacrylamide gel electrophoresis. This ^{32}P containing α could be identical with our T4 altered α. Chelala et al. (37) have reported an inactivation of E.coli RNA polymerase when the enzyme is incubated with ATP, Mg^{++} and a certain protein fraction. This reaction which was described to be reversible, may also be involved in the alteration reaction. We have not been able, however, to observe the described inactivation of RNA polymerase.

The in vivo function of the alteration reaction could be the shutting off of host transcription. This assumption is strongly supported by the results of Schachner et al. (8,38) who measured the rate of E.coli and T4 RNA synthesis within the cell before and after phage infection.

Since alteration is so rapid, the immediate early T4 genes are probably read by E_A. Is this supposition confirmed by in vitro experiments? Although E_A is only 20 % as active as E.coli enzyme on T4 DNA under the usual assay conditions it shows some specificity for T4 DNA: DNA from other sources (λ, φ X 174, calf thymus, T4 single strand) is only transcribed with an efficiency of a few per cent compared with T4 DNA (39).

Unfortunately, there exists no suitable in vitro system with E.coli DNA as template to check directly whether the shut off of host transscription occurs by altered RNA polymerase. Nevertheless, the assay of T4 altered enzyme in a cell free system with catabolite repression sensitive genes from E.coli (40,41) should enable this point to be tested.

The function of modification

The biological function of the modification reaction is not clear and seems to be rather complicated. When we summarize our findings on alteration and modification (Fig. 4) and compare it with the temporal appearance of the different RNA products, the modified core enzyme becomes a good candidate for that enzyme which transcribes the DE genes. If T4 modified core (E_M) is really involved in the transcription of the DE genes some additional conditions have to be fulfilled: one of them is the availability of one or more initiation factors because E_M transcribes T4 DNA randomly and very inefficiently (21). The problem of how the transcription of IE and DE genes is regulated seems, however, not as simple as suggested previously (42) assuming that a replacement of E.coli σ by a new T4 coded σ factor is sufficient. There is no conclusive evidence that σ_{T4e} (42,20) is, in fact, a σ like factor. Moreover, normal E.coli σ factor stimulates E_M equally as well as E.coli core enzyme in spite of its obviously lower affinity to E_M (8). Like E.coli holoenzyme, E_M + σ also transcribes the IE and DE genes in vitro (43).

The combination E_M + σ might play some role in vivo but it should be, at most, only one of several required combinations between a core enzyme and transcription factors since the transcription of IE and DE genes is subjected to manifold regulatory mechanisms during the

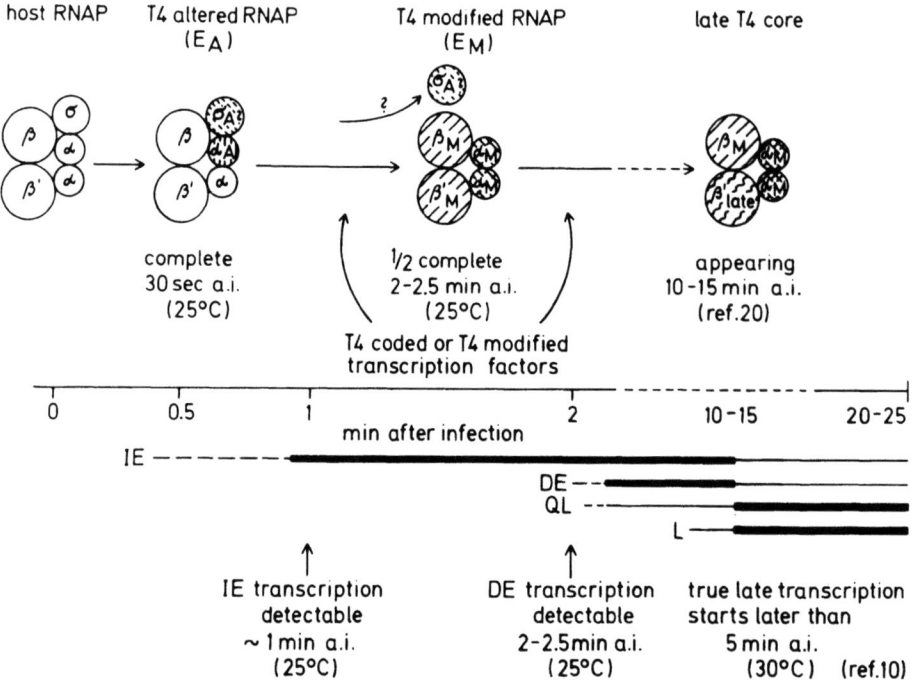

Fig. 4. Very schematic diagram representing the changes of host RNA polymerase (= RNAP) and the RNA pattern after infection with phage T4. Most time specifications for the appearance of the different RNA species are accepted from ref. 8. ω might also play a role in the specificity change of RNAP. This subunit, however, has not been included in this scheme because not enough facts are available. IE, DE, QL and L mean immediate early, delayed early, quasi-late and late RNA. a.i. = after infection

growth cycle of the phage. Included in the term transcription factor are not only σ like factors (20,21) but also additional factors acting at a higher hierarchical level than σ (like CAP (41), ψ (44)). Moreover, one has also to take into account that another function of the modification reaction could be the adaption of the host enzyme for specific interactions with newly synthesized or phage modified termination or antitermination factors.

Other modifications of host RNA polymerase

Two other changes of host RNA polymerase have been described but their functional meaning is not clear at all yet:

- Travers (20) reported the appearance of a changed ß' subunit 10 - 15 min. after infection. The first indication of this change in ß' came from Khesin at al. (45,46). The change in ß' occurred only after infection with $T4^+$ and was not visible when mutants not able to replicate DNA were used. Therefore, and because of its relatively late appearance, it cannot be identical with the modification of ß' observed in our laboratory (47,48).

- Stevens (49,50) labeled cells with ^{14}C-leucine and ^{14}C-glucose shortly after infection with phage T4. She found 3 different ^{14}C labeled small proteins (MW 9,000 - 21,000) copurifying with RNA polymerase and assumed that they could be related to the ω subunit. It is interesting that one of these small proteins was lacking in two different amber mutants in gene 55.

Some concluding remarks

T4 changes the host RNA polymerase and, unlike T7 and T3 (51,52) does most likely not synthesize its own RNA polymerase (53). The reason for this feature could be the relatively complicated regulation of T4 transcription compared to the more simple regulation of transcription of T3 and T7 whose genomes are about 5 times smaller than that of T4. Furthermore, T4 solves the problem of producing a RNA polymerase suitable for complicated regulation processes quickly and economically. The attachment of small residues (containing P or adenine or both) can be carried out much more rapidly than the synthesis of large new enzymes. Some examples of how the activity of enzymes (54) and also of ribosomes (55) can be influenced by phosphorylation, adenylylation and ADPribosylation are known. However, in the case of RNA polymerase it is possible to change or substitute selectively only one or some of its subunits. The detailed investigation of the consequence of changes and substitutions to subunits could be fruitful for learning more about the role of the single subunits in the transcription process.

The regulation of T4 transcription seems also to be a good subject for the study of principal regulatory mechanism. Its complexity lies between that of the small phages and that of E.coli or eucaryotic cells. Especially the mechanism of late transcription with its coupling to DNA replication could lead to interesting results because some animal viruses also show this coupling between DNA replication and the transcription of certain parts of the genome. Moreover, a fruitful investigation of the different regulatory elements such as σ-like initiation factors, transcription factors such as CAP and ψ, termination and antitermination factors and their hierarchical arrangement seems promising in the T4 system.

Summary

Two successive changes of the host RNA polymerase are observed after infection of E.coli by phage T4 and N82. In the first step ("alteration") one of the two α subunits is changed by covalent attachment of 2 phosphorous atoms and 1 adenine residue. σ may also be phosphorylated in this step. The function of the alteration reaction seems to be the shutting off of host transcription. In the second step ("modification"), β and β' are changed and to each α subunit 2 phosphorus atoms and 1 adenine residue are bound. The appearance of the "modified" core polymerase coincides with the beginning of delayed early transcription.

Acknowledgments

Thanks are due to Ingelore Holz and Ursula Lederer for excellent technical assistance. We thank Prof. A. Butenandt, the Deutsche Forschungsgemeinschaft und Sonderforschungsbereich 51 for generous support. Dr. E.P. Geiduschek kindly donated the T4 am B 17 Am N 120 double mutant.

References

1. TRAVERS, A., *Nature New Biology*, 229, 69 (1971).

2. SCHMIDT, D.A., MAZAITIS, A.J., KASAI, T., and BAUTZ, E.K.F., *Nature*, 225, 1012 (1970).

3. LEVINTHAL, C., HOSODA, J., and SHUB, D., in: *The Molecular Biology of Viruses* (Colter and Paranchyck, ed.) P. 71. New York.

4. KANO-SUEOKA, T., and SPIEGELMAN, S., *Proc.Nat.Acad.Sci.*, 48, 1942 (1962).

5. KHESIN, R.B., and SHEMYAKIN, M.F., *Biokhimiya*, 27, 761 (1962).

6. GRASSO, R., and BUCHANAN, J.N., *Nature*, 224, 882 (1969).

7. SALSER, W., BOLLE, A., and EPSTEIN, R.H., *J.Mol.Biol.*, 49, 271 (1970).

8. SCHACHNER, M., SEIFERT, W., and ZILLIG, W., *Europ.J.Biochem.*, in press.

9. GUHA, A., and SZYBALSKI, W., *Virology*, 34, 608 (1968).

10. GUHA, A., SZYBALSKI, W., SALSER, W., BOLLE, A., GEIDUSCHEK, E.P., and PULITZER, J.F., *J.Mol.Biol.*, 59, 329 (1971).

11. ADESNIK, M., and LEVINTHAL, C., *J.Mol.Biol.*, 48, 187 (1970).

12. BRODY, E., SEDEROFF, R., BOLLE, A. and EPSTEIN, R.H., Cold Spring Harbor Symp., 35, 203 (1970).

13. MILANESI, G., BRODY, E.N., GRAU, O., and GEIDUSCHEK, E.P., *Proc.Nat.Acad.Sci.*, 66, 181 (1970).

14. WALTER, G., SEIFERT, W., and ZILLIG, W., *Biochem.Biophys.Res.Commun.*, 30, 240 (1968).

15. SEIFERT, W., RABUSSAY, D., and ZILLIG, W., *FEBS-Letters*, 16, 175 (1971).

16. RABUSSAY, D., MAILHAMMER, R., and ZILLIG, W., Abstracts of Papers presented at the 1971 European Phage Meeting, Berlin, 1971.

17. BOLLE, A., EPSTEIN, R.H., SALSER, W., and GEIDUSCHEK, E.P., *J.Mol.Biol.* 33, 339 (1968).

18. PULITZER, J.F., and GEIDUSCHEK, E.P., *J.Mol.Biol.*, 49, 489 (1970).

19. CASCINO, A., RIVA, S., and GEIDUSCHEK, E.P., Cold Spring Harbor Symp., 35, 213 (1970).

20. TRAVERS, A., Cold Spring Harbor Symp., 35, 241 (1970).

21. HAGER, G., HALL, B.D., and FIELDS, K., Cold Spring Harbor Symp., 35, 233 (1970).

22. SEIFERT, W., QASBA, P., WALTER, G., PALM, P., SCHACHNER, M., and ZILLIG, W., *Europ.J.Biochem.*, 9, 319 (1969).

23. BURGESS, R.R., *Ann.Rev.Biochem.*, 40 (1971), in press.

24. SETHI, V.S., Progress in Biophysics and Molecular Biology, in press.

25. ZILLIG, W., ZECHEL, K., and HALBWACHS, H.-J., Hoppe Seyler's Z. Physiol.Chem., 351, 221 (1970).

26. SCHÄFER, R., MAILHAMMER, R., RABUSSAY, D., SEIFERT, W., SETHI,V.S., and ZILLIG, W., Hoppe Seyler's Z. Physiol.Chem., in press.

27. SHAPIRO, A.L., VINUELA, E., and MAIZEL, J.V., Biochem.Biophys. Res.Commun., 28, 815 (1967).

28. RABUSSAY,D., and ZILLIG, W., FEBS-Letters, 5, 104 (1969).

29. ORLANDO, J.M., WOO, S.L.C., REIMANN,E.M., and DAVIE,E.W., Biochemistry, 9, 4807 (1970).

30. HEIL, A., and ZILLIG, W., FEBS-Letters, 11, 165 (1970).

31. GOFF, Ch.G., and WEBER, K., Cold Spring Harbor Symp., 35, 101 (1970).

32. FUCHS, E., MILETTE, R.L., ZILLIG, W., and WALTER, G., Europ.J. Biochem., 3, 183 (1967).

33. SEIFERT, W., unpublished results.

34. KING, J., J.Mol.Biol., 58, 693 (1971).

35. FUCHS, E., Thesis, University of München, 1965.

36. GOFF, Ch.G., and WEBER, K., personal communication.

37. CHELALA, C.A., HIRSCHBEIN, L., and TORRES, H.N., Proc.Nat.Acad. Sci.U.S., 68, 152 (1971).

38. SCHACHNER, M., Thesis, University of Tübingen, 1970.

39. SCHACHNER, M., and SEIFERT, W., Hoppe Seyler's Z. Physiol.Chem. 352, 734 (1971).

40. CROMBRUGGHE, B. de, CHEN, B., ANDERSON, W., NISSLEY, P., GOTTESMAN, M., PASTAN, I., and PERLMAN, R., Nature New Biology, 231, 139 (1971).

41. ZUBAY, G., SCHWARTZ, D., and BECKWITH, J., Proc.Nat.Acad.Sci.U.S., 66, 104 (1970).

42. TRAVERS, A.A., Nature, 225, 1009 (1970).

43. BAUTZ, E.K.F., BAUTZ, F.A., and DUNN, J.J., Nature, 223, 1022 (1969).

44. TRAVERS, A., KAMEN, R., and CASHEL, M., Cold Spring Harbor Symp., 35, 415 (1970).

45. KHESIN, R.B., GORLENKO, Z.M., SHEMYAKIN, M.F., STROLINSKY,S.L., MINDLIN, S.Z., and ILYINA, T.S., Mol.Gen.Genet. 105, 243 (1969).

46. KHESIN, R.B., in: Lepetit Colloqu. on RNA Polymerase and Transscription (L.Silvestri, ed.), P. 167. North Holland Publ. Co., Amsterdam.

47. ZILLIG, W., ZECHEL, K., RABUSSAY. D., SCHACHNER, M., SETHI, V.S., PALM, P., HEIL, A., and SEIFERT, W., Cold Spring Harbor Symp., 35, 47 (1970).

48. SCHACHNER, M., and ZILLIG, W., Europ.J.Biochem., in press.

49. STEVENS, A., Biochem.Biophys.Res.Commun., 41, 367 (1970).

50. STEVENS, A., personal communication.

51. CHAMBERLIN, M., Mc GRATH, I., and WASKELL, L., Nature, 228, 227 (1970).

52. DUNN, J.J., BAUTZ, F.A., BAUTZ, E.K.F., Nature New Biology, 230, 94 (1971).

53. HASELKORN, R., VOGEL, M., and BROWN, R.D., Nature, 221, 836 (1969).

54. HOLZER, H., and DUNTZE, W., Ann.Rev.Biochem., 40 (1971), in press.

55. KABAT, D., Biochemistry, 9, 4160 (1970).

Discussion:

Stadtman

My recollection is that the modification of RNA polymerase is not uniquely associated with T4 infection but can also occur in response to starvation conditions. Is that right?

Rabussay

Ch. Goff reported at a recent Gordon conference that he found a ^{32}P-label in the α subunit of RNA polymerase from stationary E. coli cells. Unfortunately, it has not been decided whether this labelling corresponds to the alteration or modification described by us or to something else; he has merely separated the labelled α subunit by SDS-polyacrylamide gel electrophoresis. We have had no success in finding such a change in α of stationary cells.

Zubay

I would just like to comment that four years ago we reported conditions which, in a cell free system, support transcription and translation from T phage DNA but not from E.coli DNA. This differential transcription depended only on salt composition of the mixture and not on a modification of sigma or core.

Gancedo

At the meeting of the Spanish biochemical Society at Barcelona this month Vinuela and Salas reported that Bac.subtilis RNA polymerase is adenylylated upon infection by ⌀29 phage. They have purified the adenylylating and deadenylylating enzymes and characterized them as a single chain of about 4o,ooo and a tetramer of about 32o,ooo D, resp.

Rabussay

Does this occur only upon phage infection or also under particular physiological conditions, for example, at the beginning of the sporulation?

Gancedo

I don't think they have done such studies; they have concentrated on studies of the consequences of phage infection.

Hartmann

Have you any evidence that there are new proteases formed after phage infection?

Rabussay

We looked for them, but we couldn't find any significant new protease after the infection.

Zubay

There is some talk about new sigma factors produced by phage infection.

Rabussay

Travers (Ref. 2o) has rather crude fractions which stimulate both the immediate and delayed early transcription. Hager, Hall and Fields (Ref. 21) have purified two new sigma-like factors which stimulate the modified core enzyme in its ability to transcribe the immediate early and delayed early genes.

On the Mechanism of Action and Metabolic Control of the Multifunctional Enzyme Complex that Catalyzes Adenylylation and Deadenylylation of *Escherichia coli* Glutamine Synthetase

E. R. Stadtman, M. Brown, A. Segal, W. A. Anderson, B. Hennig, A. Ginsburg, and J. H. Mangum

National Institutes of Health, National Heart and Lung Institutes, Laboratory of Biochemistry, Bethesda, Maryland/USA

Glutamine metabolism in E. coli is modulated by enzyme catalyzed adenylylation and deadenylylation of glutamine synthetase (GS) in response to fluctuations in the intracellular concentration of glutamine, α-ketoglutarate and various nucleoside triphosphates. Adenylylation is accompanied by a decrease in catalytic activity (1,2), a change in divalent ion specificity (1), a change in sensitivity to cumulative feedback inhibition by 9 different end products of glutamine metabolism (3,4) and a substantial shift in pH optimum (3).

Adenylylation of glutamine synthetase involves attachment in phosphodiester linkage of an adenylyl group from ATP to the hydroxyl group of a specific tyrosyl residue (one of 14) on each enzyme subunit (5). Since glutamine synthetase is composed of 12 apparently identical subunits of 50,000 mol. wt. each (6), up to 12 adenylyl groups can be covalently bound to one molecule of enzyme (1). As shown in reaction [1], inorganic pyrophosphate is the other product of the reaction. Mantel and Holzer (7) showed that the adenylylation reaction is reversible and proceeds with only a slight loss in standard free energy. The free energy of hydrolysis of the adenyl-O-tyrosyl linkage is therefore of the same order of magnitude as is the free energy of hydrolysis of a pyrophosphate bond in ATP.

$$GS + 12\ ATP \xrightleftharpoons{P_I} GS\ (AMP)_{12} + 12\ PPi \qquad [1]$$

$$GS\ (AMP)_{12} + 12\ Pi \xrightarrow{P_I} GS + 12\ ADP \qquad [2]$$

Sum: 1 + 2 $12\ ATP + 12\ Pi \rightarrow 12\ ADP + 12\ PPi;\quad \Delta F_0' -3,000\ cal$ [3]

Although thermodynamically feasible, reversibility of reaction 1 is probably not a physiologically significant mechanism for the deadenylylation of the modified enzyme. Instead, as was shown by Shapiro (8,9) deadenylylation is catalyzed by a complex enzyme system consisting of 2 separable protein components, P_I and P_{II}, and is stimulated by α-ketoglutarate, UTP and ATP, Shapiro (9) detected both ADP and AMP among the products of deadenylylation, but for technical reasons, he was unable to determine which of these was the primary product. Nevertheless, for sometime it was generally assumed that deadenylylation of glutamine synthetase involved hydrolysis of the adenylyl-O-tyrosyl bond to produce AMP. This and the relatively more complex requirements for maximal deadenylylation activity compared to those needed for adenylylation, led to the belief that adenylylation and deadenylylation of glutamine synthetase were catalyzed by different enzymes. Since adenylylation involves a transfer of the adenylyl group from ATP to glutamine synthetase, the enzyme catalyzing that reaction was referred to as ATP: glutamine synthetase adenylyl transferase and was given the trival name "adenylyl transferase" (ATase) (1). In the meantime Anderson et al. (10) established that deadenylylation involves a phosphorolytic cleavage of the adenylyl-O-tyrosyl bond to produce ADP (reaction 2) and that the P_I protein catalyzing this reaction is identical with the adenylyl transferase. This identity is illustrated by the data in Fig. 1 which shows the distribution of catalytic activities in fractions obtained when a partially purified enzyme preparation capable of catalyzing both adenylylation and deadenylylation reactions (reaction [1] and [2], respectively) is filtered through a column of agarose. It can be seen, as was shown earlier by Shapiro (9), that the P_I and P_{II} protein

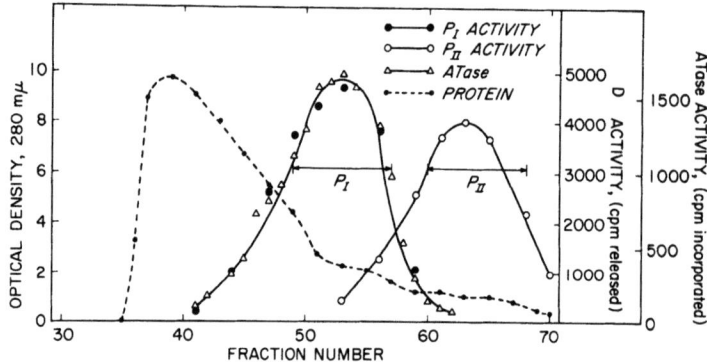

Fig. 1. Agarose gel filtration. 32 ml of a partially purified enzyme preparation was applied to a column (4 X 80 cm) of Bio-Gel A - 0.5 m agarose equilibrated with 50 mM·K phosphate buffer (pH 7.2) containing 10 mM 2-mercaptoethanol and 0.25 mM K_2·MgEDTA. 10 ml fractions were collected. Adenylyltransferase (ATase) and deadenylylation activities were measured as previously described in the legend to Fig. 1 of reference (11)

fractions which together catalyzed reaction [2] are readily separated by this procedure; moreover, it is further evident that the elution profile of the P_I protein coincides perfectly with that of the enzyme that catalyzes adenylylation of glutamine synthetase (i.e., the ATase). This and several other lines of evidence (11) not discussed here show that the ATase and P_I are the same protein.

From the chemical point of view, reaction [2] and the reverse of reaction [1] are very similar. Both involve adenylylation of a phosphoric acid group by adenylylated

glutamine synthetase but in reaction [1] inorganic pyrophosphate is the adenylyl group acceptor whereas in reaction [2] orthophosphate is the acceptor[1].
It is therefore not surprising that both reactions are catalyzed by the same enzyme. This enzyme has been isolated as nearly homogeneous protein by Ebner, et al. (12) in Holzer's laboratory and also in our laboratory in Bethesda (13). It has a molecular weight of about 130,000 (12,13) and is composed of two dissimilar subunits of about 70,000 and 60,000 mol. wt. (13). The 70,000 mol. wt. subunit has been obtained as a homogeneous protein, which by itself has the capacity to catalyze reaction [1] but not reaction [2] (14). Thus far reaction [2] has not been observed except in the presence of the intact P_I complex. The fact that reactions [1] and [2] are catalyzed by the same enzyme poses a serious problem from the standpoint of cellular regulation. It is obvious that unless the adenylylation and deadenylylation functions of the P_I protein are rigorously controlled, coupling between these two functions (reaction [1] and [2]) will occur leading to the aimless oscillation of glutamine synthetase between adenylylated and unadenylylated states; the net result would be simply a phosphorylsis of the terminal pyrophosphate bond of ATP to form ADP and inorganic pyrophosphate (reaction [3]). In addition to a loss of regulatory function, such a futile cycle is accompanied by a loss in standard free energy of approximately -3,000 calories per mole of ATP consumed. This loss in free energy derives from the fact that the standard free energy of hydrolysis of the β-γ-pyrophosphate bond of ATP to produce AMP and inorganic pyrophosphate is about -3,000 cal. greater than the free energy of hydrolysis of the terminal phosphoryl groups of either ATP or ADP (15).

Regulation of Adenylylation and Deadenylylation Functions of ATase

Haphazard coupling of reactions [1] and [2] is prevented by the reciprocal effects of glutamine, UTP and α-ketoglutarate on the two reactions. As is illustrated in Fig. 2, glutamine stimulates adenylylation but inhibits deadenylylation of glutamine synthetase, whereas both α-ketoglutarate and UTP inhibit adenylylation but stimulate deadenylylation. The teleological significance of the UTP effect is not immediately apparent, but it could be related to the fact that the glutamine is involved in the synthesis of uridine nucleotides as well as their conversion to cytidine nucleotides. The effect of α-ketoglutarate on each function is opposite to that of glutamine, and since each metabolite has an opposite effect on reactions [1] and [2], it is obvious that a high glutamine: α-ketoglutarate ratio favors adenylylation (reaction 1) of glutamine synthetase whereas a low ratio favors deadenylylation (reaction 2)

---

[1] Since reactions [1] and [2] are both adenylyltransfer reactions, ambiguity results from the use of the term adenylyl transferase to specify only the adenylylation reaction. We propose that hereafter the P_I protein be designated as adenylyl transferase (ATase), irrespective of which function it catalyzes; the deadenylylation function will be referred to as D-activity and the adenylylation function referred to as A-activity.

Moreover, this ratio is extremely sensitive to the availability of ammonia nitrogen since in the presence of ammonia α-ketoglutarate is converted to glutamine. Thus, when the supply of ammonia nitrogen is high, the glutamine: α-ketoglutarate ratio will become high with the result that glutamine synthetase will be adenylylated and thereby converted to a relatively inactive state. However, as the supply of ammonia diminishes, the glutamine: α-ketoglutarate ratio will decrease sharply due to simultaneous accumulation of α-ketoglutarate and depletion of the glutamine pool. Consequently, the adenylylation reaction will be inhibited and the deadenylylation reaction will be stimulated, resulting in an enhancement of glutamine synthetase activity.

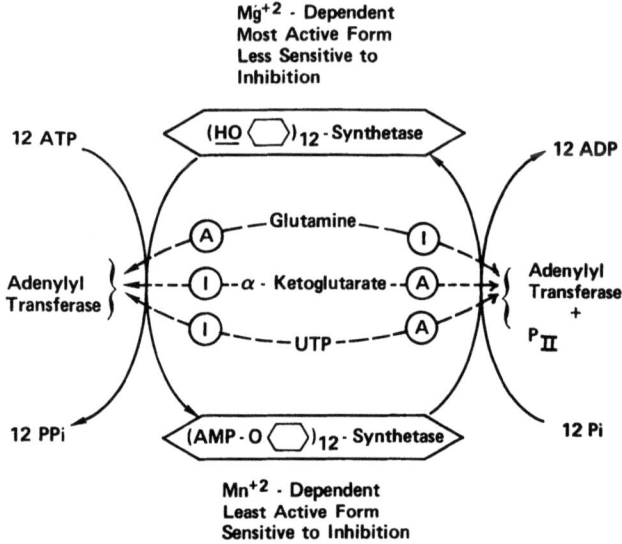

Fig. 2. Metabolite control of adenylylation and deadenylylation reactions. The encircled I and A indicate the inhibition and activation, respectively

Regulatory function of the P_{II} protein. In earlier studies, Shapiro (9) demonstrated that in addition to P_I and the above metabolites, a small protein, designated P_{II}, is required for deadenylylation of glutamine synthetase. In the meantime, the capacities of the metabolites and of P_{II} to stimulate deadenylylation have varied from one enzyme preparation to another. This variability is now partly explained by the discovery that P_{II} is a regulatory protein that can exist in two interconvertible forms; one form, $P_{II\ A}$, stimulates the adenylylation activity of P_I in the absence of glutamine and makes that activity more sensitive to inhibition by α-ketoglutarate. $P_{II\ A}$ has little influence on the deadenylylation activity of P_I either in the precence or absence of metabolite effectors. The other form of P_{II}, designated $P_{II\ D}$, stimulates markedly the deadenylylation activity of P_I in the presence of α-ketoglutarate and ATP.

Conversion of $P_{II\,A}$ to $P_{II\,D}$

As noted above, Shapiro (9) demonstrated that deadenylylation of glutamine synthetase is catalyzed by a complex system composed of P_I, P_{II}, GS·AMP and various effectors, including, α-ketoglutarate, UTP and ATP. To determine if the effectors are involved in activation of one or more of the protein components, studies were made in which each protein component by itself or in various combinations with the other protein components was incubated with α-ketoglutarate, ATP and UTP. After prior incubation the mixtures were tested for their ability to stimulate deadenylylation activity. As is shown in Table I prior incubation of either protein by itself with effectors had no effect on deadenylylation activity. However, prior incubation of mixtures of P_I and P_{II} with effectors led to a five fold increase in deadenylylation activity.

TABLE I
Requirement for both P_I and P_{II} in the activation of deadenylylation

Proteins in Prior Incubation Mixture	$\dfrac{\text{D-activity after prior incubation}}{\text{D-activity before prior incubation}}$
$P_I + P_{II}$	5.5
$P_I + P_{II} + GS$	5.4
P_I	1.1
P_{II}	0.9
$P_I + GS$	0.9
$P_{II} + GS$	1.1

As indicated, various combinations of 52 μg P_I, 27 μg P_{II}, and 36 μg glutamine synthetase (GS; $E_{\overline{8.0}}$) were incubated with 50 mM 2-methylimidazole (pH 7.2), 0.5 mM ATP, 15 mM α-ketoglutarate, 1 mM $MnCl_2$, and 1 mM dithiothreitol (DTT) in 0.06 ml at 37°C. Immediately after mixing and after 60 minutes, 0.02 ml aliquots were removed and assayed for D-activity, as described in Fig. 1. The final assay mixture was supplemented with the protein component omitted from the prior incubation mixture and in all cases contained 17.3 μg of P_I and 9 μg of P_{II}

Data in Table II show that α-ketoglutarate, UTP and ATP must all be present during the prior incubation for activation to occur. Data presented in Fig. 3 show that the activation of deadenylylation is a relatively slow process and that the extent of activation is proportional to the concentration of P_{II} present in the prior incubation mixture. This and other lines of evidence, not considered here, indicate that prior incubation leads to an increase in capacity of the P_{II} component to stimulate deadenylylation, and that the modification of P_{II} is catalyzed by P_I or by another enzyme that is present in the partially purified P_I preparations used in those studies. Results of preliminary experiments favor the latter possibility.

TABLE II

Effector requirements for the conversion of P_{II} A to P_{II} D

Prior Incubation Mixture	D-activity after prior incubation / D-activity before prior incubation
Complete	4.8
-α-ketoglutarate	1.4
-ATP	1.2
-UTP	1.1

The complete mixture contained 0.02 mM ATP, 1.0 mM UTP, 15 mM α-ketoglutarate. 20 mM $MgCl_2$, 1.0 mM $MnCl_2$, 50 mM 2-methylimidazole (pH 7.2), 1 mM DTT, 16 μg of P_I and 2.3 μg of P_{II} in a final volume of 0.03 ml. After a 30 minutes at 37°C the mixtures were assayed for D activity as in Fig. 1.

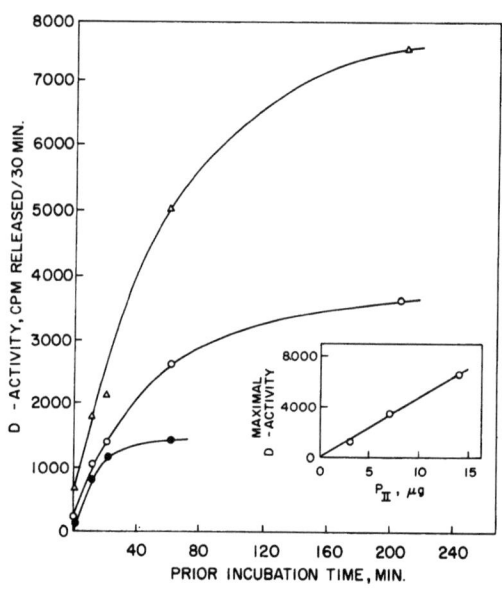

Fig. 3. Effect of P_{II} concentration on rate and extent of activation of deadenylylation. 96 μg of P_i were mixed with either 3.5 μg (-●-●-), 7 g (-o-o-) or 14 μg (-△-△-) of P_{II} in 0.16 ml. solution containing 0.02 mM ATP, 1.0 mM UTP, 15 mM α-ketoglutarate, 20 mM $MgCl_2$, 1.0 mM $MnCl_2$, 1.0 mM DTT and 50 mM 2-methylimidazole. At the times indicated on the abscissa, 0.02 ml aliquots were removed and assayed for D-activity. The assay mixture (0.1 ml) for estimation of deadenylylation activity (D-activity) contained 50 mM 2-methylimidazole·HCl (pH 7.2), 0.02 mM ATP, 1.0 mM UTP, 15 mM α-ketoglutarate, 20 mM $MgCl_2$; 5 mM Na_2Mg EDTA, 20 mM K phosphate, 1 mM DTT, 100 μg of ^{14}C-ATP adenylylated glutamine synthetase (equal to 2.0 nanomoles of adenylylated subunits, 20 X 10^3 cpm/nanomole) and enzyme. After 30 minutes at 37°C, 0.15 ml of 6% perchloric acid was added and the mixtures centrifuged. 0.15 ml of the supernatant was dissolved in 10 ml of Aquasol (New England Nuclear Corp.) and counted in a scintillation spectrometer

Whereas modification of P_{II} leads to an increase in its ability to stimulate deadenylylation activity (D-activity), its ability to stimulate adenylylation (A-activity)[2] is diminished. As is shown in Fig. 4, during prior incubation of P_{II} with α-ketoglutarate, ATP and UTP and the converting enzyme (present in the P_I preparation) its capacity to stimulate D-activity increased nearly 6-fold, whereas its capacity to stimulate A-activity declined. Unfortunately for reasons that need not be discussed here glutamine (0.3mM) and α-ketoglutarate (0.2 mM) were included in the assay mixture used to measure A-activity. An even greater loss in A-activity would have been observed if these effectors had been omitted from the assay mixture (cf., Table III).

In view of its reciprocal effects on the capacity of P_{II} to stimulate adenylylation and deadenylylation activity, it is concluded that prior incubation leads to

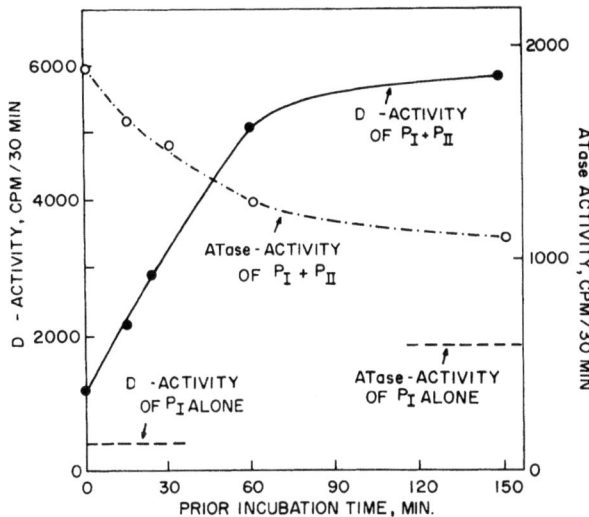

Fig. 4. Reciprocal effect of prior incubation of P_I and P_{II} on A-activity and D-activity. P_I (0.21 mg) and P_{II} (0.1 mg) were incubated together in 0.2 ml containing 0.1 mM ATP, 1.0 mM UTP, 1.0 mM α-ketoglutarate, 1.0 mM $MnCl_2$, 1.0 mM DTT and 50 mM 2-methylimidazole (pH 7.2). At times indicated, 0.02 ml aliquots were assayed for D-activity as described in the legend to Fig. 3, and for A-activity as below. A-activity was measured in mixtures (0.1 ml) containing 50 mM 2-methylimidazole (pH 7.2), 1.0 mM DTT, 100 μg of unadenylylated glutamine synthetase, 20 mM $MgCl_2$ 1.0 mM $U^{14}C$-[ATP] 0.3 mM glutamine 0.2 mM α-ketoglutarate, 0.02 mM UTP and 20 mM K-phosphate. After 30 min. at 37° C, precipitated protein was collected by filtration through nitrocellulose millipore filters of 0.45 millimicron pore size. The filters were washed with 20 ml. of 7% $Cl_3CHCOOH$ and 2 ml ethanol, dried at 60° C and counted in a liquifluor mixture (New England Nuclear Corp.)

---

[2] Whereas it has been known for some time (9) that the P_{II} protein is involved in deadenylylation of glutamine synthetase, its involvement in adenylylation was only recently recognized. The failure to demonstrate an effect of P_{II} on adenylylation in earlier experiments is due to the fact that the assay conditions used to measure adenylylation activity were different from those used here. Assay mixtures used in the earlier studies contained higher concentrations of glutamine, glutamine synthetase, and ATP and were at a higher pH than those used in the present study.

conversion of P_{II} from a form, designated $P_{II\ A}$, that stimulates A-activity to a form, designated $P_{II\ D}$, that stimulates D-activity.

TABLE III

Effects of glutamine and α-ketoglutarate on the stimulation of adenylylation of glutamine synthetase by $P_{II\ A}$ and $P_{II\ D}$

Enzyme System	SPECIFIC ACTIVITY			
	Effectors	+ GLN (1 mM)	+ α-KG (0.2 mM)	+ GLN; α-KG (1 mM; 0.2 mM)
P_I	0.25	2.9	0.25	2.9
$P_I + P_{II\ A}$	7.75	10.0	1.25	6.6
$P_I + P_{II\ D}$	0.87	9.7	0.50	7.0

Where indicated, 18 μg of either $P_{II\ A}$ or $P_{II\ D}$ were added. Standard assay conditions were used except that the amount by P_I (either 42 μg or 8.5 μg) and the times of incubation (either 15 or 30 min) were varied so that in every case adenylylation was linear with time. In all cases specific activity is expressed as pmoles of AMP bound to glutamine synthetase min/μg P_I.

Other experiments show that glutamine and orthophosphate (Pi) both inhibit the conversion of $P_{II\ A}$ to $P_{II\ D}$. When present at concentrations of 4 and 20 mM, Pi inhibits 40 and 94%, respectively. Inhibition by glutamine is strongly influenced by the levels of ATP and UTP. No inhibition is observed with 0.4 mM glutamine when the concentration of both nucleoside triphosphates is 0.1 mM, but when each is present at a concentration of 1.0 mM, inhibition by glutamine is 80%. Knowledge concerning the conversion of $P_{II\ A}$ to $P_{II\ D}$ is summarized in scheme I, which shows that this transformation is catalyzed by an enzyme (present in P_I preparations) and requires ATP, UTP and α-ketoglutarate; moreover, the conversion is inhibited by glutamine and Pi.

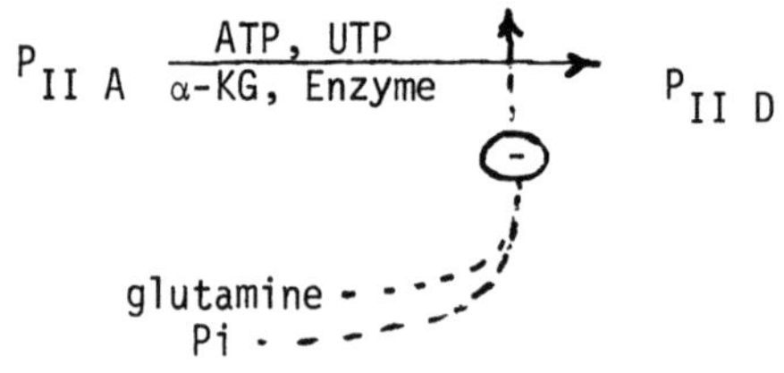

$$P_{II\ A} \xrightarrow{\text{ATP, UTP}, \alpha\text{-KG, Enzyme}} P_{II\ D}$$

glutamine, Pi (−)

Scheme I

Role of the P_{II} Regulatory Protein in Modulating Effector Responses

Data in Table III shows that the conversion of $P_{II\ A}$ to $P_{II\ D}$ not only affects its ability to stimulate the A-activity of P_I, but it also modulates the effects of α-ketoglutarate and glutamine on adenylylation. In the absence of either glutamine or P_{II}, P_I exhibits a low level of A-activity; this activity is increased 10-fold, by 1 mM glutamine, 30-fold by $P_{II\ A}$ and 40-fold by a mixture of glutamine and $P_{II\ A}$. In contrast, $P_{II\ D}$ by itself causes only slight stimulation of P_I-A-activity but it makes this activity much more sensitive to stimulation by glutamine. The P_{II} regulatory protein also sensitizes the P_I-D-activity to inhibition by α-ketoglutarate. α-Ketoglutarate (0.2 mM) does not affect the D-activity exhibited by P_I alone, nor the enhanced A-activity observed in the presence of glutamine; however, the high level of A-activity obtained in the presence of $P_{II\ A}$ is very sensitive to inhibition by α-ketoglutarate, and that obtained in the presence of glutamine when either $P_{II\ A}$ or $P_{II\ D}$ is present is partially inhibited by α-ketoglutarate.

Possible Role of UTP in the Modification of P_{II}

In an effort to establish the roles of ATP and UTP in the conversion reaction, separate experiments were carried out in which $P_{II\ A}$ was incubated with the converting enzyme in the presence of all three of the metabolites; but each reaction mixture contained a different labeled nucleoside triphosphate, either γ-^{32}P, ^{14}C-[ATP] or γ-^{32}P, ^{14}C-[UTP]. After prior incubation of P_I with P_{II} in the mixture containing γ-^{32}P, ^{14}C-[ATP], neither isotope could be detected in the P_I or P_{II} protein fractions after their subsequent separation by gel filtration. This indicates that the conversion of $P_{II\ A}$ to $P_{II\ D}$ does not involve covalent attachments of either a phosphoryl group or an adenylyl group from ATP. An allosteric role of ATP in the transformation of P_{II} is indicated by the demonstration that the β,γ-methylene phosphonic acid analog of ATP is about 50% as effective as ATP in the conversion reaction.

On the contrary, as shown in Fig. 5, when γ-^{32}P, ^{14}C-[UTP] was included in the prior incubation mixture, the P_{II} fraction that was subsequently isolated by gel filtration contained a significant amount of ^{14}C but no ^{32}P; neither isotope was recovered in the P_I fraction. These results suggest that during prior incubation the γ-phosphoryl group of UTP is cleaved and a derivative containing the uridine moiety is covalently bound to the P_{II} protein. It is therefore possible that the conversion of $P_{II\ A}$ to $P_{II\ D}$ involves covalent attachment of a UMP or a UDP group to a specific site on the P_{II} protein. This possibility is further supported by the demonstration that treatment of the reisolated $P_{II\ D}$ with snake venom phosphodiesterase leads to a release of the ^{14}C label and to complete loss of P_{II} activity; similar treatment of $P_{II\ A}$ preparations produces only a slight loss of activity. It must be emphasized that the P_{II} preparation used in these studies could not be more than 40% pure as judged by polyacrylamide gel electrophoresis. The amount of ^{14}C-[uridine derivative] bound to the P_{II} fraction corresponds to only 1 equivalent per 35 moles of enzyme of 50,000 mol. wt. Therefore the possibility that the uridine derivative is bound to a contaminating protein in the P_{II} preparation rather than to the P_{II} protein itself has not been ruled out.

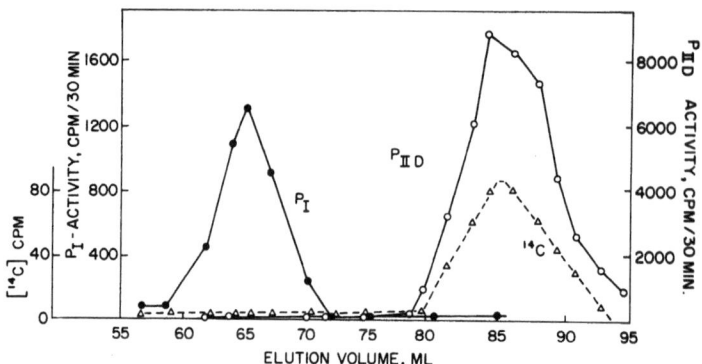

Fig. 5. Association of $[^{14}C]$-uridine with reisolated $P_{II\ D}$. 1.8 mg $P_{II\ A}$ was incubated with 0.43 mg P_I in 1.5 ml solution containing 50 mM 2-methylimidazole (pH 7.2), 1 mM α-ketoglutarate, 0.1 mM ATP, 1.0 mM $MnCl_2$, 1.0 mM DTT and 0.1 mM UTP containing a mixture of $[^{14}C]$-UTP (final specific activity 53.9 X 10^3 cpm/nanomole and γ-^{32}P UTP (final specific activity 15.8 X 10^3 cpm/nanomole). After 4 hours at 37° C the mixture was applied to a 1.5 X 100 cm. column of Sephadex G-100 equilibrated with 20 mM Tris, 0.5 mM DTT, 0.25 mM Na_2Mg EDTA, pH 7.2 at 4° C, and eluted with the same buffer. 0.05 ml aliquots of each 0.86 ml fraction were dissolved in 10 ml Aquasol and counted in a liquid scintillation spectrometer. 0.06 ml aliquots were assayed for D-activity as described in the legend to Fig. 3. Assays for P_{II} were supplemented with 21 µg P_I; the assays for P_I were supplemented with 9 µg of $P_{II\ D}$. (●-●-●)=P_I activity, (o-o-o)=P_{II} activity (--- ---)=cpm ^{14}C channel

Relationship Between the P_{II} Interconversion and the Control of Adenylylation and Deadenylylation by Metabolites

From the foregoing presentation, it is obvious that the P_{II} protein is concerned primarily with the regulation of adenylylation and deadenylylation function of ATase (i.e., the P_I protein). The complex interrelations between the interconversion of P_{II} and the control of P_I activity by metabolites is illustrated by Fig. 6. To begin with, it is important to note that P_I exhibits only slight activity for either function in the absence of P_{II} and effectors. In the absence of P_{II}, physiological concentrations of glutamine cause a significant stimulation of A-activity but have little or no effect on the D-activity of P_I; moreover, α-ketoglutarate is without effect on either activity, whether or not glutamine is present. The importance of $P_{II\ A}$ in the selective control of A-activity is evident from the fact that it is a potent activator of adenylylation in the presence or absence of glutamine but it has no demonstrable effect on D-activity; furthermore, in the presence of $P_{II\ A}$, the

A-activity is very sensitive to inhibition by α-ketoglutarate. Conversion of $P_{II\,A}$ to $P_{II\,D}$ results in a loss of its ability to stimulate A-activity in the absence of glutamine. In fact $P_{II\,D}$ does not affect A- or D-activity in the absence of effectors. However, $P_{II\,D}$ causes tremendous enhancement of the ability of glutamine to stimulate A-activity and at the same time it makes the glutamine-dependent A-activity very sensitive to inhibition by α-ketoglutarate. Furthermore, $P_{II\,D}$ makes the D-activity of P_I very sensitive to stimulation by ATP and α-ketoglutarate

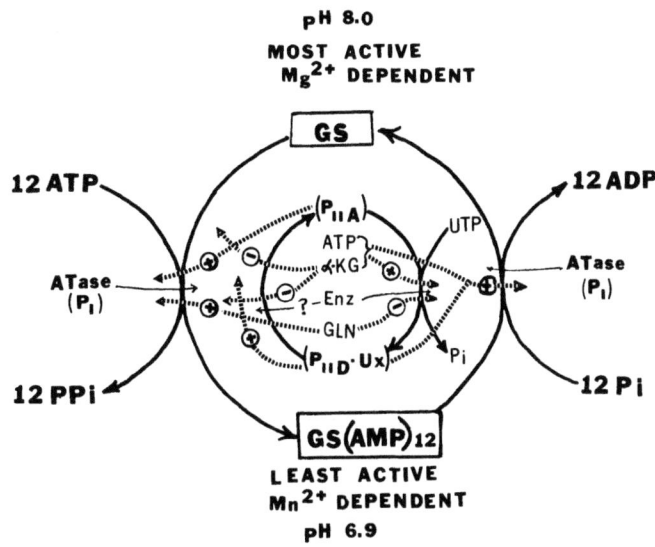

Fig. 6. Schematic representation of the interrelationship between the conversion of $P_{II\,A}$ to $P_{II\,D}$ and the effects of various metabolites on adenylylation and deadenylylation of glutamine synthetase

In addition to their direct effects on the A- and D-activities of P_I, α-ketoglutarate, ATP and glutamine appear to be allosteric effectors in the conversion of $P_{II\,A}$ to $P_{II\,D}$. This conversion requires ATP and α-ketoglutarate whereas glutamine is an inhibitor whose effectiveness is modulated by the concentrations of ATP and UTP.

The Influence of Effectors on the Average State of Adenylylation

Since glutamine synthetase is composed of 12 apparently identical subunits, each of which can be adenylylated, it is obvious that multimolecular forms of the enzyme may exist that differ from each other in the number (0 to 12) and orientation of adenylylated subunits within a single enzyme molecule. In all, 382 different forms of the enzyme are possible (M.S. Raff and W.C. Blackwelder, personal communication). Other studies that cannot be discussed here, have demonstrated that the average state of adenylylation of glutamine synthetase in E. coli varies with the availability of exogenase nitrogen, and that enzyme preparations at intermediate states of adenylylation consist of a complex mixture of hybrid forms; i.e., molecular forms containing both adenylylated and unadenylylated subunits within the same molecule (16, 19). It was further established that intramolecular heterologous interaction between adenylylated and unadenylylated subunits affects catalytic parameters (17,18,20) as well as stability of the aggregate structure (16,19).

Since adenylylation and deadenylylation are both catalyzed by one and the same enzyme, the average state of adenylylation of glutamine synthetase at any given time must reflect the relative rates of these two processes. This in turn is dictated by the state of the regulatory protein ($P_{II\ A}$ or $P_{II\ D}$) and by the relative concentrations of the metabolites, α-ketoglutarate, ATP, UTP and glutamine. In particular, as noted above, because of their opposite and reciprocal affects on adenylylation and deadenylylation, α-ketoglutarate and glutamine should regulate the state of adenylylation. To test this hypothesis, in vitro experiments were carried out to determine the effect of the α-ketoglutarate; glutamine ratio on the steady state level of adenylylation of glutamine synthetase as catalyzed by a mixture of P_I and P_{II}. Data in Fig. 7 show that when the concentrations of other components (ATP, α-ketoglutarate, glutamine synthetase, P_I and P_{II}) are held constant. The extent of adenylylation is determined by the concentration of glutamine. In the absence of glutamine, little or no adenylylation occurs; however, at each glutamine concentration there is initially a rapid increase in the state of adenylylation to a plateau value, the height of which is dependent upon the concentration of glutamine added. Presumably these plateau values are not a measure of true thermodynamic equilibria but represent

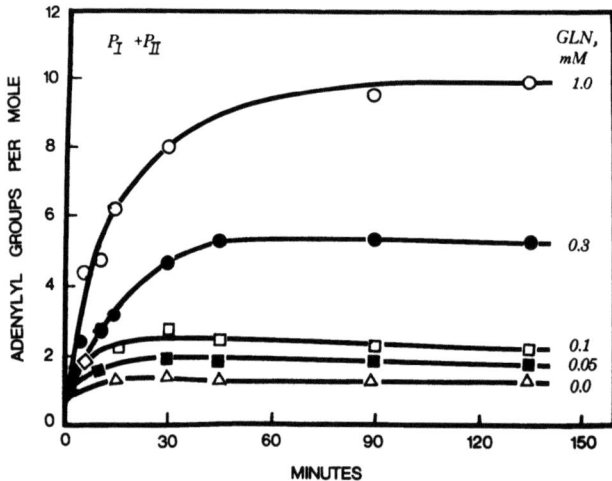

Fig. 7. Effect of glutamine concentration on the steady state level of adenylylation. Each reaction mixture (0.1 ml) contained 50 mM 2-methylimidazole buffer (OH 7.2), 20 mM K phosphate, 15 mM α-ketoglutarate, 1.0 mM UTP, 1.0 mM ATP, 20 mM, $MgCl_2$, 1.0 mM dithiothreitol, µg P_I, 83 µg $P_{II\ A}$, 195 µg glutamine synthetase ($E_{1.0}$) and varying levels of glutamine as indicated on the curves. At times indicated, aliquotes were removed as indicated on the curves. At times indicated, aliquotes were removed and the state of adenylylation of glutamine synthetase was determined (16)

steady states at which the rates of adenylylation and deadenylylation are equal for that particular α-ketoglutarate: glutamine ratio. A similar experiment in which the concentration of α-ketoglutarate is varied, while all other components including glutamine are held constant, is shown in Fig. 8. Here we see that in the absence of α-ketoglutarate the state of adenylylation increases sharply to a maximum value of about 10 adenylyl groups per mole. But, as expected the steady state level of adenylylation decreases with increasing concentrations of α-ketoglutarate. Other similar experiments show that the steady state level increases with increasing energy charge from 0 to 1.0 (21) and decreases as the ratio of Mn^{2+}/Mg^{2+} is increased. Unfortunately these experiments were all performed before it was known that P_{II} exists in two different forms and the $P_{II\ A}$ form was used throughout. Preliminary experiments show that if $P_{II\ D}$ rather than $P_{II\ A}$ is used, then the steady state level of adenylylation is much more sensitive to control by lower concentrations of α-ketoglutarate (1-2 mM). It is obvious that in further studies, the ratio of $P_{II\ A}:P_{II\ D}$ on the state of adenylylation should be investigated.

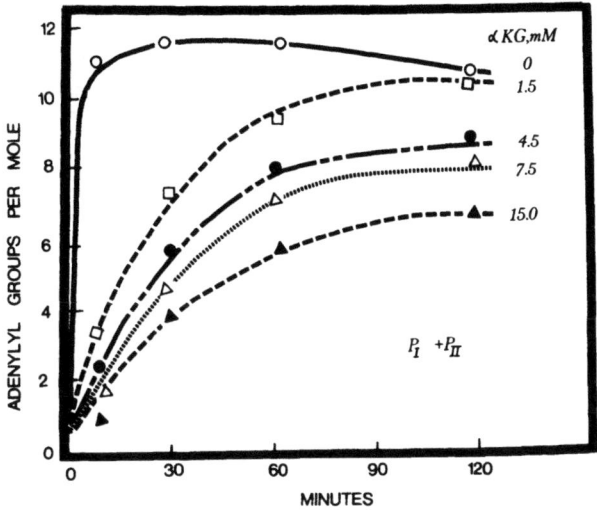

Fig. 8. Effect of α-ketoglutarate concentration on the steady state level of adenylylation. Conditions were as in Fig. 7 except that all reaction mixtures contained 1.0 mM glutamine and the concentration of α-ketoglutarate was varied as indicated

In summary, the specific activity of glutamine synthetase varies with the number of its subunits that are adenylylated. The average number of adenylylated subunits per molecule of enzyme can vary from 0 to 12 and is determined by the relative rates of adenylylation and deadenylylation as catalyzed by the P_I protein (ATase). The rates of these two processes is under strict metabolite control by α-ketoglutarate, glutamine, ATP and UTP. The effects of these metabolites are mediated by a regulatory protein, P_{II}, the activity of which is modulated by an enzyme catalyzed alteration that may involve covalent attachment of the uridylyl moiety of UTP to the P_{II} protein; this converts P_{II} from a form, $P_{II\ A}$, favoring adenylylation to a form, $P_{II\ D}$, favoring deadenylylation. This alteration of P_{II} is also under metabolite control by α-ketoglutarate, ATP, UTP and glutamine.

References

1. KINGDON, H.S., SHAPIRO, B.M. and STADTMAN, E.R., Proc. Natl. Acad. Sci. USA 58(1967)
2. MECKE, D., WULF, K., LIESS, K. and HOLZER, H., Biochem. Biophys. Res. Commun. 24 452 (1966)
3. KINGDON, H.S. and STADTMAN, E.R., J. Bacterial 94, 949 (1967).
4. KINGDON, H.S. and STADTMAN, E.R., Biochem. Biophys. Res. Commun. 27, 470 (1967)
5. SHAPIRO, B.M. and STADTMAN, E.R., J. Biol. Chem. 243, 3769 (1968).
6. WOOLFOLK, C.A., SHAPIRO, B.M. and STADTMAN, E.R., Arch. Biochem.Biophys. 116,177(1966).
7. MANTEL, M. and HOLZER, H., Proc. Nat. Acad. Sci. USA 65, 660 (1970).
8. SHAPIRO, B.M. and STADTMAN, E.R., Biochem. Biophys. Res. Communs. 30, 32 (1968).
9. SHAPIRO, B.M., Biochemistry, 8, 659 (1969).
10. ANDERSON, W.A. and STADTMAN, E.R., Biochem. Biophys. Res. Commun. 41, 704 (1970).
11. ANDERSON, W.A., HENNIG, S.B., GINSBURG, A., and STADTMAN, E.R., Proc. Natl. Acad. Sci. USA 67, 1717 (1970).
12. EBNER, E., WOLF, D., GANCEDO, C., ELASSER, S. and HOLZER, H., Europ. J. Biochem. 14, 535 (1970).
13. HENNIG, S.B., ANDERSON, W.B. and GINSBURG, A., Proc. Natl. Acad. Sci. USA 67, 1761 (1970).
14. HENNIG, S.B. and GINSBURG, A., Arch. Biochem. Biophys. 144, 611 (1971).
15. WOOD, H.G., DAVIS, J.J. and LOCHMULLER, H., J. Biol. Chem. 241, 5692 (1966).
16. STADTMAN, E.R., GINSBURG, A., CIARDI, J.E., YEH, J., HENNIG, S.B. and SHAPIRO, B.M. Adv. Enz. Regulation, 8, 99 (1970).
17. DENTON, M.D. and GINSBURG, A., Biochemistry 9, 617 (1970).
18. GINSBURG, A., YEH, J., HENNIG, S.B. and DENTON, M.D., Biochemistry 9, 633 (1970).
19. CIARDI, J.E. and CIMINO, F., Federation Proceedings 30, 1175 ABS (1971).
20. SEGAL, A., Federation Proceedings 30, 1175 ABS (1971)
21. ATKINSON, D.E., Biochemistry 7, 4030. (1965).

Inactivation of Glutamine Synthetase in Intact *E. coli* Cells

H. Holzer, H. Schutt, and P. C. Heinrich

Biochemisches Institut der Universität Freiburg, Freiburg/Germany
and Hoffmann-La Roche AG, Basel/Switzerland

INTRODUCTION

Addition of small amounts of NH_4^+ (at final concentration 10^{-4} M, for example) to suspensions of E.coli which have grown in ammonium-free medium leads to a rapid inactivation of glutamine synthetase (1, see figure 1). Pursuit of this observation led to the discovery of an ATP-

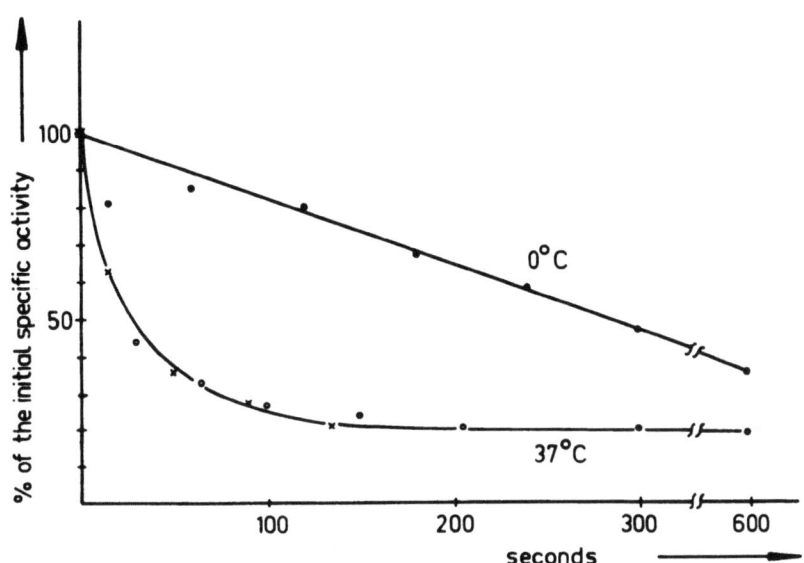

Fig. 1. Time course of the inactivation of glutamine synthetase in E.coli after addition of 10^{-4} M NH_4^+ at 0° and 37°. Cells were cultivated in glutamate-medium and at zero time ammonium sulfate was added. Samples were withdrawn at various times, sonicated 15 sec, and centrifuged. Symbols o and x denote data from two separate experiments. Taken from ref.1

and glutamine-dependent, glutamine synthetase-inactivating enzyme (2). Further research by Earl Stadtman's group in Bethesda and ours in Freiburg (3,4) elucidated a system for the regulation of glutamine synthetase by effector-controlled, enzyme-catalyzed, covalent modification schematized in figure 2. The findings collected in figure 2 are based on studies with extracts of E.coli and with the purified enzymes.

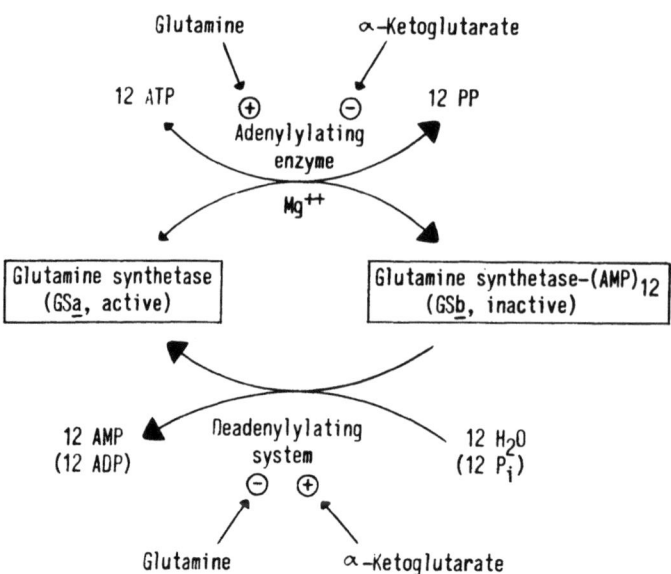

Fig. 2. <u>Regulation of glutamine synthetase from E.coli by enzyme-catalyzed chemical modification (4)</u>. ⊕ = positive effector (stimulation), ⊖ = negative effector (inhibition)

There is much evidence that this regulatory mechanism functions also in intact coli cells. There is a basic incongruency, however, between the situation <u>in vivo</u> and the picture derived from studies <u>in vitro</u>. In intact cells low concentrations of ammonium in the medium effect the inactivation of glutamine synthetase, whereas glutamine is effective only at relatively high concentrations (5, see table 1). With the isolated enzymes glutamine effects the inactivation at low concentrations, while NH_4^+ is not effective (6).

In the present work two sets of experiments are described which resolve this discrepancy. The first shows that inactivation of glutamine synthetase by adenylylation occurs also <u>in vivo</u> in consequence to the addition of NH_4^+ to intact cells. The second shows that addition of NH_4^+ to the medium leads to a rapid intracellular accumulation of glutamine, which in turn triggers the inactivation of glutamine synthetase.

ADENYLYLATION OF GLUTAMINE SYNTHETASE IN INTACT E.COLI CELLS (7)

The ATP-pool of a purine-less mutant of <u>E.coli</u> was labelled by brief incubation with ^{14}C-adenine. After inactivation of glutamine syn-

Table 1

Inactivation of glutamine synthetase in suspensions of intact E.coli cells after the addition of glutamine or NH_4^+ (5)

	Specific activity 10 min after addition of glutamine or NH_4^+ (units per mg protein)
Without addition	104
10^{-4} M NH_4^+	2.3
10^{-4} M glutamine	47
10^{-3} M glutamine	8.5
10^{-2} M glutamine	4.7

thetase by addition of NH_4^+ to the intact cells, glutamine synthetase was extracted and purified. The purified glutamine synthetase was incubated with phosphodiesterase from snake venom. As shown in figure 3, during incubation with phosphodiesterase the glutamine synthetase purified from NH_4^+-inactivated cells lost radioactivity but gained correspondingly in specific enzymatic activity. In contrast, the glutamine synthetase purified from E.coli cells incubated without addition of NH_4^+ showed neither significant change of enzymatic activity nor significant liberation of ^{14}C material during incubation with phosphodiesterase. The radioactive material lost from inactivated glutamine synthetase was soluble in trichloroacetic acid and, as shown in figure 4, consisted mainly of adenosine and adenine. These findings indicate that, when confronted with NH_4^+, E.coli cells, whose adenine nucleotide pool is labelled with ^{14}C, form ^{14}C-AMP-labelled glutamine synthetase. Phosphodiesterase hydrolyzes the glutamine synthetase-AMP bond; the acid-soluble ^{14}C-AMP is further hydrolyzed by enzymes contaminating the phosphodiesterase preparation thus giving rise to adenosine and adenine. Thus, the experiments shown in figures 3 and 4 vouch that the inactivation of glutamine synthetase by adenylylation is a physiolgical process occuring in intact E.coli cells.

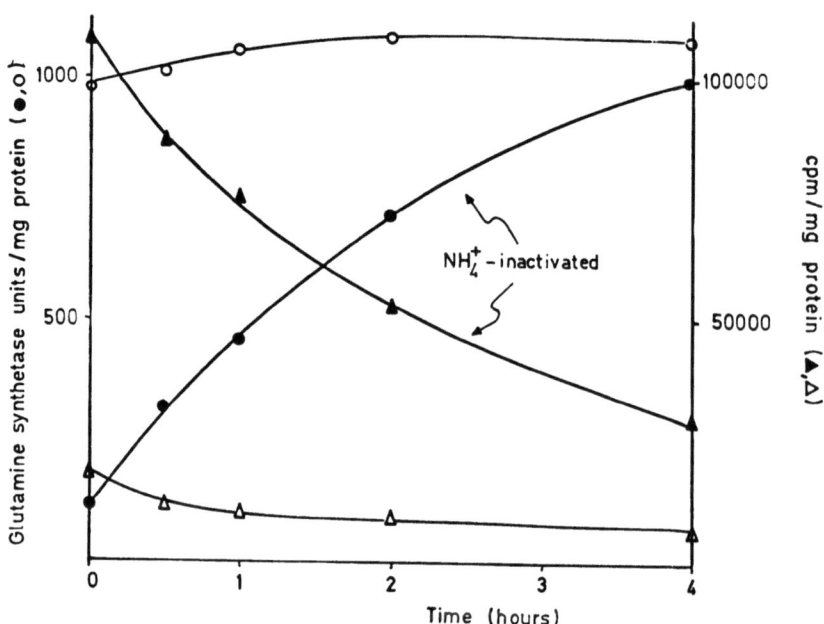

Fig. 3. <u>Incubation of glutamine synthetase preparations with snake venom phosphodiesterase (7)</u>. After streptomycin precipitation, 2 acid precipitations and 60° treatment, a specific biosynthetic activity of glutamine synthetase obtained from the NH_4^+-inactivated cell suspension (1 g moist cells) was 123 u/mg. The specific activity of glutamine synthetase isolated from the cell suspension where no NH_4^+ was added was 1390 u/mg. <u>Left ordinate</u>: synthetase activity in the incubation mixture, with non-inactivated glutamine synthetase (o), or with inactivated glutamine synthetase (●). <u>Right ordinate</u>: radioactivity of the trichloroacetic acid-insoluble material of the incubation mixture, with non-inactivated glutamine synthetase (Δ), or with inactivated glutamine synthetase (▲)

CHANGES OF CONCENTRATIONS OF METABOLITES AFTER INCUBATION OF E.COLI CELLS WITH AMMONIA (8)

Figure 5 shows changes in the concentrations of glutamate, glutamine, ATP, and of glutamine synthetase activity after treatment of <u>E.coli</u> cells with NH_4^+. It can be seen that after the addition of NH_4^+ glutamine accumulates very rapidly. As is known from <u>in vitro</u> experiments, glutamine at concentrations of 1 mM stimulates the glutamine synthetase adenylylating enzyme and thereby adenylylates, i.e. inactivates, glutamine synthetase. These findings and considerations are summarized in the scheme depicted in figure 6. According to this scheme NH_4^+ is very rapidly taken up by <u>E.coli</u> cells and used for the formation of glutamine. The accumulated glutamine then stops further formation of glutamine by stimulating the glutamine synthetase inactivating enzyme. We are dealing here with a feedback loop for control of glutamine synthesis which is

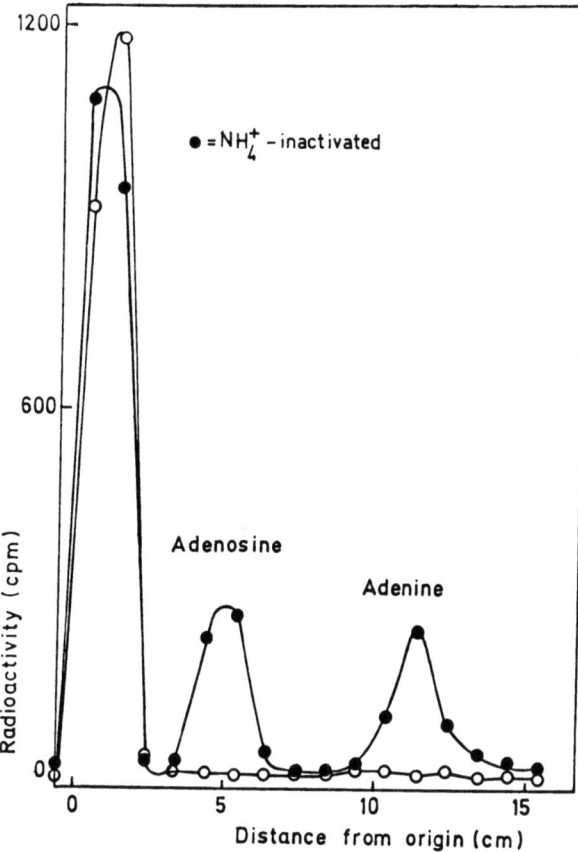

Fig. 4. Electrophoresis of the trichloroacetic acid supernatant from NH_4^+-inactivated, ^{14}C-labelled glutamine synthetase preparations before (o) and after 4 h (●) incubation with phosphodiesterase (7). The high peak of radioactivity close to the origin is attributed to trichloroacetic acid soluble material (see fig.3, right ordinate) which contaminates both glutamine synthetase preparations

similar to the allosteric feedback loops described as "endproduct inhibition" by Umbarger (9). One difference from the allosteric loops is, however, that in the case of E.coli glutamine synthetase the effector, i.e. glutamine, causes a covalent modification of the regulated enzyme and not a physical, i.e. conformational, modification.

Furthermore it can be seen from the results depicted in figure 5 that, after the addition of ammonia to E.coli cells, ATP decreases very rapidly. The magnitude of the observed changes in the concentration of glutamine suggest that the most probable reason for the rapid depletion of the ATP pool is its rapid consumption in support of ATP-dependent synthesis of glutamine from ammonia. If one considers that all important energy-consuming functions of a living cell depend on ATP, one is forced to the conclusion that the rapid throttling of the activity of glutamine

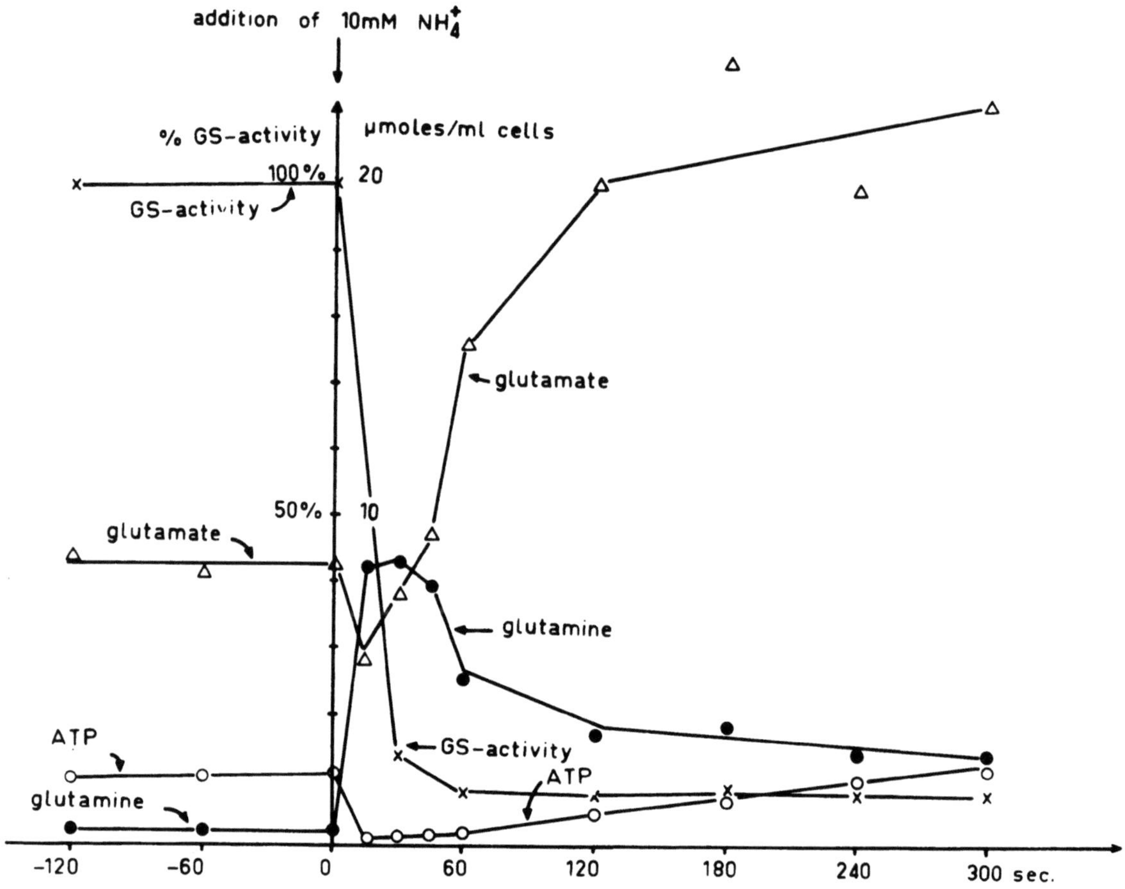

Fig. 5. <u>Glutamine synthetase activity, glutamate, glutamine and ATP after addition of NH_4^+</u>. For details see (8)

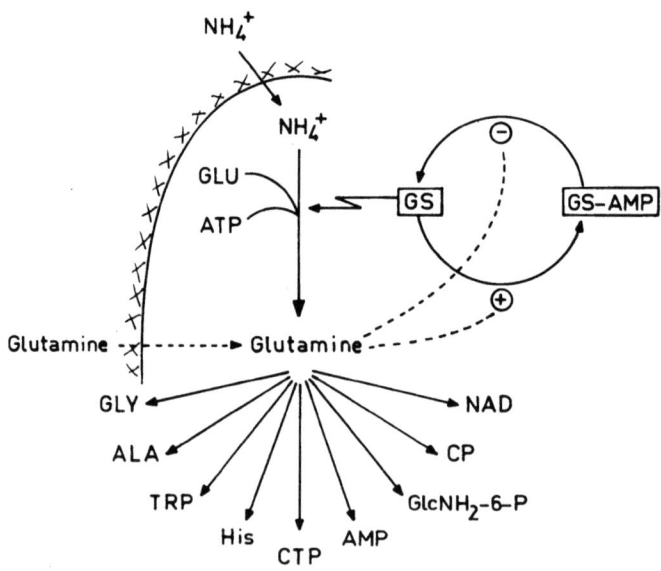

Fig. 6. <u>Scheme for the effect of NH_4^+ on glutamine synthetase in intact cells of E.coli</u>

synthetase by the accumulating glutamine is a "life-saving mechanism" which preserves the cells from the deleterious consequences of ATP-deficiency.

SUMMARY

Experiments with crude extracts and with purified enzymes from E.coli have shown that glutamine synthetase is inactivated in the presence of the effector glutamine by way of an ATP-dependent enzymatic reaction, which involves the adenylylation of the enzyme protein. In intact cells of E.coli inactivation of glutamine synthetase takes place upon addition of NH_4^+ to the growth medium, but not by glutamine itself. The question was raised whether or not the NH_4^+-induced inactivation of glutamine synthetase still involves adenylylation of the enzyme. In order to settle this question the ATP pool of a purine-less mutant of E.coli was labelled by brief incubation with ^{14}C-adenine. After inactivation of glutamine synthetase by addition of NH_4^+ to the intact cells, glutamine synthetase was extracted and purified. It was shown that the purified enzyme preparation contained radioactivity originating from adenine which could be liberated by treatment with phosphodiesterase. It was of importance that simultaneously with the liberation of ^{14}C-labelled material reappearance of enzymatic activity was also observed.

The question why NH_4^+ and not glutamine effects in intact cells the adenylylation of glutamine synthetase was answered with the aid of analyses of the steady state concentrations of relevant metabolites. From the observed changes of concentrations of ATP, glutamine, and glutamate, and the activity of glutamine synthetase, three main conclusions can be drawn: 1) addition of NH_4^+ causes a rapid accumulation of glutamine and thereby stimulates the adenylylation (i.e. inactivation) of glutamine synthetase, 2) the glutamine-stimulated covalent modification (i.e. adenylylation) of glutamine synthetase represents a feedback loop, similar to the allosterically effected feedback loops described as "endproduct inhibition" (9), 3) the very fast, ATP-dependent synthesis of glutamine from ammonia and glutamate or α-ketoglutarate causes a rapid depletion of the ATP pool of the cells. The rapid turning off of the activity of glutamine synthetase by the accumulating glutamine is therefore a "life-saving mechanism" preserving the cells from the deleterious consequences of ATP-deficiency.

REFERENCES

1. MECKE, D. und HOLZER, H., Biochim.Biophys.Acta, 122, 341 (1966).
2. MECKE, D., WULFF, K., LIESS, K., and HOLZER, H., Biochem.Biophys. Res.Commun., 24, 452 (1966).
3. STADTMAN, E.R., SHAPIRO, B.M., GINSBURG, A., KINGDON, H.S., and DENTON, M.D., Brookhaven Symposium in Biology, 21, 378 (1969).
4. HOLZER, H., Adv.Enzymol., 32, 297 (1969).
5. HOLZER, H., MECKE, D., WULFF, K., LIESS, K., and HEILMEYER, L.,Jr., in "Advances in Enzyme Regulation", Vol.5 (Ed. G.Weber), Pergamon-Press-Oxford & New York 1967, p.211.
6. MECKE, D., WULFF, K., und HOLZER, H., Biochim.Biophys.Acta, 128, 559 (1966).
7. HEINRICH, C.P. and HOLZER, H., Arch.Mikrobiol., 73, 97 (1970).
8. SCHUTT, H. and HOLZER, H., manuscript submitted for publication (1971).
9. UMBARGER, H.E., Cold Spring Harbor Symposia on Quantitative Biology, 26, 301 (1961).

Studies on the Mechanism of the
ATP: Glutamine Synthetase Adenylyltransferase Reaction*

D. Wolf, R. Wohlhueter, and E. Ebner

Biochemisches Institut der Universität Freiburg, Freiburg and Gesellschaft für Strahlen- und Umweltforschung, München/Germany

ATP:glutamine synthetase adenylyltransferase (ATase) catalyzes the adenylylation of glutamine synthetase (GS), yielding (glutamine synthetase)-tyrosyl-O-AMP, and thereby modifies the catalytic and allosteric properties of glutamine synthetase (1,2). The transferase reaction is magnesium-dependent and modulated by several effectors, notably the activator glutamine and inhibitor α-ketoglutarate (3). All experiments reported here were carried out with a 115,000 dalton species of adenylyltransferase (4), purified from E.coli B according to Ebner et al. (3), and possessing potential deadenylylating activity in the sense of Anderson and Stadtman (5).

Since the substrate of this reaction is itself an allosteric protein, it is necessary to distinguish whether the effectors operate on catalyst or substrate. By titration of sulfhydryl groups with 5,5'-dithiobis(2-nitrobenzoic acid) we can detect conformational changes in adenylyltransferase as illustrated in figure 1. We see that addition of Mg^{2+} or of Mg^{2+} plus ATP (at a concentration providing optimal reaction velocity) increases the number of titratable sulfhydryl groups. Inhibitory concentrations of ATP or UTP cause a disappearance of SH-groups. Besides MgATP, glutamine synthetase (not shown here) also leads to an exposure of SH-groups in adenylyltransferase. Thus apparently both substrates bind independently to ATase.

---

*Taken in part from the doctoral dissertation of D. Wolf, Faculty of Biology, University of Freiburg, 1971. A detailed publication of methods and results is in preparation.

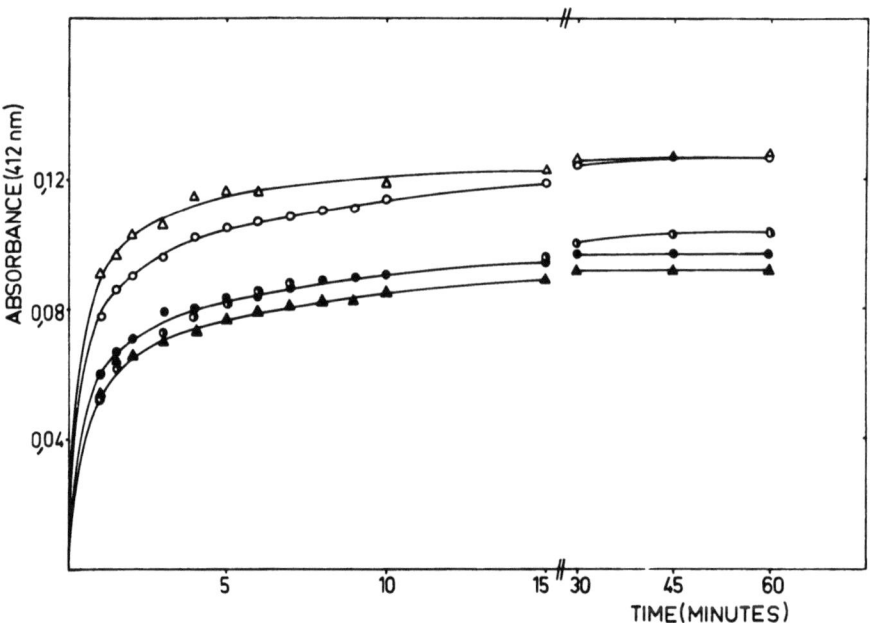

Fig. 1. Example of DTNB titration of adenylyltransferase. Change in extinction at 412 nm of a solution of 0.216 mg protein/ml in 0.1 M Tris/HCl, pH 7.6, containing initially 0.5 mM DTNB was followed (25°). Reference curve without addition (●-●); plus 25 mM $MgCl_2$ (o-o); plus 25 mM $MgCl_2$ and 5 mM ATP (Δ-Δ); plus 25 mM $MgCl_2$ and 20 mM ATP (▲-▲); plus 25 mM $MgCl_2$ and 10 mM UTP (●-●)

The results of several similar titrations are condensed in table 1.

Table 1

Change in the number of titratable thiol groups of adenylyltransferase by effectors of the adenylylation reaction. The values were taken after 60 minutes when the reaction was completed. All effectors were added at 20 mM. Experimental details will be published elsewhere.

Effector Added	Reference Conditions	Δ SH with Respect to Reference
L-Glutamine	without Effector	0
α-Ketoglutarate	without Effector	-0.05
3-Phosphoglycerate	without Effector	-0.25
L-Glutamine	25 mM Mg^{2+}	0
α-Ketoglutarate	25 mM Mg^{2+}	-0.3 to -0.5
3-Phosphoglycerate	25 mM Mg^{2+}	-0.5 to -1.0
L-Glutamine	25 mM Mg^{2+} + 1 mM ATP	+0.2 to +0.3
L-Glutamine	25 mM Mg^{2+} + 20 mM α-Ketoglutarate	+0.3 to +0.5

20 mM α-ketoglutarate or 3-phosphoglycerate, which inhibit the transferase reaction 24 and 91 %, respectively (3), diminish the number of titratable sulfhydryls. Glutamine alone has no discernable effect, but was found to expose additional sulfhydryl to DTNB titration in the presence of ATP (at concentrations suboptimal for the transfer reaction) or α-ketoglutarate. Thus we conclude that these effectors do influence the state of adenylyltransferase, and in a way which may be correlated qualitatively to their influences on the rate of the transferase reaction.

Furthermore, the negative effectors of adenylyltransferase (UTP, α-ketoglutarate, 3-phosphoglycerate), which promote a "closed" conformation of adenylyltransferase, are simultaneously positive effectors of the deadenylylating complex (5, and unpublished observations). This suggests that the regulation of this complex, too, depends upon the modulation of the conformation of its ATase component.

Our second line of inquiry concerns the course of the transfer reaction. We would like to know if the adenylyl group is transferred directly to glutamine synthetase or through the mediacy of a stable, adenylylated transferase. This decision bears on our understanding of two aspects of glutamine synthetase modification. First, it appears likely from the work of Stadtman and his colleagues (5,6,7) that the pyrophosphorolytic activity of adenylyltransferase is metamorphosed to a phosphorolytic activity by means of interaction with a second protein. Assuming that the two reactions proceed analogously, the question arises whether GS-AMP or ATase-AMP is the (pyro)phosphorolyzed species. Second, an ATase-AMP bond would necessarily have a free energy of hydrolysis comparable to that of GS-AMP, i.e. about 9 kcal/mole (8,9). Existence of such a bond would provide another example of a "high energy" protein-adenylylate and prompt elucidation of its chemistry.

We have attempted to form and isolate an adenylylated adenylyltransferase by the two routes corresponding to the two hypothetical partial reactions:

(1a)   ATP + ATase ⇌ ATase-AMP + PP_i

(1b)   ATase-AMP + GS ⇌ ATase + GS-AMP

$[U-^{14}C]$ATP was incubated with enzyme and the reaction mixture filtered on Sephadex G-25. The distribution of radio- and enzymatic-activity is shown in figure 2, expressed as molarity of adenylyltransferase (MW = 115,000) or ATP in each fraction. The slight radioactivity found in enzyme-containing fractions was only about 2 % of that expected for the 1:1 stoichiometry postulated by equation 1a.

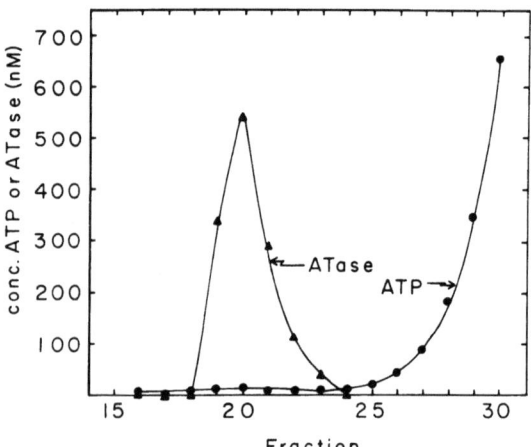

Fig. 2. Sephadex filtration of reaction mixture ATase plus {^{14}C}ATP. 270 mg adenylyltransferase was incubated 5 min at 25° with 750 μM {U-^{14}C}ATP, 30 Ci/mole, 20 mM L-glutamine, 25 mM MgCl$_2$, and 0.1 M imidazole/HCl, pH 7.6, in a total volume of 100 μl. The reaction mixture was subsequently passed through Sephadex G-25 (0.9 x 30 cm). Concentrations were calculated from the specific radioactivity of ATP, and a specific enzymatic activity of 1300 u/mg, MW = 115,000 daltons

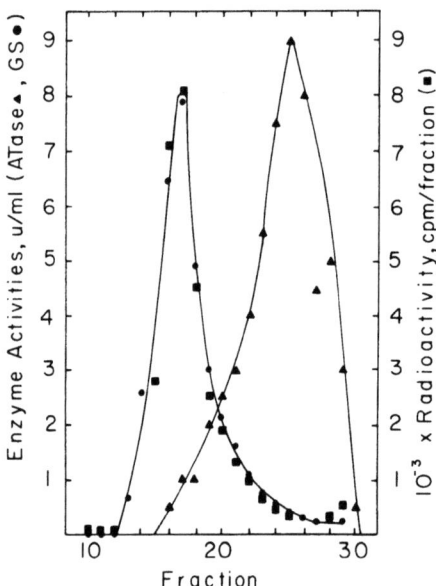

Fig. 3. Density-gradient fractionation of reaction mixture ATase plus ({^{14}C}AMP)$_{12}$-glutamine synthetase. 1.74 nmoles adenylyltransferase was incubated with 0.167 nmoles ({^{14}C}AMP)$_{12}$-glutamine synthetase (300 Ci/mole) 30 min at 4° in 20 mM L-glutamine, 10 mM MgSO$_4$, 0.1 M imidazole/HCl, pH 7.0, in a total volume of 50 μl. The mixture was then placed on a 5 to 20 % sucrose gradient (buffer, glutamine and MgSO$_4$ as above) and centrifuged at 39000 rpm for 3.5 h in rotor SW-39 of Beckman L2/65B centrifuge. 60 μl fractions were collected

Similarly $(\{^{14}C\}AMP)_{12}$-glutamine synthetase was incubated with adenylyltransferase and the reaction products separated by sucrose density-gradient centrifugation (figure 3). Again there was no evidence for an attachment of the radioactive adenylyl moiety to adenylyltransferase.

We tested further the hypothesis represented by equation 1a by looking for the expected pyrophosphate:ATP exchange. Purified adenylyltransferase was incubated with $\{^{32}P\}$pyrophosphate and cold ATP and the appearance of radioactivity in charcoal adsorbable form measured. As seen from table 2, there was a small measurable exchange. Troublesome, however, was the fact that this activity was variable from preparation to preparation and unresponsive to the effector glutamine, which stimulates the overall reaction 20-fold. Glutamate also failed to accelerate the exchange(cf.6). The third experiment of table 2 shows, however, that addition of adenylylated glutamine synthetase increases the exchange rate. That is, conditions supporting the overall reaction,

(2) $\quad$ GS-(AMP)$_{12}$ + PP$_i$ $\underset{\text{gln}}{\overset{\text{ATase}}{\rightleftharpoons}}$ GS-(AMP)$_{11}$ + ATP ,

provide a more rapid PP$_i$/ATP exchange than those supporting reaction 1a. This situation is inconsistent with reaction 1a's being a part of the overall reaction. The data in table 3 verify the stoichiometry of equation 2. The rate of $\{^{32}P\}$PP$_i$ incorporation into ATP observed with the overall reaction is similar, with and without effector, to the rate of incorporation of adenylyl groups originating from adenylylated glutamine synthetase.

The foregoing experiments provided negative evidence for the existence of ATase-AMP. We sought with the help of kinetic experiments to make a positive distinction between the mechanism represented by equations 1a and 1b and the concerted mechanism implied by equation 2.

Equation 1a + 1b is of a "ping-pong" reaction type, which is characterized kinetically by a family of parallel reciprocal curves generated by varying one substrate at several fixed concentrations of the other (10). An "ordered" type mechanism, where both substrates build a tertiary complex together with enzyme, is characterized by a family of converging reciprocal curves.

Figure 4 presents the results of such a kinetic analysis. Adenylyltransferase activity was measured as the appearance of protein-bound ^{14}C when $\{U-^{14}C\}$ATP was used as substrate. The substrate inhibition reported by Ebner et al. (3) is visible here also, but the linear portions clearly converge. We should note one technical problem with the kinetic analysis. ATP is a substrate for glutamine synthetase, as well

Table 2

PP$_i$: ATP exchange with adenylyltransferase. Reactions were carried out in 0.1 M Tris/HCl, pH 7.6, 20 mM MgCl$_2$, 2 mM ATP, and 2.5 mM {^{32}P}Na$_2$PP$_i$ at 30° for 30 min in a total volume of 100 µl. Where indicated 100 µg of glutamine synthetase containing 12 moles ATP per mole ("GS$_{12}$"), ca. 50 µg (145 units) of adenylyltransferase ("ATase"), or 20 mM effector were added

Component Added	pmole PP$_i$ exchanged / 30 min ATase unit
expt. 1	
ATase	4.5
ATase + glutamine	3.7
expt. 2	
ATase	2.7
ATase + glutamine	0.7
expt. 3	
ATase	2.7
ATase + glutamine	3.2
GS	0.5
ATase + GS	19.7
ATase + GS + glutamine	233
ATase + GS + 3-phosphoglycerate	7.6

as for the adenylyltransferase. The ATP concentrations must, therefore, be corrected for the amount bound to the glutamine synthetase catalytic site. We have estimated this binding constant (under conditions of the kinetic measurements) to be about 10^4 M^{-1}, and have corrected the ATP concentrations accordingly. This value may need refinement; adjustments will alter the values of the kinetic constants somewhat, though not the overall pattern.

Secondary plots of slopes and intercepts versus the reciprocal of glutamine synthetase concentration are linear as required by the rate equation for an ordered bimolecular reaction.

Table 3

Comparison of velocities of PP:ATP exchange and release of ^{14}C ATP from adenylylated glutamine synthetase. Parallel reaction mixtures contained 0.1 M Tris/HCl, pH 7.6, 4 mM ATP, 2.5 mM Na_4PP_i, 40 mM $MgCl_2$, 1 mM EDTA and 0.52 mg glutamine synthetase in a total volume of 100 μl. For exchange reaction $\{^{32}P\}PP_i$, $16 \cdot 10^6$ dpm, was used; for the ATP-release reaction ($\{^{14}C\}AMP)_{11}$-glutamine synthetase, $67 \cdot 10^5$ dpm, was used. Incubations were at 30° for 30 min

Reaction Measured	- glutamine	+. 20 mM glutamine
PP:ATP exchange (nmoles/30 min)	8.1	54
$\{^{14}C\}$ ATP release (nmoles/30 min)	6.0	63

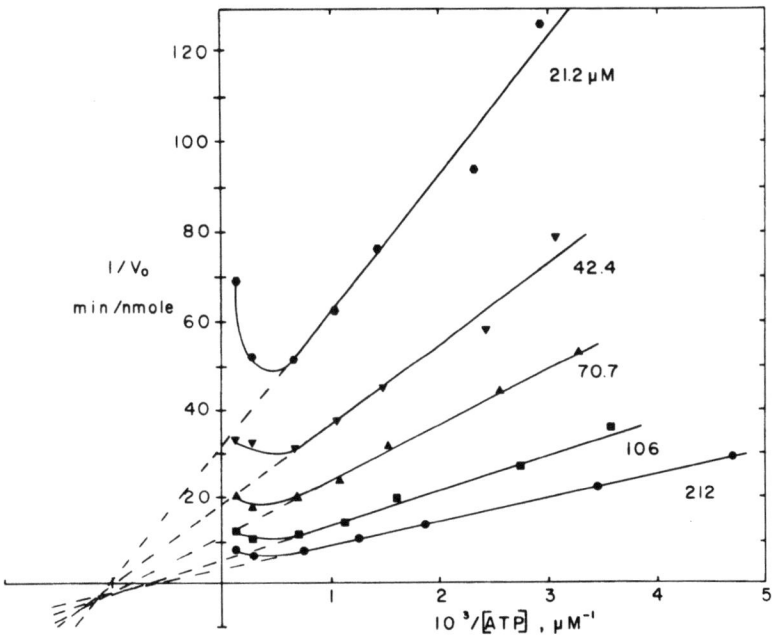

Fig. 4. Kinetics of adenylylation of glutamine synthetase. Initial velocities of incorporation of ^{14}C into protein were measured over a range of concentrations of $\{U-^{14}C\}MgATP$ (14.8×10^6 dpm/μmole) and GS_0 (calculated as subunit, MW = 50000 daltons), in 0.1 M Tris/HCl, pH 7.6, 20 mM L-glutamine, and 20 mM $MgCl_2$. Concentrations of glutamine synthetase subunit are given for each curve (μM). [ATP] was corrected for binding to glutamine synthetase, $k_a = \dfrac{[GS:ATP]}{[GS][ATP]} = 10^4 \text{ M}^{-1}$

We would like to draw then the following general conclusions.
1) The conformational changes observed in adenylyltransferase by DTNB titration indicate that this enzyme is target for effector control of the adenylylation of glutamine synthetase. 2) The ability of both substrates to bind independently, the failure to find a stable ATase-AMP form, the slowness of PP_i-ATP exchange in comparison to the overall reaction, and the kinetic analysis all support the idea of a concerted transfer of the adenylyl group, not involving an ATase-AMP intermediate.

References

1. HOLZER, H. and DUNTZE, W., Ann.Rev.Biochem., 40, 345 (1971).
2. STADTMAN, E.R., SHAPIRO, B. M., GINSBURG, A., KINGDON, H.S., and DENTON, M.D., Brookhaven Symposium in Biology, 21, 378 (1969).
3. EBNER, E., WOLF, D., GANCEDO, C., ELSÄSSER, S., and HOLZER, H., European J. Biochem., 14, 535 (1970).
4. WOLF, D., EBNER, E., and HINZE, H., manuscript submitted for publication (1971).
5. ANDERSON, W.B. and STADTMAN, E.R., Arch.Biochem.Biophys., 143, 428 (1971).
6. ANDERSON, W.B., HENNIG, S.B., GINSBURG, A., and STADTMAN, E.R., Proc.Nat.Acad.Sci. (U.S.), 67, 1417 (1970).
7. HENNIG, S.B., ANDERSON, W.B., and GINSBURG, A., Proc.Nat.Acad.Sci. (U.S.), 67, 1761 (1970).
8. MANTEL, M. and HOLZER, H., Proc.Nat.Acad.Sci.(U.S.), 65, 660 (1970).
9. WOHLHUETER, R.M., European J.Biochem., 21, 575 (1971).
10. CLELAND, W.W., Biochim.Biophys.Acta, 67, 104 (1963).

The Role of Enzyme Inactivation in the Regulation of Glutamine Synthesis in Yeast: *in vivo* Studies Using $^{15}NH_3$

A. P. Sims and A. R. Ferguson

School of Biological Sciences, University of East Anglia, Norwich/England

Studies using isotopically-labelled ammonia have shown that in the yeast, Candida utilis, almost all the ammonia assimilated enters into two compounds, glutamate and glutamine (Figure 1).

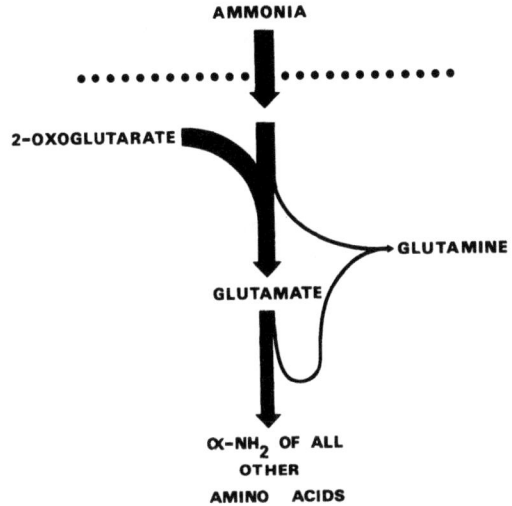

Figure 1.

The primary pathways of nitrogen assimilation in Candida utilis.

The thickness of the arrows is in proportion to the amount of nitrogen (and carbon) passing through the pools of intermediates

Balanced growth requires that nearly 75% of the ammonia is incorporated into glutamate, and between 10-15% into the amide group of glutamine (Sims and Folkes, 1964). Clearly controls must operate to ensure that there is an appropriate flow of ammonia into both glutamate and glutamine, and also that the rates of glutamate synthesis are sufficient to meet the demands made on it by the synthesis of glutamine and by the transamination reactions leading to the synthesis of all other amino acids.

Some aspects of the regulation of glutamate synthesis have already been discussed by Sims, Folkes and Bussey (1966). Regulation can be achieved through changes in the level of the assimilatory (NADP-dependent) glutamate dehydrogenase, and by modulation of the feedback restraint imposed on the activity of this enzyme by the total pool of soluble amino acids, the endproducts of glutamate metabolism. We now discuss our reasons for believing that in yeast the synthesis of glutamine is not subjected to endproduct inhibition; rather it seems likely that rapid and reversible changes in enzyme level, apparently brought about by changes in the relative concentrations of glutamate and glutamine, constitute the principle method of regulation and allow yeast to adjust rapidly to changes in nutritional conditions.

Regulation of glutamine synthesis by control of enzyme level

It is possible to determine in vivo rates of glutamine synthesis by supplying yeast with ^{15}N-labelled ammonia and then evaluating the kinetics of the incorporation of isotope into the amide group. We have established that in yeast growing under steady-state conditions with ammonia, there is a close match between the in vivo rate of synthesis and that predicted from measurements of glutamine synthetase specific activity (Table 1). Hence it would appear that there can be little increase in the rate of amide synthesis unless there is a corresponding increase in enzyme level. Although the level of glutamine synthetase is much higher in glutamate-grown cells, it can be seen (Table 1) that when the limited availability of ammonia is taken into account, the predicted rate of amide synthesis is comparable to that in yeast growing on ammonia.

Our belief that amide synthesis is regulated by control of enzyme level is supported by other experiments. If yeast, previously adapted to growth on a constant level of ammonia, is subjected to a brief period of nitrogen depletion, the pool of glutamine falls as soon as the concentration of ammonia in the

TABLE 1

Comparison of in vivo rates of glutamine synthesis with rates predicted from measurements of glutamine synthetase specific activity and ammonia availability

For cells growing with ammonia (25°C)

Measured rate of amide synthesis	= 0.0641 mg amide N/g DW/min
Specific activity of glutamine synthetase	= 0.140 mg amide N/g DW/min
But mean cellular concentration of ammonia	= 1.45×10^{-3} M
∴ Enzyme is 92% saturated w.r.t. ammonia	
∴ Predicted rate of amide synthesis	= 0.129 mg amide N/g DW/min
(Assuming all other substrates non-limiting)	

For cells growing with glutamate (25°C)

Probable rate of amide synthesis	= 0.0640 mg amide N/g DW/min
Specific activity of glutamine synthetase	= 0.634 mg amide N/g DW/min
But mean cellular concentration of ammonia	= 1.2×10^{-4} M
∴ Enzyme is 14% saturated w.r.t. ammonia	
∴ Predicted rate of amide synthesis	= 0.089 mg amide N/g DW/min
(Assuming all other substrates non-limiting)	

medium drops below the steady-state value (Figure 2). In contrast, the pools of glutamate and of other amino acids are not affected until the medium is almost completely exhausted of ammonia. This suggests that the level of glutamine synthetase must have been carefully adjusted to the availability of ammonia, and once the ammonia level falls the steady-state rate of amide synthesis can no longer be sustained. The results support the idea that the rate of glutamine synthesis in vivo is normally determined by enzyme level, and also indicate that a decrease in the ammonia supply is not compensated for by removal of any feedback restraint imposed on the enzyme, as is the case in the synthesis of glutamate (Sims, Folkes and Bussey, 1968; Folkes and Sims, in preparation). In similar experiments we have found that within a few minutes of the fall in the ammonia concentration in the medium, there is a sustained rise in glutamine synthetase (Figure 3), which results

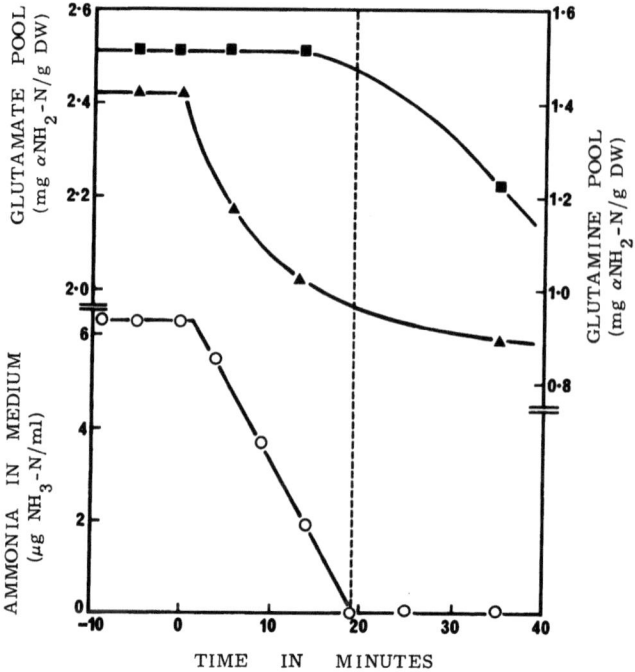

Figure 2.

Changes in the pools of glutamate and glutamine following a reduction of ammonia concentration in the culture medium.

Cells were grown in a turbidostat on a slight excess of ammonia. At time 0 mins the inflow medium was switched to one without nitrogen.

- ■ glutamate in the yeast
- ▲ glutamine in the yeast
- ○ residual ammonia in the culture medium.

The dashed line indicates the time at which ammonia in the culture medium was exhausted

in an increased capacity to synthesize glutamine once ammonia is again supplied. Measurements of the rate of amide synthesis show that this increased capacity is directly related to the increase in enzyme level (Sims and Ferguson, in preparation), yet another indication that rates of synthesis are determined by the level of glutamine synthetase.

The level of glutamine synthetase is, in turn, probably determined by the glutamine pool. We have found a good correlation between the rate of net enzyme formation and the glutamine content (but not of ammonia) of yeast growing on a number of different nitrogen sources (Figure 4). These results indicate that the control of amide synthesis, even under steady-state conditions, does not result in a glutamine pool of constant size. In contrast to this however, the level of many other amino acids are more nearly constant.

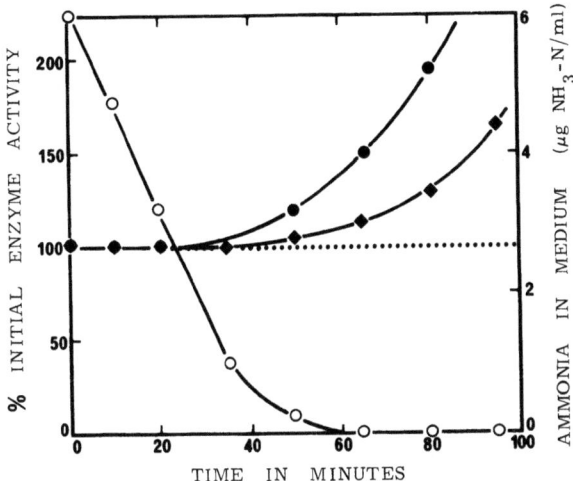

Figure 3.

Changes in the level of glutamine synthetase and the glutamate dehydrogenases in yeast cells following a reduction of ammonia concentration in the culture medium.

The specific activities of the enzymes are shown relative to the level prevailing at time 0 mins.

- ● glutamine synthetase
- ◆ the dissimilatory (NAD) glutamate dehydrogenase
- ○ residual ammonia in the culture medium.

The dotted line shows there is no change in the level of the assimilatory (NADP) glutamate dehydrogenase during this time (measured in a separate experiment)

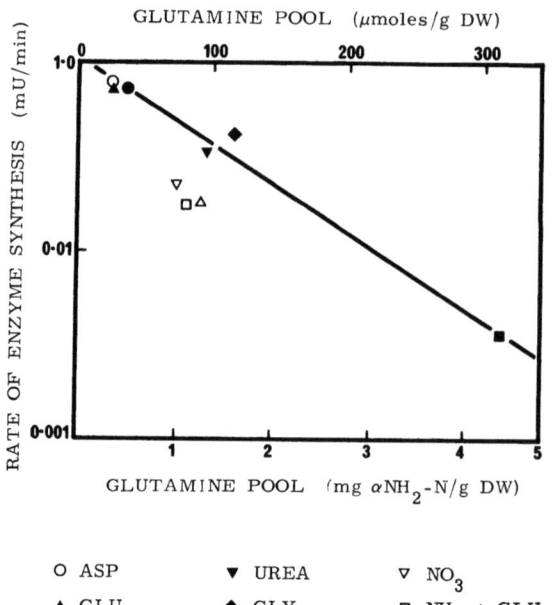

Figure 4.

Correlation between the rate of glutamine synthetase formation and the concentration of glutamine in the metabolic pool.

Yeast was grown on a number of compounds each as sole source of nitrogen. The cells were harvested during the exponential phase of growth and measurements of the specific activity of glutamine synthetase, growth rate and protein content of the cells were obtained to calculate the steady-state rate of enzyme synthesis. Samples were also analysed for their glutamine content.

Source of nitrogen for growth

O	asparate	▼	urea	▽	nitrate
▲	glutamate	◆	glycine	□	ammonia
●	alanine	■	glutamine	△	ammonia and glutamate

Inactivation of glutamine synthetase

When yeast is grown with glutamate or some other amino acids, as sole nitrogen source there is an increase in the levels of glutamine synthetase in response to the limited availability of ammonia. The addition of either glutamine or ammonia to such cells results in a rapid and extensive fall in enzyme level, a fall that is much faster than can be accounted for by the immediate cessation of enzyme synthesis followed by dilution of existing enzyme by growth. On resuspension of the culture into glutamate there is then an equally rapid increase in enzyme level (Ferguson and Sims, 1971). Enzyme inactivation appears to occur only when changed growth conditions have resulted in a marked and rapid increase in the pool of glutamine, an increase that is usually associated with a fall in the pool of glutamate.

Effects of enzyme inactivation

The addition of isotopically-labelled ammonia to glutamate-grown cells allows measurement of the actual rates of amide synthesis, and in this way, we can determine the effects of enzyme inactivation on the capacity of the organism to synthesize glutamine. Our results show that on the transition from glutamate to ammonia the rate of amide synthesis cannot be determined solely by the level of glutamine synthetase. Following the addition of ammonia there is an immediate increase in the rate of amide synthesis (Figure 5), such that, after seven minutes, rates are at least six times those found in steady-state cells growing on ammonia. This initial increase in the rate of synthesis occurs at a time when the level of glutamine synthetase is falling rapidly (Figure 6), and although subsequent changes in amide synthesis reflect closely changes in enzyme level, the relationship between in vivo rate and enzyme level is not linear (Figure 7). The observed changes in the rate of amide synthesis per unit enzyme (Figure 8) indicate that some restraint is being removed and then imposed on enzyme activity.

(a) Endproduct inhibition. Such restraint could have resulted from fluctuations in the concentration of glutamine and the endproducts of its metabolism, if these compounds were inhibitors of enzyme activity. We have measured changes in the pools of soluble amino acids following the addition of ammonia to glutamate-grown cells. There is a very large increase in glutamine and an equally dramatic

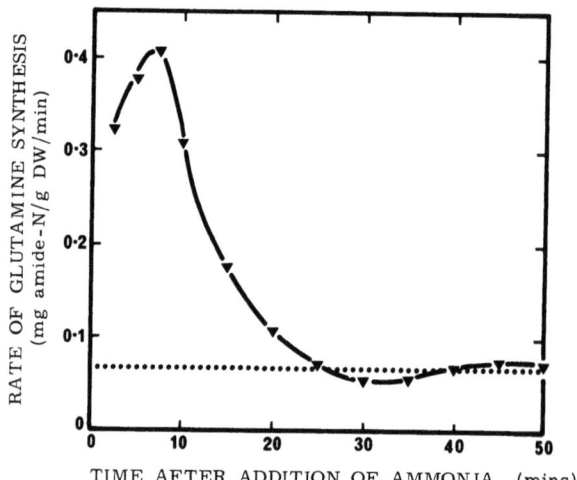

Figure 5.

The regulation of glutamine synthesis in vivo as revealed by isotope incorporation into the amide group of glutamine following the addition of $^{15}NH_3$ to cells growing on glutamate.

The dotted line indicates the rate of amide synthesis measured in steady-state cells growing on ammonia

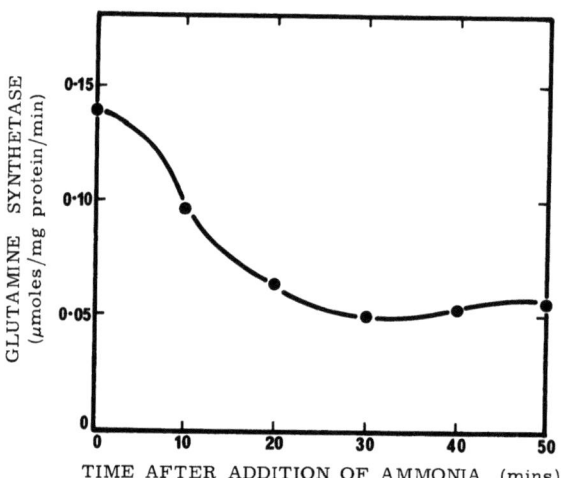

Figure 6.

Changes in the specific activity of glutamine synthetase following the addition of ammonia to steady-state cells growing on glutamate.

These measurements were carried out at the same time as the determinations of the in vivo rate of amide synthesis shown in Figure 5

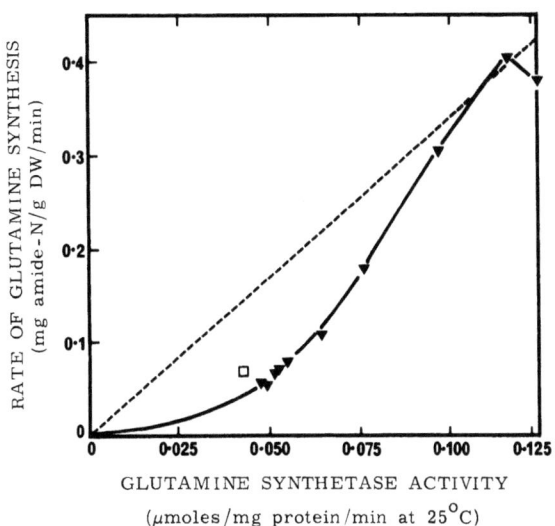

Figure 7.

The correlation between the rate of glutamine synthesis and the level of glutamine synthetase.

▼   data derived from a glutamate-ammonia transition experiment at 25°C.

□   figure derived from a separate ammonia steady-state experiment at 25°C.

The dotted line represents the maximal rate of glutamine synthesis possible as calculated from determinations of the cellular concentration of glutamine synthetase. The enzyme was assayed using saturating levels of all substrates at 25°C. The protein content is 325 mgs/g dry wt. of cells

fall in the pool of glutamate, but there is little change in the levels of most other amino acids (Figure 9) and the soluble nucleotides. Over the period that the rate per unit enzyme fell most rapidly, there was actually a decrease in the concentration of alanine, glycine, and histidine which, on the basis of <u>in vitro</u> studies, Hubbard and Stadtman (1967) has suggested as possible inhibitors of glutamine synthetase in the living cell. Our own studies have shown that glutamine synthetase activity, when measured by the synthetase assay with Mg^{++} as metal cofactor, is not significantly inhibited by glutamine itself, or by amino acids or nucleotides either separately or together. Moreover we have found in other experiments (Sims and Ferguson, in preparation) that depletion

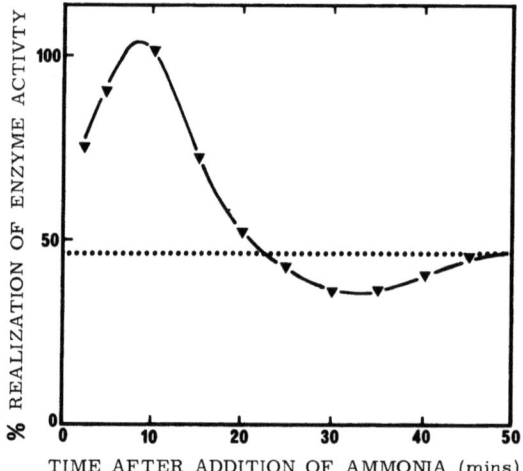

Figure 8.

Changes in the rate of glutamine synthesis per unit of enzyme with time.

The in vivo rates of synthesis were expressed as a percentage of those predicted from measurements of enzyme specific activity made at the same time.

The dashed line represents the percentage found in steady-state cells growing on ammonia as a sole source of nitrogen

of the soluble intermediates by a brief period of nitrogen-starvation, does not result in any significant increase in the potential rate of amide synthesis that cannot be attributed to increases in enzyme level. We have therefore concluded that it is most unlikely that feedback inhibition plays an important part in the regulation of glutamine synthesis in yeast.

(b) Changes in enzyme properties. There is no evidence to indicate that changes in the rate of amide synthesis were due to modification of the properties of glutamine synthetase on inactivation. We have compared the behaviour of enzyme preparations from cells containing de-repressed levels of glutamine synthetase with those from cells in which most activity has been lost because of enzyme inactivation. No differences were observed in the affinities of these enzyme preparations for ammonia (Figure 10), glutamate (Figure 11), or magnesium, in their metal specificity, or in their pH profiles.

(c) Changes in substrate availability. Measurement of the cellular

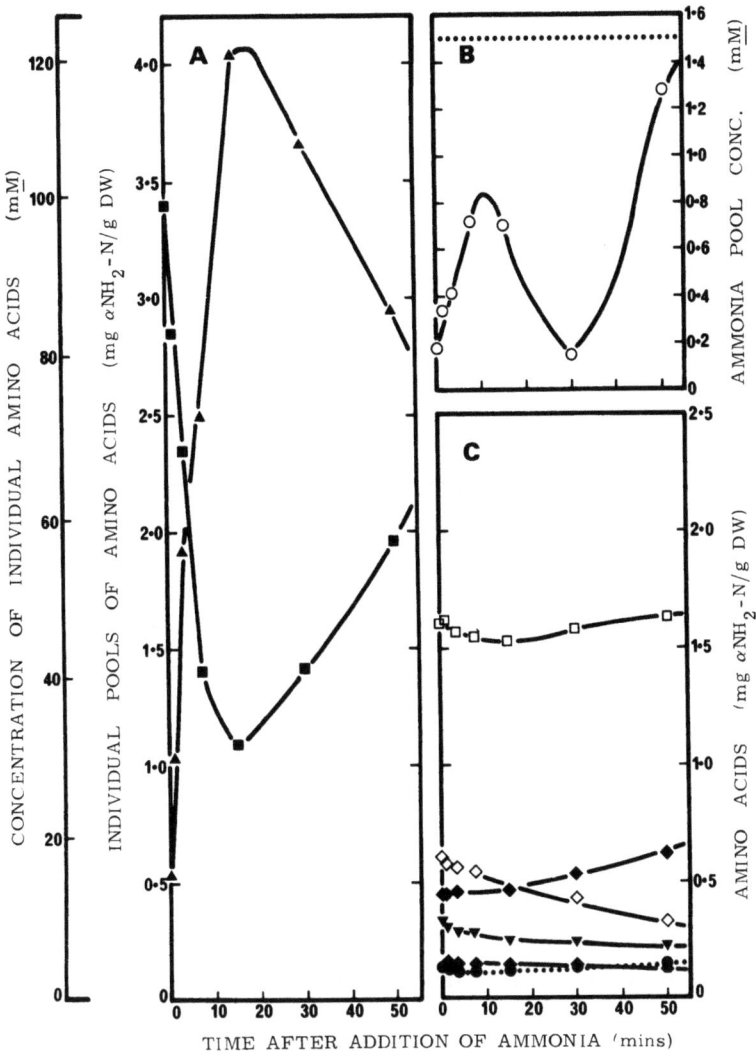

Figure 9.

Changes in the pools of amino acids in yeast following the addition of ammonia to a culture growing on glutamate as a sole source of nitrogen.

All the results, with the exception of the ammonia data, are expressed as mgs α NH$_2$-N/g dry wt. of cells. The ammonia data have been converted to a mean cellular concentration by assuming that the water content of 1 g dry wt. of cells is 2.448 mls. (A. Wiemken: unpublished data using ^{14}C inulin method) The dotted line indicates the steady-state concentration of ammonia in yeast cells growing on this as a sole source of nitrogen.

A.	▲ glutamine	C.	□ alanine	▼ ornithine
	■ glutamate		◇ lysine	⬟ histidine
B.	○ ammonia		◆ arginine	●··● glycine

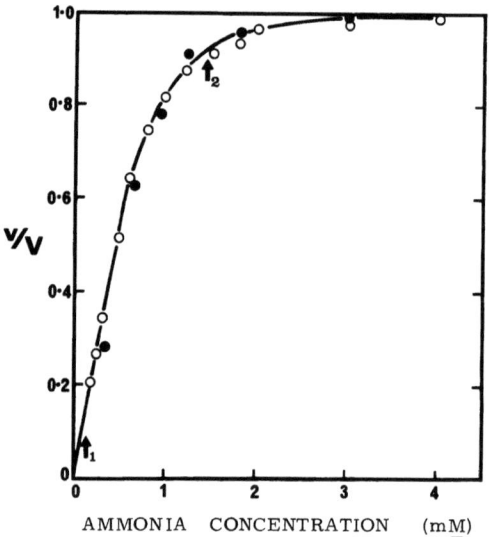

Figure 10.

The effects of varying the concentration of ammonia, expressed relative to the maximal rate, on glutamine synthetase from yeast.

- ● enzyme prepared from cells growing on glutamate as sole source of nitrogen.
- ○ enzyme prepared from cells containing inactivated enzyme obtained by transferring glutamate grown cells to ammonia for 1 hour.

Values were calculated for the mean cellular concentration of ammonia in yeast: the 2 arrows indicate the values obtained.

1. growth on glutamate
2. growth on ammonia

concentrations of glutamate and ammonia indicate that whereas glutamine synthetase is likely to be saturated with respect to glutamate, the level of ammonia could limit enzyme activity. In such circumstances, the synthesis of glutamine would be particularly sensitive to small changes in the concentration of ammonia since not only does the enzyme exhibit co-operative responses towards this substrate (Figure 10) but also the greatest changes in rate might be expected to occur over the range of ammonia concentrations found in the cell. Thus on the addition of ammonia to glutamate-grown cells, a marked increase in the

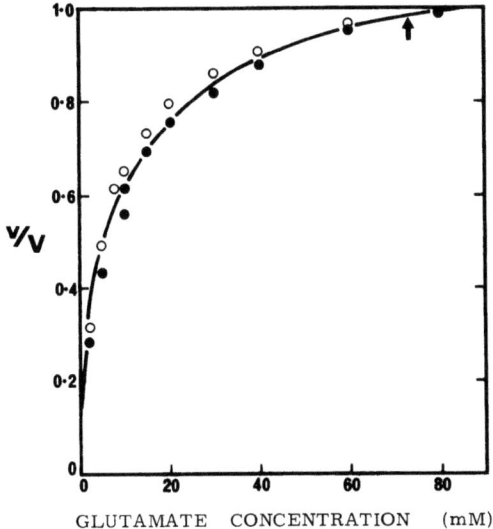

GLUTAMATE CONCENTRATION (mM)

Figure 11.

The effects of varying the concentration of glutamate, expressed relative to the maximal rate, on glutamine synthetase from yeast.

- ● enzyme prepared from cells grown on glutamate as sole source of nitrogen.
- ○ enzyme prepared from cells containing inactivated enzyme obtained by transferring glutamate grown cells to ammonia for 1 hour.

A value was calculated for the mean cellular concentration of glutamate in yeast grown on ammonia: the arrow indicates the value obtained

cellular concentration of ammonia (hence in the degree of saturation of the enzyme) could account for the initial rise in the rate of amide synthesis per unit enzyme (Figure 8). We have established that on the addition of ammonia, there is, in fact, a very rapid increase in the cellular concentration of ammonia (Figure 9) and measurements of the levels of ammonia and glutamate have been used to calculate the probable extent to which glutamine synthetase had, at different times, been saturated by these two substrates. These results were then used in conjunction with measurements of enzyme specific activity, to predict rates of glutamine formation during the transition from glutamate to ammonia. When the calculated rate for time 7.5 minutes was matched with that measured in vivo at that time, we found that there was very good agreement between the subsequent rates predicted and those actually measured in the living cell (Fig. 12). Thus provided we take into consideration both the changes in the level of enzyme and in the availability of substrates we can fully account for the changes observed in the in vivo rate of amide synthesis during the glutamate ammonia transition experiment.

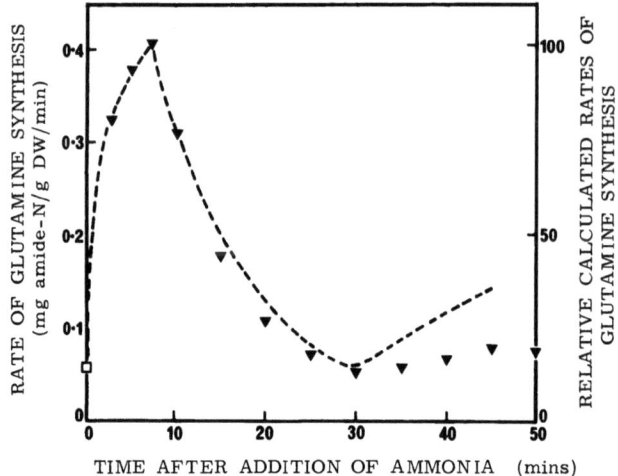

Figure 12.

A comparison of the relative rates of amide synthesis with values calculated from measurements of enzyme concentrations after allowing for changes in substrate availability.

Measurements of the mean cellular concentrations of ammonia and glutamate shown in Fig. 9 were used to estimate the extent to which the enzyme was saturated with these substrates at different times (also Figures 10 and 11). The enzyme specific activity (Figure 6) were corrected using the appropriate saturation factors and the rate calculated for 7.5 mins was matched to the in vivo rate. All the other values are shown relative to this point.

▼ Rate of amide synthesis
-- Relative calculated values

Summary

In vivo studies using isotopically labelled ammonia have established that in yeast the rate of amide synthesis is determined primarily by the level of glutamine synthetase the full activity of which is limited only by the availability of substrates. Changes in the concentration of these substrates, particularly that of ammonia, result in compensatory changes in the level of enzyme and so ensure that appropriate rates of glutamine synthesis are maintained. Glutamine and the endproducts of its metabolism do not cause any inhibition of enzyme activity in vitro, and feedback inhibition does not appear to play an important part in the regulation of glutamine synthesis in the living cell. Glutamine can, however, effectively control its own synthesis since changes in the pool of glutamine, usually accompanied by changes in the pool of glutamate, can bring about a rapid and reversible modulation of enzyme level.

Acknowledgements

This work was supported by the Science Research Council of Great Britain. We are very much indebted to Mrs. J. Toone for her skilled technical assistance, and to Mr. S. Howitt for the determinations of ^{15}N abundance. We wish also to thank Professor B. F. Folkes for his discussion and helpful advice.

References

FERGUSON, A.R. and SIMS, A.P. (1971). In vivo inactivation of glutamine synthetase and NAD-specific glutamate dehydrogenase: its role in the regulation of glutamine synthesis in yeasts. Journal of General Microbiology (in press)

HUBBARD, J.S. and STADTMAN, E.R. (1967). Regulation of glutamine synthetase. II. Patterns of feedback inhibition in microorganisms. Journal of Bacteriology 93, 1045-1055.

SIMS, A.P. and FOLKES, B.F. (1964). A kinetic study of the assimilation of (^{15}N)-ammonia and the synthesis of amino acids in an exponentially growing culture of Candida utilis. Proceedings of the Royal Society, B. 159, 479-502.

SIMS, A.P., FOLKES, B.F. and BUSSEY, A.H. (1968). Mechanisms involved in the regulation of nitrogen assimilation in microorganisms and plants. In Recent Aspects of Nitrogen Metabolism in Plants (First Long Ashton Symposium, 1967) pp 91-114. Edited by E.J. Hewitt and C.V. Cutting. London and New York: Academic Press.

Discussion:

Segal

Is there something known about the control of glutamine-synthetase, avoiding a futile cycle?

Stadtman

In E. coli there are two glutaminases. We found that the constitutive one is inhibited by ATP. According to Holzer the ATP level is low when glutamine is present. Reversely the loss of glutamine is followed by an increase of ATP and this might have an effect on glutaminase.

Segal

You showed that, if glutamine is present, it doesn't matter whether P_2 is in the a or b form. Under these circumstances the control of P_2 would be as important as the level of glutamine.

Stadtman

The system's sensitivity to α-ketoglutarate is very different. Fraction P_2 already has an activated enzyme. It is only slightly sensitive to glutamine and has a low sensitivity to α-ketoglutarate. This causes an effect on catabolite susceptibility depending on whether the a or b form is present.

Buc

I was struck by the similarity of the enzyme forms P_1 and P_2 with the CTP synthesizing system. Did you look for the CTP synthetase activity in P_1 and P_2?

Stadtman

Yes, we did.

Helmreich

Is it possible that the effectors could act on the converting enzyme rather than on P_2?

Stadtman

For conversion you need ATP and α-ketoglutarate in addition to UTP, whose function I think is probably intimately concerned with the reaction itself.

In addition, after conversion there is no deadenylation capacity of the b form unless ATP and α-ketoglutarate are present. This means that α-ketoglutarate must interact either with the P_2 protein or the P_1 protein or with glutaminesynthetase. I don't think that we can distinguish between these possibilities, but from the experiments in today's second paper there is no reason to believe that interaction of the effectors and P_1 protein is not the site of action.

Fischer

Do you know anything about this strong inorganic phosphate inhibition? Is it possible that it acts on a phosphorolysis of the uridinylgroup?

Stadtman

That is not the explanation. Crude preparations contain activity that catalyses the hydrolysis of the uridinylmoiety from whatever it is bound. We don't know whether it is physiologically important or whether it is simply the manifestation of a non-specific phosphodiesterase. This point has to be looked at.

N.N.

Dr. Holzer, could you clarify your interpretation of the rapid decrease in ATP?

Holzer

The ATP is not used up to adenylate glutamine synthetase. It is used for the synthesis of glutamine from glutamate. Therefore the "life-saving" mechanism of glutamine dependent inactivation of glutamine synthetase occurs only in organisms with very high glutamine synthetase activity, because here all available ATP is used up very rapidly. And you even have to regenerate two to three times the amount of initially available ATP during the first 15 seconds from elsewhere.

Sols

Was an energy source supplied when you added ammonia to the intact cells?

Holzer

Yes, the energy source was glycerol.

Hartmann

Couldn't it be, that a large amount of ATP is used for the entrance of ammonia into the cell or is there any control at this point.

Holzer

Perhaps. We know only, that it enters very fast.

Lynen

There are two forms of ammonia, NH_3 and NH_4^+. NH_3 penetrates very easy.

Holzer

The pH of the growth medium was 7.6 to 8.0.

Lynen

Then I think the penetration is no problem.

Lynen

I understand the drainage in ATP after addition of NH_4^+ but I do not understand why the ATP regeneration takes so long.

Holzer

ATP does recover in five minutes.

Lynen

Well, why does it take so long?

Holzer

After addition of NH_4^+ there is still some activity of glutamine synthetase because it is not completely inactivated, and perhaps the glutaminase mentioned by Stadtman is operating and so causes this slow recovery of the ATP level.

Lynen

So you think that this is evidence for the futile cycle that was mentioned before?

Stadtman

I only wanted to comment that from your data. I think that the glutamine level continues to rise quite considerably during this period. We now know that probably the major mechanism for glutamate formation in E. coli involves glutamine as a precursor. This would, of course, require ATP as it is an ATP-dependent process. So that as long as glutamate is being synthesized, ATP is going to be consumed. I believe that part of the explanation for the persistence of the low level of ATP is that you are consuming ATP for the synthesis of glutamate from glutamine.

Molecular Aspects of the Regulation of the Mammalian Pyruvate Dehydrogenase Complex

Lester J. Reed, Tracy C. Linn, Ferdinand Hucho, Genshin Namihira,
Cecilio R. Barrera, Thomas E. Roche, John W. Pelley, and Douglas D. Randall

Clayton Foundation Biochemical Institute and Department of Chemistry,
The University of Texas at Austin, Austin, Texas/USA

Pyruvate dehydrogenase systems have been isolated from <u>Escherichia coli</u>, avian and mammalian tissues, and <u>Neurospora crassa</u> as functional units with molecular weights in the millions (1). The bacterial, avian, and mammalian pyruvate dehydrogenase complexes have been separated into three enzymes - pyruvate dehydrogenase, dihydrolipoyl transacetylase, and dihydrolipoyl dehydrogenase (a flavoprotein), and functional units resembling the native complexes have been reassembled from the individual enzymes. These three enzymes act in a coordinated manner as indicated in Fig. 1 (1,2). The discussion will be limited to the structure, function, and regulation of the mammalian pyruvate dehydrogenase complex.

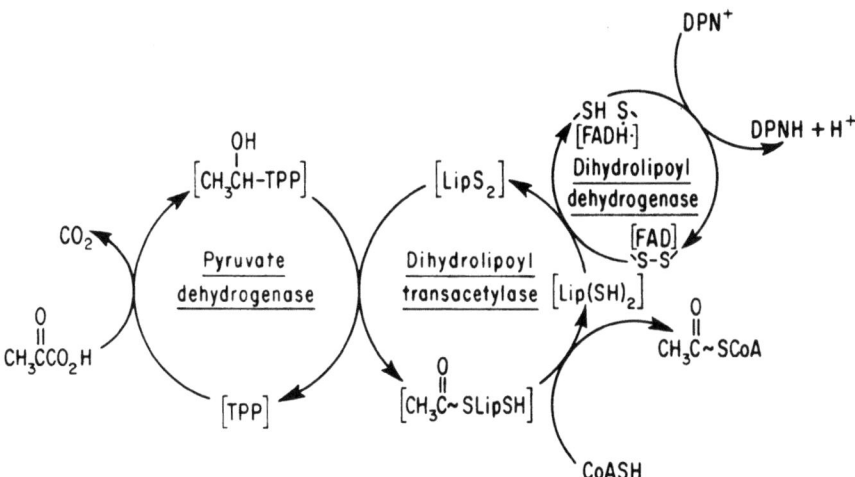

Fig. 1. Reaction sequence in pyruvate oxidation

<u>Subunit Structure of Mammalian Pyruvate Dehydrogenase Complex</u>

Since the mammalian pyruvate dehydrogenase complex and its regulatory enzymes are localized in mitochondria, we developed procedures for preparing bovine kidney and heart mitochondria on a large scale and for isolating the desired enzymes from the mitochondria (3). Since relatively mild conditions, i.e. freezing and thawing, were effective in extracting the pyruvate dehydrogenase complex and its regulatory enzymes from kidney and heart mitochondria, it appears that these enzymes are bound loosely, if at all, to the mitochondrial membranes. Although we have made no attempt to localize these enzymes within mitochondria, results reported from other laboratories (4) suggest that the pyruvate and α-ketoglutarate dehydrogenase complexes are localized within the mitochondrial matrix.

Major features of the structure of the mammalian pyruvate dehydrogenase complex have been elucidated by biochemical and electron microscopic studies (5-7). The complex contains a core, consisting of the dihydrolipoyl transacetylase, to which the pyruvate dehydrogenase and dihydrolipoyl dehydrogenase are joined. Thus the core enzyme plays both a catalytic and a structural role. The appearance of the transacetylase in the electron microscope is that of a pentagonal dodecahedron (Fig. 2).

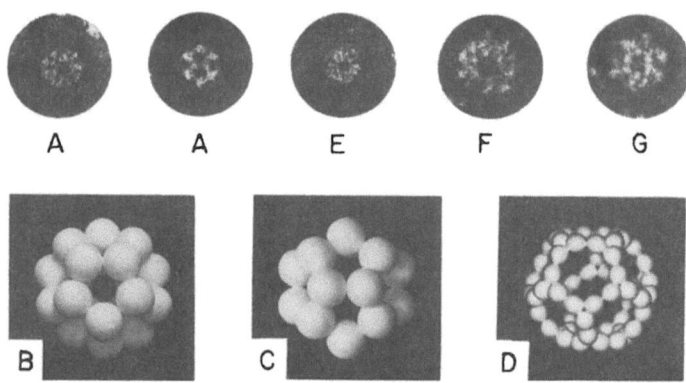

Fig. 2. Electron micrograph images of the mammalian pyruvate dehydrogenase complex and its transacetylase component, and interpretative models of the transacetylase. (A) Individual images (X250,000) showing two orientations of the bovine kidney dihydrolipoyl transacetylase. (B,C) Corresponding views of a model of the transacetylase photographed down a 5-fold axis and a 2-fold axis, respectively. The model consists of 20 spheres placed at the vertices of a pentagonal dodecahedron. Each sphere represents a group of 3 polypeptide chains. (D) Expanded model of the transacetylase showing trimer clustering at the 20 vertices. (E) Image (X250,000) of the bovine heart dihydrolipoyl transacetylase viewed down a 5-fold axis. (F,G) Individual images (X250,000) of the pyruvate dehydrogenase complex from bovine kidney and heart, respectively. The electron micrographs were taken by Robert M. Oliver. The samples were negatively stained with phosphotungstate

This structure would be expected to contain 60 very similar, if not identical, polypeptide chains organized into 20 groups of three chains clustered about the three-fold axes of symmetry. We have found that the bovine kidney and heart dihydrolipoyl transacetylases are very similar and that each enzyme does indeed consist of 60 apparently identical polypeptide chains of molecular weight about 52,000 (7). Each chain apparently contains one molecule of covalently bound lipoic acid.

The pyruvate dehydrogenase components of the bovine kidney and heart pyruvate dehydrogenase complexes are also very similar. Both enzymes have been crystallized

(3). The uncomplexed pyruvate dehydrogenases have a molecular weight of about 154,000 and possess the subunit compostion $\alpha_2\beta_2$ (7). The molecular weights of the two chains are about 41,000 and 36,000, respectively. Pyruvate dehydrogenase is a major site of regulation of the activity of the pyruvate dehydrogenase complex (8). This enzyme undergoes phosphorylation and concomitant inactivation in the presence of a regulatory enzyme, pyruvate dehydrogenase (PDH) kinase, and ATP. Reactivation and concomitant dephosphorylation occur in the presence of a second regulatory enzyme, PDH phosphatase, and millimolar concentrations of Mg^{2+}. Only the α-chain undergoes phosphorylation and dephosphorylation (7).

In view of the apparent icosahedral (532) symmetry of the transacetylase, we should expect 60 equivalent binding sites on the transacetylase for pyruvate dehydrogenase and 60 sites for dihydrolipoyl dehydrogenase. It appears that the bovine kidney and heart pyruvate dehydrogenase complexes do indeed contain about 60 pyruvate dehydrogenase units ($\alpha\beta$), but only 10-12 flavoprotein chains (7). The simplest organization of the pyruvate dehydrogenase complex would involve the attachment of one pyruvate dehydrogenase unit ($\alpha\beta$) to each of the 60 transacetylase chains. On the other hand, if pyruvate dehydrogenase is present in the complex as tetramers ($\alpha_2\beta_2$), we should expect the complex to contain 30 such units, and they presumably would be located on the 30 two-fold axes of the transacetylase pentagonal dodecahedron. The available data do not permit a decision between these two alternatives.

The uncomplexed flavoprotein contains two apparently identical polypeptide chains and two molecules of FAD per molecule of enzyme of molecular weight about 110,000 (9). The available data do not permit a decision as to whether the flavoprotein is present in the complex as monomers or dimers. From the FAD content of the bovine kidney and heart pyruvate dehydrogenase complexes (3), we estimate that there are only 5-6 molecules of flavoprotein or 10-12 flavoprotein chains per molecule of complex. It is difficult to reconcile these numbers with the apparent icosahedral (532) symmetry of the transacetylase and the apparent 1:1 ratio of pyruvate dehydrogenase chains and transacetylase chains. A possible explanation of this discrepancy is that the flavoprotein dissociates from the pyruvate dehydrogenase complex during its extraction and purification from mitochondria. Consistent with this possibility is our observation that substantial amounts of uncomplexed flavoprotein are present at certain stages of the purification procedure (3). Other investigators (10) have also reported the existence of an uncomplexed form of dihydrolipoyl dehydrogenase in extracts of mammalian tissues. However, we have observed that uncomplexed flavoprotein (i.e. the dimer) has little, if any, effect on the rate of pyruvate oxidation by the purified pyruvate dehydrogenase complex. Another possibility is that a single flavoprotein chain or molecule, bound to the transacetylase, can interact, either intermolecularly or intramolecularly, with more than one lipoyl moiety. Until these problems are resolved, the stoichiometry of the complex with respect to the flavoprotein will remain uncertain.

The regulatory enzyme, PDH kinase, is copurified with the bovine kidney and heart pyruvate dehydrogenase complexes (3,8). When the complexes were resolved, the kinase was found to be tightly bound to the transacetylase (3,11). We have separated and purified the kinase from the kidney transacetylase. The kinase comprises only about 10% by weight of the transacetylase. We estimate that there are at most about 5 kinase molecules per molecule of the bovine kidney transacetylase (3). It is difficult to rationalize these data in terms of specific binding sites for the kinase on the transacetylase. It is possible that binding of the kinase to the transacetylase is adventitious.

A second regulatory enzyme, PDH phosphatase, appears to be loosely associated with the bovine kidney and heart pyruvate dehydrogenase complexes (3,11). We have purified the kidney and heart PDH phosphatases 400- to 1000-fold from mitochondrial extracts (3). Gel electrophoresis indicates that the heart PDH phosphatase is almost homogeneous. The molecular weight of the two phosphatases was estimated by gel filtration to be about 100,000. We estimate that there are only about 5 molecules of PDH phosphatase per molecule of pyruvate dehydrogenase complex in bovine heart and kidney mitochondrial extracts (3). Although we have not eliminated the possibility that there is a nonactive form of the PDH phosphatase or the PDH kinase, or both enzymes, in bovine kidney and heart mitochondrial extracts, it appears that the molar concentration of the phosphatase, and probably the kinase as well, is about an order of magnitude less than that of its protein substrate, pyruvate dehydrogenase.

Regulation of the Mammalian Pyruvate Dehydrogenase Complex

The activity of the mammalian pyruvate dehydrogenase complex is inhibited by the products of pyruvate oxidation, acetyl-CoA and DPNH, and these inhibitions are reversed by CoA and DPN, respectively (12-14). The site of DPNH inhibition appears to be the flavoprotein component of the complex. Although the site of acetyl-CoA inhibition has not been established, it is probably the transacetylase. Although there has been considerable speculation that acetyl-CoA/CoA and DPNH/DPN ratios may regulate the activity of the mammalian pyruvate dehydrogenase complex in vivo, definitive evidence in support of this possibility remains to be obtained.

Another regulatory mechanism, involving phosphorylation and dephosphorylation of the pyruvate dehydrogenase component of the mammalian pyruvate dehydrogenase complex (Fig. 3), was uncovered in this laboratory (8,11). We found that highly purified preparations of the bovine kidney pyruvate dehydrogenase complex were inactivated by incubation with 1-10 μM ATP. AMP, ADP, CTP, GTP, and UTP were ineffective. Using radioactive ATP labeled with ^{32}P in either the α-, $\alpha\beta$-, or γ-phosphoryl moieties, we established that the terminal phosphoryl moiety of ATP is transferred to the pyruvate dehydrogenase complex. When the phosphorylated (inactivated) pyruvate dehydrogenase complex was resolved, essentially all of the protein-bound radioactivity was found in the pyruvate dehydrogenase component. Further investigation revealed that the phosphorylated, inactivated pyruvate dehydrogenase complex was reactivated by incubation with millimolar concentrations of Mg^{2+}. Restoration of activity was accompanied by release of inorganic orthophosphate. The data presented in Fig. 4C illustrate the time course of reciprocal changes in enzymic activity and protein-bound phosphoryl groups observed with the bovine kidney pyruvate dehydrogenase complex (8).

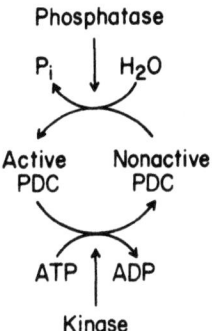

Fig. 3. Interconversion of active and nonactive (phosphorylated) forms of the mammalian pyruvate dehydrogenase complex (PDC)

These data established that the bovine kidney pyruvate dehydrogenase complex is subject to regulation by phosphorylation and dephosphorylation and that the site of this regulation is the pyruvate dehydrogenase component of the complex, which catalyzes the first step in pyruvate oxidation. Subsequent studies provided evidence that phosphorylation and concomitant inactivation of pyruvate dehydrogenase are catalyzed by a kinase (i.e. pyruvate dehydrogenase kinase), and dephosphorylation and concomitant reactivation are catalyzed by a phosphatase (i.e. pyruvate dehydrogenase phosphatase) (8,11). Wieland and co-workers (15,16) extended these observations to the pyruvate dehydrogenase complex from porcine heart muscle. Jungas (17), Siess, Wittmann, and Wieland (18), and Randle and co-workers (19) have made similar observations with crude preparations of the pyruvate dehydrogenase complex from rat epididymal adipose tissue and porcine brain. We have demonstrated that preparations of the pyruvate dehydrogenase complex isolated from mitochondria of bovine heart, liver, and brain, porcine liver, and rat kidney and heart are also subject to regulation by phosphorylation and dephosphorylation (11,20). The time course of reciprocal changes in enzymic activity and protein-bound phosphoryl groups obtained with preparations of the pyruvate dehydrogenase complex from mitochondria of bovine heart and brain and porcine liver are shown in Fig. 4. The differences in rates of inactivation and reactivation are apparently due to differences in the amounts and possibly the activities of the kinase and the phosphatase.

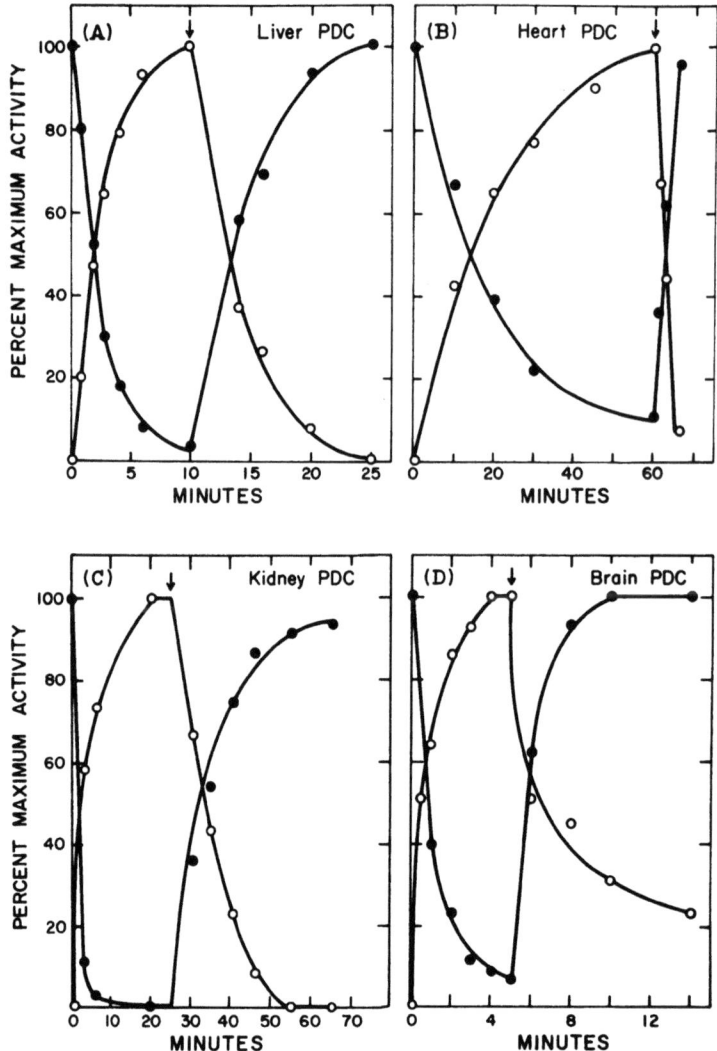

Fig. 4. Time course of phosphorylation and dephosphorylation of purified mammalian pyruvate dehydrogenase complexes (PDC). The reaction mixtures contained 20 mM phosphate buffer, pH 7.0-7.5, 0.5 or 1.0 mM $MgCl_2$, 2 mM dithiothreitol, 0.01-0.03 mM (A,B,C) or 0.5 mM (D) $[\gamma-{}^{32}P]ATP$, and enzyme complex in a total volume of 1.0 ml. The mixtures were incubated at 25° (A,C) or 30° (B,D), and aliquots were removed at the indicated times and assayed for DPN-reduction activity (●) and for protein-bound radioactivity (○). At the time interval indicated by the vertical arrow, sufficient $MgCl_2$ was added to give a final concentration of 10 mM (A,B,C) or 20 mM (D)

Subunit Structure and Function of Pyruvate Dehydrogenase

As indicated above, we have shown that it is the pyruvate dehydrogenase component of the mammalian pyruvate dehydrogenase complex which undergoes phosphorylation and dephosphorylation (3,8). We have investigated the molecular basis of this control mechanism. The bovine kidney and heart pyruvate dehydrogenases contain two different types of subunits with molecular weights of about 41,000 (α-subunit) and 36,000 (β-subunit), respectively (7). We have separated the α- and β-chains of the kidney pyruvate dehydrogenase by chromatography on phosphocellulose in the presence of 8 M urea. To our knowledge the mammalian pyruvate dehydrogenase represents the first example of a thiamine pyrophosphate-dependent enzyme that consists of two nonidentical subunits. Only the α-subunit undergoes phosphorylation when the pyruvate dehydrogenase complex or the uncomplexed pyruvate dehydrogenase is incubated with PDH kinase and ATP. A radioactive tetradecapeptide has been isolated from tryptic digests of the ^{32}P-labeled kidney pyruvate dehydrogenase, and its amino acid sequence has been determined (21). The phosphoryl moieties are attached to seryl residues. The first

Tyr-His-Gly-His-Ser(P)-Met-Ser-Asn-Pro-Gly-Val-Ser(P)-Tyr-Arg

seryl residue in this sequence is rapidly phosphorylated, and this phosphorylation results in inactivation of the pyruvate dehydrogenase complex. The third seryl residue in this sequence is slowly phosphorylated. The physiological significance, if any, of this latter phosphorylation site remains to be determined.

Pyruvate dehydrogenase catalyzes a decarboxylation of [1-^{14}C]pyruvate. When the enzyme was incubated with [2-^{14}C]pyruvate, thiamine pyrophosphate, Mg^{2+}, and purified dihydrolipoyl transacetylase, [^{14}C]acetyl groups were rapidly incorporated into the transacetylase (7). The protein-bound acetyl groups were quantitatively released by exposure of the protein to performic acid vapor. When [1-^{14}C]pyruvate was used, no radioactivity was incorporated into protein. These observations are consistent with Reactions 1 and 2, i.e. a decarboxylation of pyruvate to form hydroxyethyl-thiamine pyrophosphate, followed by a reductive acetylation of the lipoyl moieties which are covalently bound to the transacetylase.

$$CH_3COCO_2H + \text{thiamine-PP} \longrightarrow CH_3CHOH\text{-thiamine-PP} + CO_2 \quad (1)$$

$$CH_3CHOH\text{-thiamine-PP} + lipS_2 \longrightarrow CH_3CO\text{-S-lip-SH} + \text{thiamine-PP} \quad (2)$$

We have considered several possible functions for the two nonidentical subunits of the mammalian pyruvate dehydrogenase. One possibility is that the α-subunit, which undergoes phosphorylation, is a regulatory subunit and that the β-subunit is a catalytic subunit. Another possibility is that both subunits have a catalytic function. One subunit might catalyze Reaction 1 and the other subunit might catalyze Reaction 2. We have recently obtained evidence for this latter hypothesis (22). Pyruvate dehydrogenase catalyzes a reductive acetylation of the lipoyl moieties of the transacetylase with either [2-^{14}C]pyruvate (Reactions 1 and 2) or [^{14}C]hydroxyethyl-thiamine-PP (Reaction 2) as substrate. Phosphorylation of pyruvate dehydrogenase with PDH kinase and ATP markedly inhibits the reaction with [2-^{14}C]pyruvate, but does not inhibit the reaction with [^{14}C]hydroxyethyl-thiamine-PP. Phosphorylation of pyruvate dehydrogenase also markedly inhibits its ability to catalyze the decarboxylation of [1-^{14}C]pyruvate (Reaction 1). These observations are compatible with the interpretation that the α-subunit catalyzes Reaction 1 and that the β-subunit catalyzes Reaction 2.

It is interesting to note that the pyruvate dehydrogenase component of the Escherichia coli pyruvate dehydrogenase complex, in contrast to the mammalian pyruvate dehydrogenase, consists of two apparently identical polypeptide chains of molecular weight about 96,000 (23,24). Moreover, the E. coli pyruvate dehydrogenase is not subject to regulation by phosphorylation and dephosphorylation (25). It appears that the E. coli pyruvate dehydrogenase has the two catalytic sites corresponding to Reactions 1 and 2 in a large polypeptide chain, whereas the mammalian version of pyruvate dehydrogenase has the two sites segregated on two smaller, nonidentical chains.

Properties of PDH Kinase and PDH Phosphatase

The PDH phosphatases isolated from mitochondria of bovine kidney, bovine heart, and porcine liver are functionally interchangeable (11). The bovine kidney PDH kinase is functional with the bovine heart and porcine liver pyruvate dehydrogenase complexes (11). Rabbit antiserum to the bovine kidney PDH kinase crossreacts with the bovine heart and bovine brain PDH kinases (T. E. Roche and L. J. Reed, unpublished observations).

In our investigations, no effect of cyclic 3',5'-AMP was found on the activity of either the PDH kinase or the PDH phosphatase, even though preparations were examined at various stages of purification, including crude mitochondrial extracts (F. Hucho, T. C. Linn, and L. J. Reed, unpublished observations). Our results are at variance with the data of Wieland and Siess (16), who reported a cyclic AMP stimulation of the activity of PDH phosphatase from porcine heart muscle. These investigators attributed this effect to a hypothetical cyclic AMP-dependent PDH phosphatase kinase.

Preliminary studies (T. E. Roche and L. J. Reed, unpublished observations) on the protein specificity of PDH kinase indicate that this kinase does not catalyze a phosphorylation of histone, in the presence or absence of cyclic AMP. A sample of skeletal muscle cyclic AMP-dependent protein kinase, kindly furnished by Dr. Edwin Krebs, did not inactivate preparations of the bovine kidney or heart pyruvate dehydrogenase complexes in the presence of ATP and cyclic AMP. The protein kinase exhibited little, if any, ability to phosphorylate the bovine kidney or heart pyruvate dehydrogenase complexes or the crystalline pyruvate dehydrogenase from bovine kidney.

Some of the kinetic parameters of PDH kinase and PDH phosphatase from bovine kidney and heart mitochondria have been determined (11,26). The true substrate for the kinase is $MgATP^{2-}$, and the apparent K_m is about 0.02 mM. ADP is competitive with ATP, and the apparent K_i value for ADP is about 0.1 mM. Magnesium ion is required for PDH phosphatase activity. Recent measurements with highly purified preparations of the kidney PDH phosphatase indicate that the apparent K_m for Mg^{2+} is about 3 mM. The phosphatase is also active with Mn^{2+} (apparent K_m about 1 mM), whereas Ca^{2+} and Zn^{2+} are inhibitory.

Pyruvate protects the pyruvate dehydrogenase complex against inactivation by ATP, and this effect appears to be more pronounced with the bovine heart pyruvate dehydrogenase complex than with the bovine kidney complex (11). The apparent K_i values for pyruvate are about 0.08 mM and 0.9 mM, respectively (26). Pyruvate is noncompetitive with ATP. It has been difficult to determine whether pyruvate exerts its effect on the kinase or on the pyruvate dehydrogenase. We have recently obtained evidence for the former possibility (26). The kidney PDH kinase catalyzes a relatively slow phosphorylation of casein, and this reaction is inhibited by pyruvate. The apparent K_i for pyruvate is about 0.07 mM. The molecular basis of these apparent differences in the K_i value for pyruvate is under investigation.

Our findings suggest that the intramitochondrial concentrations of pyruvate and uncomplexed Mg^{2+} play important roles in regulating the activity of the PDH kinase and the PDH phosphatase, respectively. The total content of magnesium in rat liver mitochondria is 20-25 nmoles/mg protein (27). About one-half of this magnesium is in the matrix. This amount is approximately equal to the amount of mitochondrial adenine nucleotides (10-15 nmoles/mg protein). The concentration of free Mg^{2+} in the matrix may be determined, at least in part, by the ATP/ADP ratio, since ADP forms a much weaker complex with Mg^{2+} than does ATP. Obviously, we need to know more about changes in the concentration of uncomplexed Mg^{2+} in the matrix under physiological conditions.

Interconversion of Active and Nonactive Forms of Pyruvate Dehydrogenase in vivo

There have been recent reports from the laboratories of Wieland (28,29), Söling (30), Jungas (17), and Randle (19) that interconversion of the active (nonphosphorylated) and nonactive (phosphorylated) forms of the pyruvate dehydrogenase complex (PDC) occurs under in vivo conditions and that this interconversion is under metabolic and hormonal control. PDC activity decreased markedly in rat heart and kidney during starvation, and the activity was restored to normal on refeeding glucose or fructose (28). PDC activity fell to less than 15% of the total activity in heart and kidney of alloxan diabetic rats on withdrawal of insulin, and the activity was restored to normal following insulin treatment (28). Intravenous injection of fructose led to a significant increase in PDC activity in rat liver within a few minutes without affecting total PDC activity (30). Significant increases were observed in PDC activity extracted from rat epididymal fat-pads briefly exposed to insulin (17,19,29). In all of these investigations PDC activity was measured in homogenates of the tissues. In certain cases PDC activity was determined in the presence and absence of 10-15 mM Mg^{2+}. An increase in PDC activity in the presence of Mg^{2+} was attributed to dephosphorylation and concomitant reactivation of the phosphorylated (nonactive) complex.

Although these results are highly suggestive, the need for caution in interpreting these data should be emphasized. In experiments with purified preparations of the pyruvate dehydrogenase complex and its regulatory enzymes, we have observed that under optimal conditions the rates of phosphorylation or dephosphorylation are relatively fast. It seems that the possibility has not been ruled out that interconversion of the active and nonactive forms of the pyruvate dehydrogenase complex occurs during the course of isolating tissues from animals or of extracting the complex for assay. A more direct demonstration of the interconversion of the active and nonactive forms of the pyruvate dehydrogenase complex in vivo is desirable. There are technical problems to be overcome before this objective can be achieved. In any event, the relationship between the effects produced by starvation, intravenous injection of fructose, exposure to insulin, etc. and the phosphorylation-dephosphorylation cycle remains to be established. It would appear that some, if not all, of the effects are indirect.

Acknowledgments

This investigation was supported in part by Grant GM06590 from the U. S. Public Health Service.

References

1. REED, L.J., in "Current Topics in Cellular Regulation" (B.L. Horecker and E.R. Stadtman, eds.), Vol. 1, p. 233, Academic Press, New York, 1969.
2. GUNSALUS, I.C., in "The Mechanism of Enzyme Action" (W.B. McElroy and H.B. Glass, eds.), p. 545, Johns Hopkins Press, Baltimore, 1954.
3. LINN, T.C., PELLEY, J.W., PETTIT, F.H., HUCHO, F., RANDALL, D.D., and REED, L.J., Arch. Biochem. Biophys., in press.
4. SMOLY, J.M., KUYLENSTIERNA, B., and ERNSTER, L., Proc. Nat. Acad. Sci. U.S.A., 66, 125 (1970).
5. ISHIKAWA, E., OLIVER, R.M., and REED, L.J., Proc. Nat. Acad. Sci. U.S.A., 56, 534 (1966).
6. HAYAKAWA, T., KANZAKI, T., KITAMURA, T., FUKUYOSHI, Y., SAKURAI, Y., KOIKE, K., SUEMATSU, T., and KOIKE, M., J. Biol. Chem., 244, 3660 (1969).
7. BARRERA, C.R., NAMIHIRA, G., HAMILTON, L., MUNK, P., ELEY, M.H., LINN, T.C., and REED, L.J., Arch. Biochem. Biophys., in press.
8. LINN, T.C., PETTIT, F.H., and REED, L.J., Proc. Nat. Acad. Sci. U.S.A., 62, 234 (1969).
9. MASSEY, V., HOFMANN, T., and PALMER, G., J. Biol. Chem., 237, 3820 (1962).
10. SAKURAI, Y., FUKUYOSHI, Y., HAMADA, M., HAYAKAWA, T., and KOIKE, M., J. Biol. Chem., 245, 4453 (1970).
11. LINN, T.C., PETTIT, F.H., HUCHO, F., and REED, L.J., Proc. Nat. Acad. Sci. U.S.A., 64, 227 (1969).
12. GARLAND, P.B., and RANDLE, P.J., Biochem. J., 91, 6c (1964).
13. BREMER, J., Eur. J. Biochem., 8, 535 (1969).
14. WIELAND, O., von JAGOW-WESTERMANN, B., and STUKOWSKI, B., Hoppe-Seyler's Z. Physiol. Chem., 350, 329 (1969).
15. WIELAND, O., and von JAGOW-WESTERMANN, B., FEBS Letters, 3, 271 (1969).
16. WIELAND, O., and SIESS, E., Proc. Nat. Acad. Sci. U.S.A., 65, 947 (1970).

17. JUNGAS, R.L., *Metabolism*, 20, 43 (1971).
18. SIESS, E., WITTMANN, J., and WIELAND, O., *Hoppe-Seyler's Z. Physiol. Chem.*, 352, 447 (1971).
19. DENTON, R.M. COORE, H.G., MARTIN, B.R., and RANDLE, P.J., *Nature New Biology*, 231, 115 (1971).
20. REED, L.J., LINN, T.C., PETTIT, F.H., OLIVER, R.M., HUCHO, F., PELLEY, J.W., RANDALL, D.D., and ROCHE, T.E., in "Energy Metabolism and the Regulation of Metabolic Processes in Mitochondria" (M.A. Mehlman and R.W. Hanson, eds.), Academic Press, New York, in press.
21. HUTCHESON, E.T., BROWN, J.R., and REED, L.J., in preparation.
22. ROCHE, T.E., and REED, L.J., in preparation.
23. VOGEL, O., and HENNING, U., *Eur. J. Biochem.*, 18, 103 (1971).
24. NAMIHIRA, G., HAMILTON, L., MUNK, P., COX, D.J., and REED, L.J., in preparation.
25. SCHWARTZ, E.R., and REED, L.J., *Biochemistry*, 9, 1434 (1970).
26. HUCHO, F., RANDALL, D.D., ROCHE, T.E., BURGETT, M.W., and REED, L.J., in preparation.
27. BOGUCKA, K., and WOJTCZAK, L., *Biochem. Biophys. Res. Commun.*, 44, 1330 (1971).
28. WIELAND, O., SIESS, E., SCHULZE-WETHMAR, F.H., von FUNCKE, H.G., and WINTON, B., *Arch. Biochem. Biophys.*, 143, 593 (1971).
29. WEISS, L., LOFFLER, G., SCHIRMANN, A., and WIELAND, O., *FEBS Letters*, 15, 229, (1971).
30. SÖLING, H.D., and BERNHARD, G., *FEBS Letters*, 13, 201 (1971).

Regulation of the Mammalian Pyruvate Dehydrogenase Complex: Physiological Aspects and Characterization of PDH-Phosphatase from Pig Heart

O. Wieland, E. Siess, H. J. v. Funcke, C. Patzelt, A. Schirmann, G. Löffler, and L. Weiss

Klinisch-chemisches Institut und Forschergruppe Diabetes, Städtisches Krankenhaus, München-Schwabing/Germany

Abstract

Studies on intact rats as well as on perfused rat liver and rat heart indicate that the PDH-complex underlies reversible interconversion by phosphorylation and dephosphorylation, in vivo. With the exception of the brain the proportion of the a and b-forms is low under situations of an increased supply of fatty acid, and vice versa. These changes are in accordance with the expected change of pyruvate utilization under these conditions. However, as indicated from heart perfusion experiments with labeled pyruvate feedback control of PDH most probably by acetyl-CoA has to be considered as an important control mechanism beyond PDH interconversion.

Insulin which stimulates conversion of carbohydrate to fatty acids in liver and adipose tissue also leads to increased PDHa-formation in these tissues. Although other mechanisms cannot definitely be excluded there is evidence that in liver the effect is due to the lowering of free fatty acid levels resulting from the antilipolytic action of the hormone. Whether insulin acts on PDH-interconversion by some other mechanism, especially in adipose tissue, remains to be clarified. Inorganic pyrophosphate, in vitro, is a powerful inhibitor of ATP-dependent inactivation of purified heart muscle PDH exceeding the effectiveness of ADP or pyruvate. The physiological significance of this finding remains to be established.

PDH phosphatase was purified over 8000-fold from pig heart muscle and some of the properties of the enzyme are presented. From studies with chelating agents it is suggested that beyond Mg^{++}, another metal, probably Ca^{++} is essential for phosphatase activity.

The pyruvate dehydrogenase (PDH) reaction occupies a strategic position between the reactions of pyruvate generation and of pyruvate utilization. In animal tissues this step needs special regulation to meet the different metabolic requirements. This may be achieved by at least two mechanisms: First, by interconversion of the two molecular forms of the PDH-complex, and second, by direct control of the active form of the enzyme by acetyl-CoA or some other metabolic effectors. In the first part of this paper the possible participation of the interconversion mechanism in the regulation of pyruvate metabolism, in vivo, will be discussed. The second part deals with the purification and some of the properties of PDH-phosphatase, one of the enzymes of the interconversion cycle.

Physiological aspects of PDH-interconversion, in vivo

In these studies it seemed first of all of interest to establish the normal distribution pattern of the active and inactive forms of PDH in various rat tissues. Our method for determining the two forms of the enzyme in tissue homogenates is based on the fact that the PDH-phosphatase converts the inactive (phosphorylated) form to the active (dephosphorylated) form in presence of high concentrations of Mg^{++}. Thus, after incubation at 10 mM Mg^{++} active plus inactive, i.e. total PDH activity is obtained whereas the activity of the samples incubated without Mg^{++} is taken to represent the a-form of the complex (1).

The validity of this procedure has been substantiated by the following experiments in which ^{32}P-incorporation and release was measured in parallel to the enzymatic activity of the PDH-complex. For these studies the enzyme was purified to some extent from the homogenates without changing the ratio of the two molecular forms originally present. As shown in Table 1, the purification started from two different homogenates from pooled rat hearts: One was prepared from the hearts of fasted rats and contained a low percentage of the a-form. Another, with a high percentage of the a-form was obtained by nicotinic acid treatment of fasted donor rats prior to sacrifice. This treatment had been shown previously to normalize the lowered PDHa levels due to fasting (1). As may be seen the proportion of PDHa to PDHb remained fairly constant during the purification procedure.

In the labelling experiments illustrated in Figs. 1 and 2, a Mg^{++}-treated, activated sample and a non activated sample of the enzyme preparations were incubated with γ-^{32}P-ATP. As may be seen the enzyme aliquot which - due to phosphatase action - contained the active, i.e. dephospho form only, was able to incorporate a correspondingly higher amount of ^{32}P as compared to the non activated sample. After 40 min. as indicated by the arrow reactivation of both samples was induced by addition of phosphatase and Mg^{++}. As may be seen this resulted in an increase of PDH activities to a value approximating the original level of the activated sample before ATP-inactivation. Simultaneously, the incorporated ^{32}P was released completely and at the same rate.

Table 1

RATIO OF ACTIVE TO INACTIVE FORM OF PDH DURING PURIFICATION FROM RAT HEART.

Animal treatment		step of purification					
		crude extract		pH 5.2 precipitate		glycerol gradient fraction	
		mU/mg	$PDH_a:PDH_b$	mU/mg	$PDH_a:PDH_b$	mU/mg	$PDH_a:PDH_b$
Rats starved for 64 hrs	active form (PDH_a)	7.0	0.14	114	0.20	216	0.17
	inactive form (PDH_b)	49.5		557		1284	
Starved rats (20 hrs) treated with nicotinic acid (250 mg/kg s.c.) 2 hrs before sacrifice	active form (PDH_a)	38.8	2.4	624	2.4	769	1.5
	inactive form (PDH_b)	16.4		258		503	

Tab. 1. Homogenates from 9-10 hearts from rats pretreated as indicated were prepared as described previously (1) except that glycerol was omitted from the phosphate buffer. Solid $(NH_4)_2SO_4$, 24.5 g per 100ml crude extract, was added and the precipitate after centrifugation (5min at 15000rpm) in the Sorvall rotor type SS-34 was dissolved in a volume of 20 mM potassium phosphate buffer pH 7.0 (=buffer A) equal to 1/4 of the volume of the crude extract. After dialysis overnight against buffer A, the enzyme solution was brought to pH 5.8 by addition of 1% acetic acid and centrifuged as above. The supernatant was further acidified to reach pH 5.2. After centrifugation, the yellow precipitate was dissolved in 0.4 - 0.5 ml buffer A, and adjusted to pH 7 by adding a small volume of 1 M K_2HPO_4 solution. Prior to separation of the phosphatase from the PDH-complex by density gradient centrifugation, a portion of the enzyme solution was activated: the activation mixture containing 220 μl PDH solution and 10 μl of a 230 mM $MgCl_2$ solution was incubated at 25°C for 15 min. The control mixture contained water in place of $MgCl_2$. For further purification the activated and the non activated enzyme preparation were separately layered on top of linear glycerol gradients (10%-50%) in buffer A and centrifuged in the Spinco rotor type SW 50 for 90 min at 45 000 rpm (+4°C). Gradient fractions of 0.4 - 0.5 ml were collected and assayed spectrophotometrically (17) for PDH activity

Fig. 2 represents an inactivation-reactivation cycle when the same experiment was carried out with the preparation obtained from the starved animals containing a low percentage of the a form. Again ^{32}P-incorporation corresponded rather closely to the portion of the active form with both the non-activated and the activated enzyme preparations. These studies lend appreciable support for the conclusion that our assay procedure in fact allows to differentiate between phospho- and dephosphoforms of PDH in tissue homogenates.

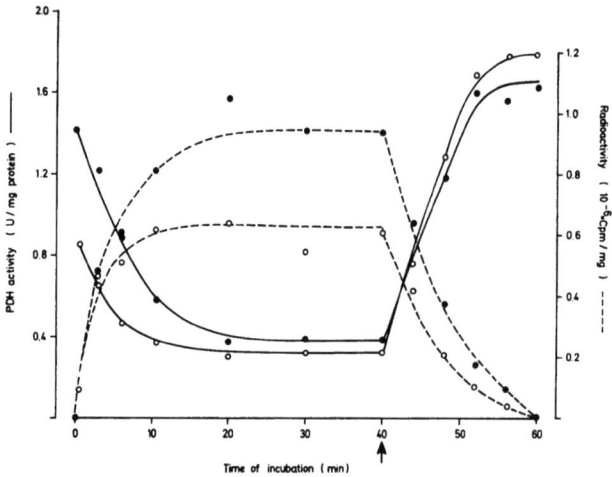

Fig. 1. Interconversion of purified rat heart PDH. PDH was purified from the hearts of 10 male Sprague-Dawley rats (240-280g) as described in the legend to Table 1. The animals were starved for 20 hrs and injected s.c. with 250 mg/kg body wt. nicotinic acid in saline (neutral pH) two hrs prior to sacrifice. The inactivation mixture contained 270 µl of activated (●) or non-activated (o) PDH in glycerol containing buffer A, respectively, 20 µl γ-^{32}P-ATP solution (corresponding to 19.8 x 10^6 cpm) to give a final concentration of 1x10^{-4}M ATP, and 10 µl of MgCl$_2$ in the same concentration. Incubation was carried out at 25°C. At the times indicated, PDH activity (——) and protein-bound radioactivity (----) of 10 µl-aliquots were determined. At the time indicated by arrow 20 µl of PDH-phosphatase purified from pig heart muscle (corresponding to 68 µg protein) and 10 µl of a 200 mM MgCl$_2$ solution were added to the remaining 160 µl of the incubation mixtures

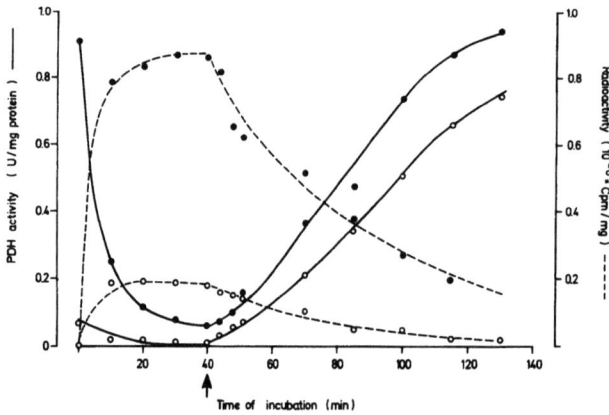

Fig. 2. Interconversion of purified rat heart PDH. PDH was purified from the hearts of 9 starved (64 hrs) Sprague-Dawley rats (240-280 g) as described in the legend to Table 1. The inactivation mixture contained 270 µl of activated (●) or non-activated (o) PDH in glycerol containing buffer A, respectively, 20 µl γ-^{32}P-ATP solution (corresponding to 16.8 x 10^6 cpm) to give a final concentration of 1x10^{-4}M ATP and 10 µl of MgCl$_2$ in the same concentration. Incubation was performed at 25°C. At the times indicated, PDH activity (——) and protein-bound radioactivity (----) of 10 µl aliquots were determined. At the time indicated by arrow 10 µl of PDH phosphatase purified from pig heart muscle (corresponding to 17 µg protein) and 10 µl of a 200 mM MgCl$_2$ solution were added to the remaining 180 µl of the incubation mixtures

The distribution of PDH in various rat tissues is illustrated in Fig. 3. As may be seen about 60% of total PDH is in the a-form in heart muscle, kidney and brain, somewhat less in adipose tissue, and only about 15% in the liver. In these studies the organs were taken from rats fed a standard laboratory chow ad libitum. The enzyme pattern is markedly changed when the animals were fasted for 24 hours prior to sacrifice. As also shown in Fig. 3 there was a pronounced drop of PDHa-levels in heart muscle and kidney, and-to a much smaller but significant extent-in liver, whereas the enzyme from brain remained essentially unchanged (2). Upon refeeding the fasted rats glucose by stomach tube PDH-activities returned to the normal values in heart muscle, kidney and liver, indicating that reversible interconversion of PDHa to b had taken place in these organs during starvation and refeeding.

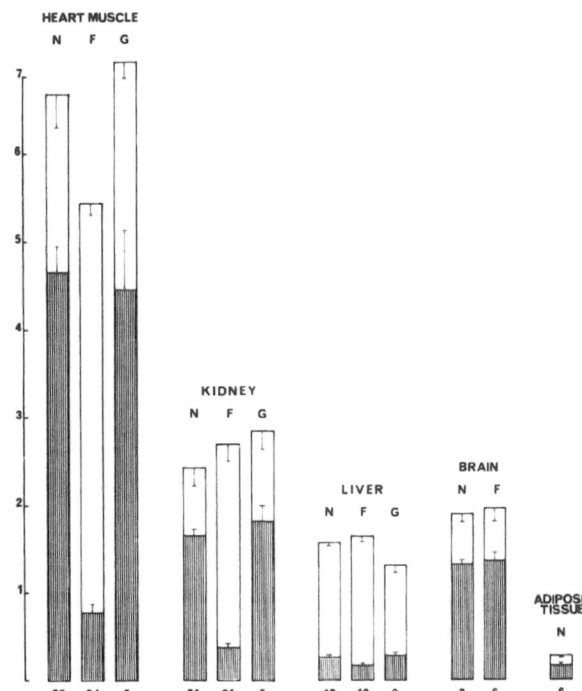

Fig. 3. Distribution of active and inactive forms of pyruvate dehydrogenase in tissues from adult rats under various nutritional conditions. Mean values ± S.E.M. are given as U/g fr. wt/min at 37°C. N=normal fed, F=24 hrs fasted, G=24 hrs fasted and refed with 3 ml of a 75% glucose solution by stomach tube. Data for heart muscle and kidney are taken from (1), for liver from (16), for brain from (2), and for adipose tissue from (12)

Control of pyruvate decarboxylation in the perfused rat heart

From previous studies on intact rats we had come to the conclusion that fatty acid oxidation, in some way, promotes the conversion of the active form of PDH to the inactive form (1). This could be demonstrated more directly in the isolated perfused rat heart upon addition of palmitate, ketone bodies, and acetate to the medium (3). Using 1-^{14}C-pyruvate, the perfusion technique seemed to open the possibility to follow the rate of pyruvate decarboxylation by the beating heart and to compare it with the activity of PDH extracted from the same organ. The results are shown in Table 2. The first finding that may be noted from the control perfusion with glucose is that PDH$\underline{a}$-activities determined in the extracts were several-fold higher than the rates of pyruvate decarboxylation as derived from $^{14}CO_2$-formation. This is not surprising since the enzyme is assayed $\underline{in}$ $\underline{vitro}$ at saturating substrate concentrations which probably do not exist in the intact myocardium. Addition of albumin-bound palmitate to the medium resulted in a decrease of PDH-activity by about 50%. However, as judged from the rate of $^{14}CO_2$-production pyruvate decarboxylation was depressed by more than 80%. Similar though somewhat smaller effects were produced by β-hydroxybutyrate instead of palmitate. We therefore suggest, that, in the beating heart, apart from PDH-interconversion another mechanism is involved in the control of the PDH-reaction.

Control of dephospho-PDH by acetyl-CoA

As to the nature of this control it seems very likely that acetyl-CoA is involved. Acetyl-CoA has been shown by Garland and Randle to inhibit PDH from heart muscle, $\underline{in}$ $\underline{vitro}$ (4). This inhibition is of the competitive type with Coenzyme A and has been verified in this laboratory to occur also with the purified enzymes from brain and liver (Fig. 4) (2,5). Acetyl-CoA concentrations required for half maximal inhibition are indicated in Table 3. These concentrations fall in the range of acetyl-CoA levels measured in heart muscle and liver. Therefore this inhibition might be an important mechanism for control of the active portion of the PDH-complex especially during an increased supply of fatty acid where an elevation of acetyl-CoA in these tissues is observed (6,7).

Regulation of PDH-interconversion in liver

In the following studies the regulation of PDH-interconversion in the liver was investigated. As already shown in Fig. 3, in the liver a much lower percentage of PDH occurs in the $\underline{a}$ form as compared to the other tissues. This may be related to the function of the liver

Table 2

EFFECT OF PALMITATE AND OF β-HYDROXYBUTYRATE ON PDH-ACTIVITY AND ON PYRUVATE DECARBOXYLATION IN THE PERFUSED RAT HEART

No. of exp.	additions to the medium	PDH-activity total	PDH_a	%	Pyruvate decarboxylation $^{14}CO_2$ formed	Lac + Pyr utilized	percent decrease of PDH_a	$^{14}CO_2$ formed	Lac + Pyr utilized
6	glucose, 5 mM	2.93±0.24	1.96±0.2	67	0.58±0.05	0.5±0.04	0	0	0
5	glucose, 5 mM palmitate 1mM	2.47±0.12	1.02±0.4$^+$	41	0.1±0.01^x	0.09±0.01^x	48	83	82
4	glucose, 5 mM D,L-β-HOB 10mM	2.80±0.25	1.22±0.23^x	44	0.13±0.01^x	0.13±0.02	38	78	74

Mean values ± S.E.M. are given as μMoles per g wet weight x min. Tracer amounts of 1-^{14}C-pyruvate (1 μCi corresponding to 85 nMol) were added for measurement of pyruvate decarboxylation. Perfusion was performed for 20 min. at 37°C. The medium (15 ml) consisted of Krebs-Henseleit-Buffer containing 2% bovine serum albumine with the additions indicated. Perfusion pressure was maintained at 60 mm Hg.

^{x} p < 0.001; $^+$ p < 0.01; ^{x} p < 0.05

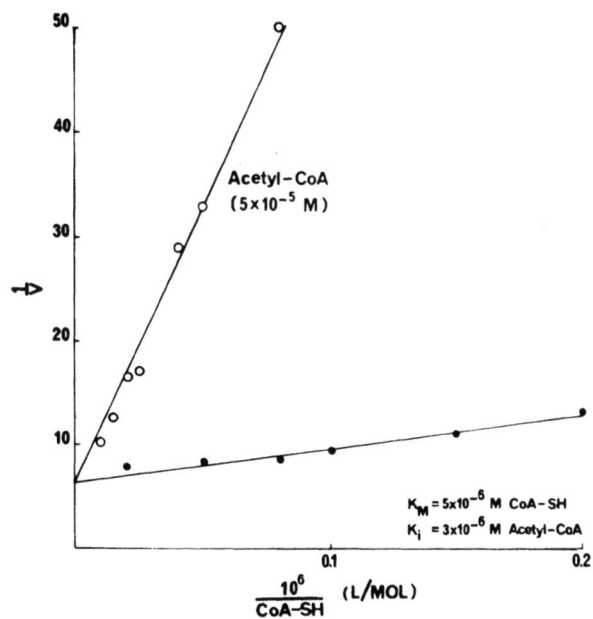

Fig. 4. Lineweaver and Burk plot of the kinetics of the inhibition of purified pyruvate dehydrogenase from rat liver by acetyl-CoA as a function of CoA-SH concentration. Micro assays (0.2ml) containing 57 μg protein were performed with pyruvate as the substrate (17). Final concentrations of acetyl-CoA and CoA-SH as indicated. Ordinate: 1/v (v=ΔE_{334} per 24 sec.)

to synthesize rather than to oxidize glucose. On the other hand, the liver can convert effectively glucose into fatty acids and it seemed therefore of interest to examine the state of the PDH-system under conditions of increased lipogenesis. In the following experiments livers were taken from fed rats after treatment with a high dose of insulin which is known to lower gluconeogenesis and to favour fatty acid synthesis. The results are illustrated in Fig. 5. As may be seen there was a marked increase of PDH$\underline{a}$ which accounted for about 50% of total activity 20 min after insulin injection. Total enzyme activity remained rather constant. These changes could explain the increased flux of pyruvate to acetyl-CoA during hyperlipogenesis.

Table 3

INHIBITION OF MAMMALIAN PYRUVATE DEHYDROGENASES BY ACETYL-CoA

SOURCE OF PDH	CoA-SH K_m (M)	ACETYL-CoA K_i (M)
HEART MUSCLE	6×10^{-6}	6×10^{-6}
LIVER	5×10^{-6}	3×10^{-6}
BRAIN	9×10^{-6}	1×10^{-5}

As to the mechanism of this $\underline{b}$ to $\underline{a}$-interconversion the possibility exists that the action of insulin is not a direct one but may rather be mediated by the lowering of plasma FFA concentrations due to the antilipolytic effect of the hormone. If so one would expect that, in turn the insulin-effect should be antagonized by exogenous supply of FFA. This could indeed be observed in the following studies where albumin-bound oleate was infused into the tail veins of the insulinized rats in order to produce high plasma FFA-levels. As may be seen in Fig. 6 PDH$\underline{b}$ to $\underline{a}$ interconversion was completely abolished under these conditions

Further investigations on that point were carried out on the isolated perfused rat liver. As illustrated in Fig. 7 the state of the PDH-system remains fairly constant in the livers from fed rats during 100 min of control perfusions without addition of substrate. Addition of glucose in concentrations of 5 -20 mM remained without significant effects (not shown here). In contrast there was a marked increase of PDH$\underline{a}$ after addition of fructose or of pyruvate. Fig. 8 illustrates the results obtained with fructose. After 80 min of perfusion with fructose about 60% of PDH existed in the $\underline{a}$-form whereas total activity was not grossly changed. Similar, yet much smaller effects were observed in livers from rats treated with fructose by Söling (8). It is well known that fructose is more rapidly utilized for lipogenesis than glucose in the liver. This could be explained by the "opening" of the PDH reaction as demonstrated here.

Fig. 5. Effect of insulin treatment on liver PDH activities. Normal fed rats were injected 10 IU of insulin each intraperitoneally. The controls received 0.9% NaCl-solution. Total height of the bars represents the total PDH activity (after activation) and the shaded areas represent PDHa. Half-shaded columns: controls; fully-shaded columns: insulin treated rats. Each value represents the mean of three experiments

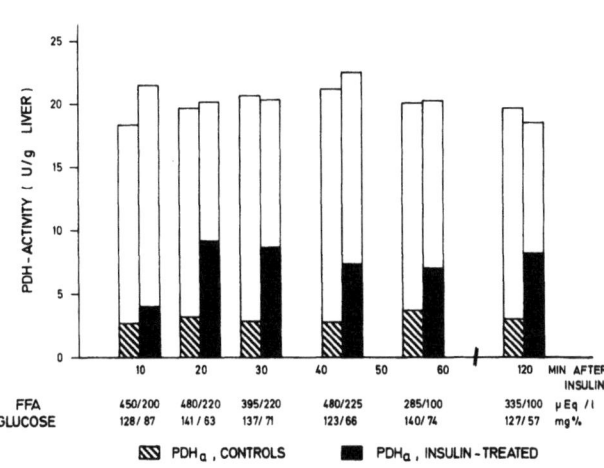

Fig. 6. Effect of oleate infusion on liver PDH in normal fed, insulin treated rats. Animals were treated with insulin as described in the legend to Fig. 5. 20 min thereafter they were anesthetized with ether and infused, into tail vein, the following preparations: Controls (C) 3 ml of a 11.5% solution of crystalline BSA; I. 3 ml of a 6.5% solution of BSA containing 20 µMol of oleate; II. 3 ml of a 10.1% solution of BSA containing 34.2 µMol of oleate; III. 3 ml of a

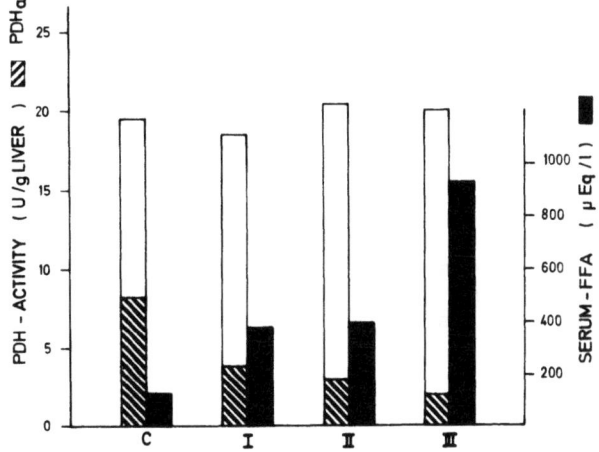

19.3% solution of BSA containing 60.9 µMol of oleate. Duration of infusion was 20 min. The fully shaded columns represent the levels of serum free fatty acids and the half-shaded columns the PDH activity as described in Fig. 5. Each value represents the mean of three experiments

In the liver perfusion system the effect of free fatty acids on PDH was also studied. Perfusion was again started with fructose in order to elevate the a-form. 40 min later oleate was added in albumin-bound form. As shown in Fig. 8 the conversion from PDHb to a due to fructose was completely reversed 20 min after addition of the fatty acid. The same effect of oleate was observed when PDHa was increased by addition of pyruvate instead of fructose to the perfusion medium.

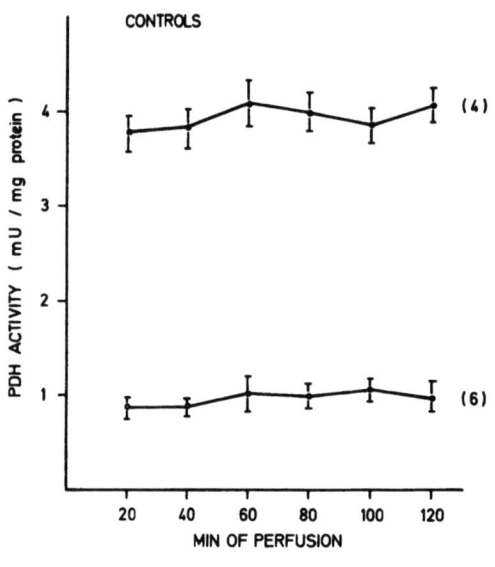 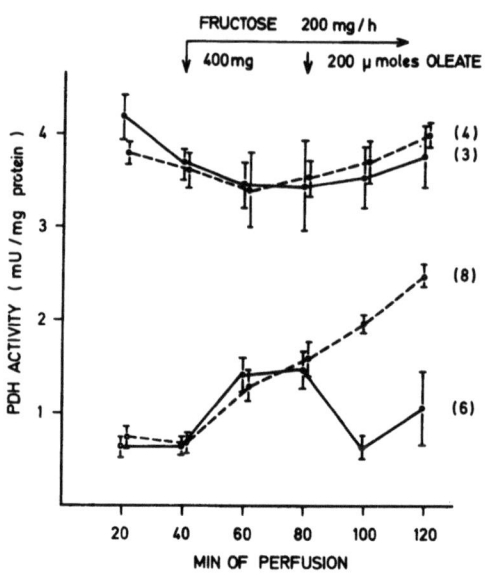

Fig. 7. PDHa and total PDH activity in perfused rat liver in the absence of substrate. Vertical bars represent S.E.M.. In the brackets the numbers of experiments are given. For methodological details see (18)

Fig. 8. Influence of fructose and fructose together with oleate on PDHa and total PDH in the perfused rat liver. Fructose and oleate were added as indicated at the time given by the arrows. Vertical bars represent S.E.M.. In the brackets the numbers of experiments are given. For methodological details see (18)

Attempts to demonstrate an effect of insulin or glucagon on the regulation of PDH in the perfused rat liver have so far been unsuccessful. It seems possible therefore that the proportion of the two molecular forms of the enzyme is controlled by metabolic intermediates rather than by hormones.

Interconversion of pyruvate dehydrogenase in adipose tissue

More recently we directed our interest on the regulation of the PDH-complex in adipose tissue. This tissue is most specialized in converting glucose to fatty acids, and hormonal effects upon this process can be readily demonstrated in vitro. Since fat cells contain more than 90% of triglyceride, a special technique had to be developed for preparing the extracts without gross dilution for enzyme measurements. This was achieved by homogenization in the presence of silicone oil (12). As illustrated in Fig. 9 the fat layer is entirely separated on top by centrifugation whereas the proteins are solubilized in the aqueous phase.

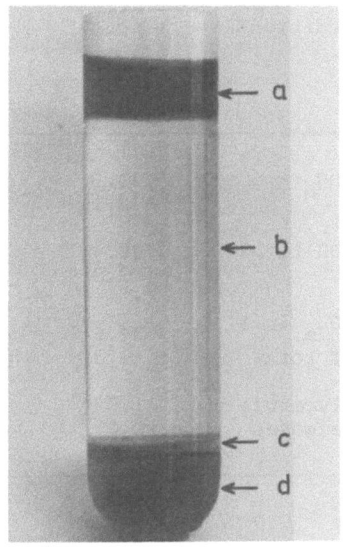

Fig. 9. Pattern of the fractions of a homogenate from adipose tissue prepared by the silicone oil technique (12). a = triglyceride; b = silicone oil; c = interphase; d = aqueous phase

Marked changes of the PDH-system were observed on incubation of adipose tissue, in vitro in the presence of insulin. In these experiments the fat pads from fed rats were preincubated, for 30 min in a basal medium, and then after addition of insulin, a second incubation of 30 min was carried out. As shown in Table 4 the percentage of a-form was raised 3-fold in the presence of insulin. The diminished glycerol release indicates that insulin had inhibited lipolysis. In order to investigate the specificity of the insulin effect other antilipolytic agents such as nicotinic acid and agmatine were used. As may be seen in Table 4 these compounds where also effective by about doubling PDHa-activities and reducing glycerol output. This opens the possibility that, similar as discussed for the liver the control of adipose tissue-PDH by insulin is also a consequence of its antilipolytic action. Similar effects of insulin on adipose tissue-PDH as described here have been observed by Jungas (13), and by Randle and his group (4). Due to the failure in obtaining an effect by the antilipolytic agent prostaglandin E_1 the latter authors suggested that insulin might interact by a special messenger with the PDH-system. This point deserves further investigation.

Table 4

EFFECT OF INSULIN, AGMATINE AND NICOTINATE ON PDH INTERCONVERSION IN ADIPOSE TISSUE IN VITRO

(Enzyme activities are expressed as mµmoles acetyl-CoA formed per g wet tissue per min at 25°C $\pm$ S.E.M.. Glycerol formation is given in µmoles per g wet tissue in 30 min at 37°C $\pm$ S.E.M.)

additions	none n=8	insulin 1 mU/ml n=8	agmatine 5×10^{-3}M n=5	nicotinate 10^{-4}M n=6
PDH_a	30.0 $\pm$ 3.1	102.4** $\pm$ 9.5	61.3*** $\pm$ 9.6	81.3** $\pm$ 7.3
PDH total	125.4 $\pm$ 13.0	145.1 $\pm$ 12.0	116.4 $\pm$ 8.4	163.4 $\pm$ 14.0
PDH_a in % of total	24.6 $\pm$ 2.3	70.8** $\pm$ 4.0	51.8** $\pm$ 5.0	49.8** $\pm$ 0.7
Glycerol released	1.03 $\pm$0.09	0.42 $\pm$ 0.05	0.72 $\pm$0.05	0.47 $\pm$0.13

** $p < 0.001$

*** $p < 0.005$

Effectors acting on the PDH-interconversion system

As observed by Reed and his group (9) and also in this laboratory (2,11) pyruvate and ADP, in vitro, protect PDH from inactivation by ATP probably by interfering with the kinase reaction. The observed increase in PDH_a in our liver perfusion studies resulting from pyruvate or fructose (the latter leading to increased levels of pyruvate) could be explained by such a protecting mechanism.

We have recently observed that inorganic pyrophosphate is a powerful inhibitor of PDH-kinase. The effect of increasing concentrations of pyrophosphate is illustrated in Fig. 10. Halfmaximal inhibition of the inactivation reaction can be derived to occur at about 20 µM pyrophosphate. Owing to the lack of information about the occurrence of pyrophosphate in animal tissues nothing can be said at the moment on the physiological significance of this phenomenon.

Likewise, the mechanism by which fatty acids promote the transition of dephospho-PDH to the phosphorylated form, is not yet understood. Since in the perfused rat heart acetoacetate, ß-hydroxybutyrate, and acetate produced the same effects as palmitate (3) it would appear, at least, that ß-oxidation is not essentially involved. In view of the Mg^{++}-dependency of the phosphatase, regulation of this enzyme by changes of the concentrations of free Mg^{++} have also to be considered. Experimental verification of this point, however, has as yet not been achieved.

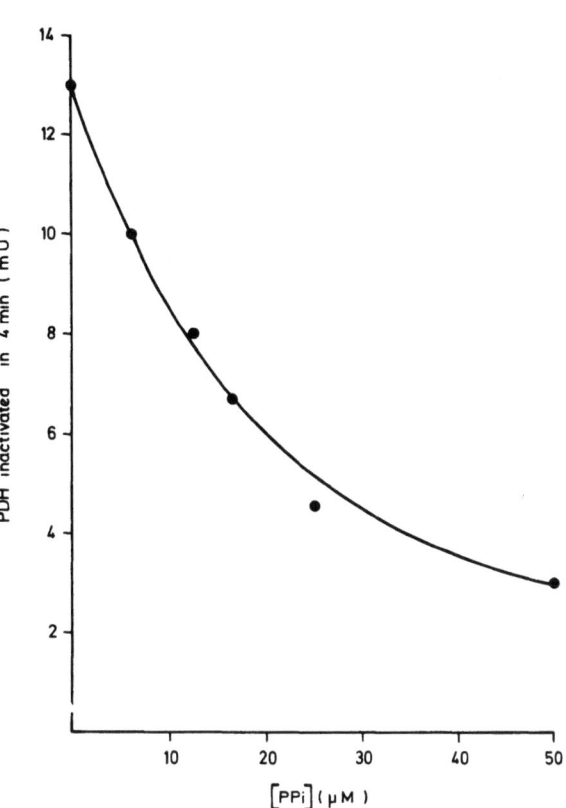

Fig. 10. Protective effect of inorganic pyrophosphate on PDH inactivation by ATP. A mixture containing 20 µl of purified pig heart PDH, corresponding to 25 mU, 5 µl of a solution of inorganic pyrophosphate in 20 mM potassium phosphate buffer, pH 7.0 to give the concentrations indicated, and 5 µl of a 1:1 mixture of 0.3 mM ATP and 12.5mM $MgCl_2$ was analyzed for PDH activity before and after incubation at 25^oC for 4 min. After hydrolysis of pyrophosphate with pyrophosphatase which yielded two moles of inorganic phosphate per mole of pyrophosphate the protective effect was abolished

Pyruvate dehydrogenase phosphatase: Purification and properties (10,15)

PDH-phosphatase was purified about 8 500-fold from pig heart muscle until apparent homogeneity as judged from polyacrylamide gel electrophoresis (Fig. 11).

The molecular weight was determined by gel-filtration on Sephadex G-100 (Fig. 12) and by SDS-gel electrophoresis to lie between 92 000 and 95 000 Daltons. No indication of a subunit structure was obtained.

The purified phosphatase is dependent on Mg^{++} or Mn^{++} with half-saturation concentrations (in histidine buffer) of 2.5 mM, and 1.8 mM, respectively. 3'5'- cyclic AMP and 5'-AMP showed no effect on the purified enzyme. Fluoride is a strong inhibitor of the phosphatase producing half

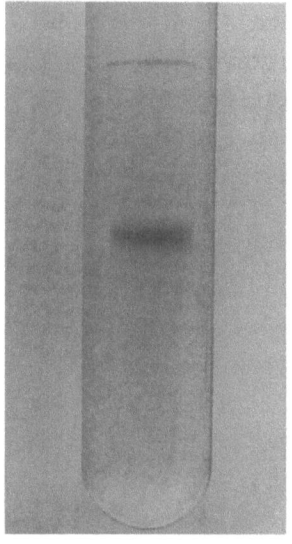

Fig. 11. Polyacrylamide gel (7.5%) electrophoresis of PDH-phosphatase purified from pig heart muscle

maximal inhibition at 0.6 mM concentration. Interestingly, fluoride acts much less in the presence of Mn^{++} instead of Mg^{++} as the activator.

Saturating kinetics were studied at increasing concentrations of ^{32}P-labelled phospho-PDH as the substrate. As shown in Fig. 13,

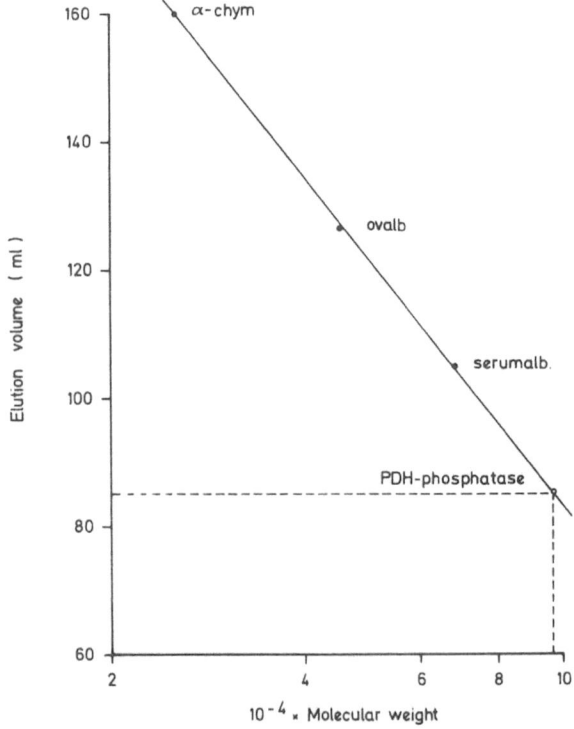

Fig. 12. Molecular weight estimation of PDH-phosphatase by gel filtration. PDH-phosphatase (o) and proteins of known molecular weight (●) were eluted from a Sephadex G-100 column (2x80 cm) with 20 mM potassium phosphate buffer containing 1 mM $MgCl_2$ and 2 mM 2-mercaptoethanol. α-chym = α-chymotrypsinogen, ovalb = ovalbumin, serumalb = bovine serum albumin

Fig. 13. Effect of substrate concentration on PDH-phosphatase activity. Substrate concentration is expressed as the concentration of the phosphate moiety calculated from the ^{32}P content of ^{32}P labelled PDH. This was prepared by incubating, at 25°C, a mixture of 460 µl of a PDH solution, corresponding to 6.9 mg protein (spec. act. = 3.7 U/mg), 30 µl γ-^{32}P-ATP, 3.3 mM, corresponding to 10.76 x 10^6 cpm, and 10 µl of MgCl$_2$, 5 mM. After complete inactivation of the enzyme, excess ATP was removed by dialysis against several changes of 20 mM potassium phosphate buffer pH 7.0 two liters each, and finally against two liters of 50 mM histidine-Cl buffer, pH 6.9, overnight. After dialysis the PDH preparation contained 1.74 nmoles ^{32}P per mg protein. Reactivation was assayed by incubating a mixture of 5 µl of PDH-phosphatase, corresponding to 0.15 µg protein, 5 µl of MgCl$_2$, 40 mM, and 10 µl of ^{32}P-PDH in the concentrations indicated at 25°C for 5 min. Re-

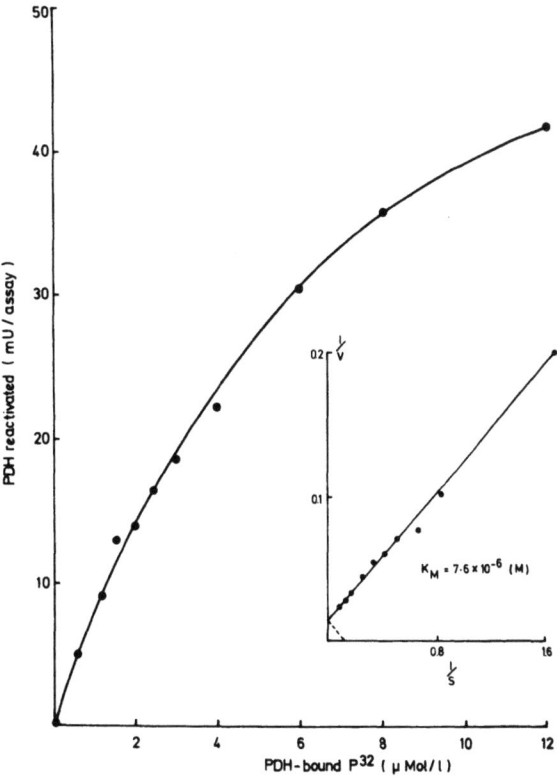

activated PDH was measured spectrophotometrically (17). Reactivation proceeded at a constant rate for at least 7.5 min

a hyperbolic curve was obtained from which an apparent $K_M = 7.6 \times 10^{-6} M$ is derived. There is also a high specificity of the reaction phospho-PDH representing so far the only substrate for the phosphatase. No dephosphorylation was observed with the following compounds: Acetylphosphate, inorganic pyrophosphate, L-α-glycerophosphate, D,L-o-phosphothreonine, o-phosphoryl-ethanolamine, p-nitrophenylphosphate, G-6-p, F-6-p, F-1,6-dp, AMP, ADP, ATP, 3'5'-cAMP, NAD$^+$, casein, phosvitin, glycogen phosphorylase a.

Finally, studies on the effect of chelating agents on PDH-phosphatase may deserve some interest. This is shown in Table 5. As may be seen phosphatase activity was reduced by more than 80% in the presence of EGTA and this was completely prevented by Ca^{++}, but not by Mg^{++}. EDTA was also effective but only after preincubation with the phosphatase. These findings could be explained by assuming that PDH-phosphatase contains a metal which, beyond Mg^{++} or Mn^{++}, is essential for its function. It seems not unreasonable to suggest that calcium might represent the essential factor.

Table 5

EFFECT OF EGTA AND EDTA ON PDH-PHOSPHATASE

Additions	Time of preincubation (min)	PDH reactivated (mU/assay)	PDH-phosphatase activity (%)
none	0	90.0	100
EDTA, 0.5 mM	0	90.0	100
EGTA, 0.5 mM	0	16.7	18.6
Ca^{++}, 0.66 mM plus EGTA, 0.5 mM	0	101.7	113
EGTA, 0.5 mM plus Ca^{++}, 0.66 mM	0	95.0	105.6
Mg^{++}, 0.63 mM plus EGTA, 0.5 mM	0	16.7	18.6
Ba^{++}, 0.65 mM plus EGTA, 0.5 mM	0	27.0	30.0
none	15	61.7	68.6
EDTA, 0.5 mM	15	26.7	29.7
EGTA, 0.5 mM	15	18.3	20.3

Acknowledgement: The authors wish to thank Mrs. R.Milfull, Mrs. U.Zechmeister, Mrs. B. v.Jagow-Westermann, and Mrs. E.Sincini, Miss J.Mayr, Miss G.Stejskal and Miss E.Hungbauer for their skillful technical assistance.

This work was supported by the Deutsche Forschungsgemeinschaft, Bonn Bad Godesberg, Germany.

References

1. Wieland, O., Siess, E., Schulze-Wethmar, F.H., v.Funcke, H.J., and Winton, B., Arch.Biochem.Biophys. 143, 593 (1971).
2. Siess, E., Wittmann, J., and Wieland, O., Hoppe-Seyler's Z. Physiol.Chem. 352, 447 (1971).
3. Wieland, O., v.Funcke, H.J., and Löffler, G., FEBS-Letters 15, 295 (1971).
4. Garland, P.B., and Randle, P.J., Biochem. J. 91, 6C(1964).
5. Wieland, O., Proceedings of the Seventh Congress of the International Diabetes Federation, Buenos Aires, August 1970, in press.
6. Randle, P.J., Garland, P.B., Hales, C.N., Newsholme, E.A., Denton, R.M., and Pogson, C.I., Recent Progr. Horm. Res. 22, 1 (1966).
7. Wieland, O., and Weiss, L., Biochem.Biophys.Res.Comm. 10, 333 (1963).
8. Söling, H.D., and Bernhard, G., FEBS-Letters 13, 201 (1971).

9. Linn, T.C., Pettit, F.H., Hucho, F., and Reed, L.J., Proc. Nat. Acad.Sci. U.S.A. 64, 227 (1969).
10. Siess, E., presented at the Symposium on "Chemical and Physical Studies on Enzymes", Reisensburg Castle (Germany) 28.-30. Sept. 1971.
11. Wieland, O., and v.Jagow-Westermann, B., FEBS-Letters 3, 271 (1969).
12. Weiss, L., Löffler, G., Schirmann, A., and Wieland, O., FEBS-Letters 15, 229 (1971).
13. Jungas, R.L., Metabolism 20, 43 (1971).
14. Denton, R.M., Coore, H.G., Martin, B.R., and Randle, P.J., Nature New Biology 231, 115 (1971).
15. Siess, E., and Wieland, O., Eur. J.Biochem., submitted.
16. Wieland, O., Patzelt, C., and Löffler, G., Eur.J.Biochem., submitted.
17. Wieland, O., v.Jagow-Westermann, B., and Stukowski, B., Hoppe-Seyler's Z. Physiol.Chem. 350, 329 (1969).
18. Patzelt, C., Löffler, G., and Wieland, O., Proceedings of the 1st European Meeting on the Technique of Perfusion of Isolated Liver and its Application. Milan, 15.-17. July 1971, Raven Press, New York, in press.

Discussion:

Sols

Dr. Wieland, the carbon for gluconeogenesis in the fasting animal comes from muscle proteins by way of alanine. What is the state of the pyruvate dehydrogenase complex in skeletal muscle under conditions of gluconeogenesis?

O. Wieland

We have not yet studied pyruvate dehydrogenase in skeletal muscle.

Dixon

Dr. Reed, in view of the marked difference between the kinase involved in the control of pyruvate dehydrogenase and Ed. Krebs's 3',5'-cyclic-AMP dependent protein kinase just a general question: Is there any process in mitochondria which is affected by cyclic-AMP?

Reed

I really can't answer, may be Ed. Krebs can. I think I asked Earl Sutherland and he was not aware of any cyclic-AMP effect in mitochondria.

Dixon

With histones, besides phosphorylation there is also acetylation especially of lysines. Could pyruvate dehydrogenase kinase be controlled by acetylation in view of the feedback inhibition by acetyl-CoA?

Reed

Acetyl-CoA does not affect the kinase.

Reinauer

We found a sedimentation coefficient of 78 S and in the electron microscope a diameter of the pyruvate dehydrogenase complex of 420 Å in agreement with low angle x-ray scattering measurements. From comparison with models of the acetyltransferase and the whole complex it appears that there are two protein shells around the pentagonal dodecahedron of the lipoyl-transferase. The question arises as to the real size of the whole complex? What is the molecular weight of the complex?

Reed

We have not determined the M.W. directly, but Koike did with the pig heart complex. I think $7,4 \times 10^6$ is the latest value. I would estimate, from various data, a value probably somewhat over 8×10^6. But there is a hole in the transacetylase. This hole you also see in the complex.

Reinauer

With the perfused guinea pig heart we found a rapid conversion of the pyruvate dehydrogenase b to a by epinephrine, whereas in your experiments, Dr. Wieland, the b to a conversion proceeds more slowly. Have you seen an effect of epinephrine on pyruvate dehydrogenase?

O. Wieland

We have not yet studied the effect of epinephrine.

Krebs

I have a question either for Dr. Wieland or Dr. Reed.
Can you calculate whether or not the amount of pyruvate dehydrogenase in the mitochondria is sufficient to account for all of the phosphate incorporated in vivo?

Lynen

Nobody seems to want to answer that.

Segal

Prof. Wieland I wonder if you could clarify the EGTA and EDTA experiments. Are you suggesting that Ca^{++} is required as well as Mg^{++}?

O. Wieland

These experiments were done as follows: Phosphatase and EGTA were first mixed and then, after dilution, aliquots were removed for measurements of phosphatase activity. Thus EGTA was diluted out so that it could not interfere with the high concentration of Mg^{++} present in the phosphatase assay. Therefore the inhibitory effect of EGTA cannot be explained by complexing Mg^{++}. Since EGTA could be knocked out by Ca^{++} it would seem that the enzyme requires both Mg^{++} and Ca^{++} (or some other divalent metal).

Seubert

Dr. Wieland, does the correlation between pyruvate dehydrogenase a and fatty acids only hold for heart muscle or does it also apply to liver?

O. Wieland

It is the same in kidney and in liver, although quantitative differences exist.

Seubert

Then this could explain the stimulatory effect of fatty acids on gluconeogenesis. Your K_i for acetyl-CoA and pyruvate dehydrogenase is low, thus pyruvate dehydrogenase should be completely inhibited.

O. Wieland

Inhibition of pyruvate dehydrogenase by acetyl-CoA is reversed by CoA-SH. Thus the rate of pyruvate decarboxylation, in vivo, will depend on the ratio of acetyl-CoA to CoA-SH rather than on acetyl-CoA alone. Another relevant point is of course compartmentation. I think that the stimulatory effect of fatty acids on gluconeogenesis might be explained by a decrease of pyruvate dehydrogenase a in the liver perfused with high concentrations of pyruvate or lactate.
Under these conditions pyruvate dehydrogenase b is converted to a. (C. Patzelt, G. Löffler and O. Wieland, Proc. of the 1st Eur. Meeting on the Technique of Perfusion of Isolated Liver and its Application, Raven Press, New York, in press). On addition of fatty acid pyruvate dehydrogenase a decays to the basal low level and preserves pyruvate for glucose synthesis. At lower concentrations of pyruvate where the activity of pyruvate dehydrogenase is low anyway, this control mechanism would be less effective.

Sols

With regard to the sparing of pyruvate for gluconeogenesis, I should like to emphasize that most of the pyruvate comes to the liver from other tissues. Therefore sparing of pyruvate through control of pyruvate dehydrogenase in peripheral tissues could be decisive regardless whether liver pyruvate dehydrogenase is controlled or not.

O. Wieland

Yes, if pyruvate in peripheral tissues is spared, more is made available to the liver. However, in order to channel pyruvate into gluconeogenesis, control of pyruvate dehydrogenase in liver and kidney is also required.

Graves

Dr. Reed, did I understand you correctly that the β-subunit of pyruvate dehydrogenase catalyzes the transfer of the hydroxyethyl intermediate but not the formation of it. If that is true then how does it get there?

Reed

The second step is not a transport step, it is a reductive acetylation. The hydroxyethyl moiety can not come off during catalysis. I didn't go into that.

NN

Will the phosphatase dephosphorylate the tetradekapeptide?

Reed

Yes, very slowly.

Cori

Dr. Reed, with 60 subunits in the complex you should have phosphodephosphohybrids present all the time. Kinetically it is unlikely that you have complete phosphorylation of one complex before another one is phosphorylated or dephosphorylated.

Reed

We have not looked at that but I would assume that you have an equilibrium ratio depending on the binding constant for the kinase and the phosphatase and pyruvate dehydrogenase.

Pette

Dr. Wieland, did you study the influence of fatty acids or carnitine esters on isolated mitochondria?

O. Wieland

We are just in the process of studying pyruvate dehydrogenase in isolated mitochondria. From earlier work by Garland, Bremer, and Dr. v. Jagow in our laboratory it is known that fatty acids and carnitine esters inhibit pyruvate oxidation in isolated liver mitochondria. Therefore we expect that pyruvate dehydrogenase interconversion in mitochondria can be triggered by fatty acids or their carnitine esters.

Hasilik

Dr. Reed mentioned that the pyruvate dehydrogenase complex in bacteria is not interconvertible. We have looked at interconversion of pyruvate dehydrogenase in bakers yeast and S. carlsbergensis, likewise with negative results.

Krebs

I have a question for either Dr. Reed or Dr. Wieland. Can all the phosphatase and kinase in the mitochondria be accounted for by what is bound to the pyruvate dehydrogenase or is there an excess of either enzyme?

Reed

The kinase is tightly bound to the complex. Of course, whether there is a non-active form we can't say because there is no way for testing. The phosphatase is loosely bound and you can get it off easily. It does not seem that there is an excess of either kinase or phosphatase. There are probably about 15 molecules of pyruvate dehydrogenase complex per mitochondrium. There would be five kinases and five phosphatases molecules for each molecule of complex.

Krebs

How do you know that you dont have a situation similar to that of phosphorylase kinase from muscle in which a trace amount of kinase is acting catalytically in the phosphorylation reaction. What is the evidence for the stoichiometry?

Reed

The kinase that is attached to the complex can be removed. Based on the activity of the purified kinase and assuming a M.W. of about 1oo.ooo based on a S value of 5,5 you can calculate the number of kinase molecules bound to the complex. The same calculation may be done with the phosphatase.

Hess

In an attempt to localize the consequences of the ATP inhibition of the pyruvate dehydrogenase I have analyzed recently the states of NAD/NADH as well as of bound FAD/FADH in a preparation of pyruvate dehydrogenase from heart, which was prepared by Professor Wieland and his collaborators. Both components were continuously analyzed with a double-beam technique as shown on the first figure, in which a record of an experiment is represented. The transmission change of NADH is recorded at 340 mµ, the fluorescence emission of bound FAD is simultaneously measured with an excitation beam selected with a filter of 420 mµ and a fluorescence emission filtered at 530 - 3000mµ.

The purified preparation of pyruvate dehydrogenase was incubated in the presence of a small amount of lactate dehydrogenase for recycling of NADH being formed during pyruvate dehydrogenation. On addition of pyruvate to the system a rapid formation of NADH as well as FADH (a decrease of fluorescence) is observed, which cycles back as soon as the pyruvate falls below the Michaelis constant of $2 \times 10^{-5}M$ for the pyruvate dehydrogenase reaction, but still serves as a substrate for the recycling of NADH in the lactate dehydrogenase reaction. (experiments see B. Hess et al. Hoppe-Seyler's Z. Physiol. Chem. 351, 515 (1971) Fig. 19).

After addition of ATP in excess to a second sample, the system is inhibited and the addition of pyruvate does not change anymore the steady-state of FAD and only a very slow cycle of NADH is induced, indicating the inhibition of the enzyme complex by phosphorylation in accord with the experimental results of O. Wieland and his collaborators as well as L. Reed and his collaborators (see fig. 2, unpublished observations). A comparison of fig. 1 and 2 shows that the initial level of FAD fluorescence is appreciably lower in case of the inhibited enzyme compared to the high initial fluorescence of the uninhibited preparation. We therefore analyzed the reaction of pyruvate dehydrogenase to the addition of ATP in more detail and

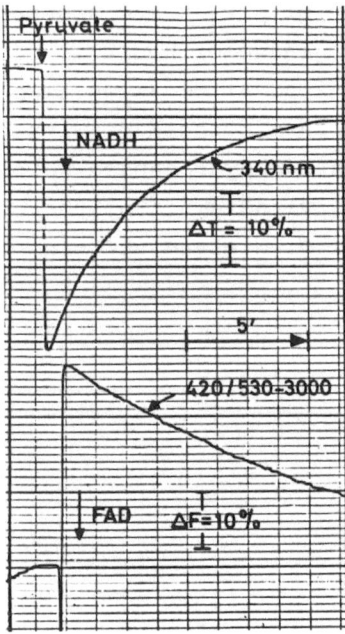

Fig. 1. Pyruvate dehydrogenase of pig heart (specific activity 0.6 U/mg protein, 11 mg protein/1.5 ml buffer, + CoA SH, thiamine pyrophosphate, Mg^{++}, NAD (each 7×10^{-5}M pyruvate added as indicated.) Fluorescence unit: 10% of total fluorescence (titrated with dithionite). Fluorescence as absorbancy increase indicated by a downward deflection (d = 1 cm, 22°C)

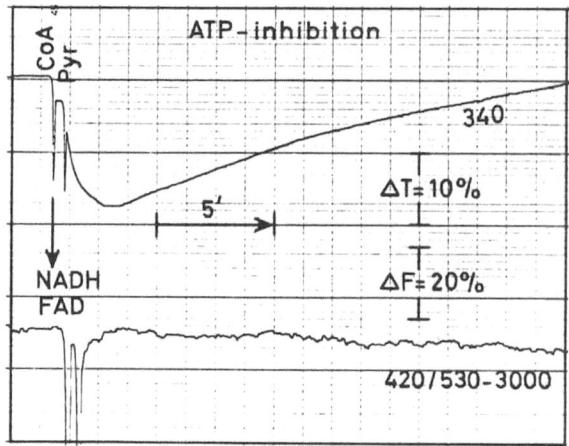

Fig. 2.   Conditions as given in fig. 1, ATP 0.33 mM (experiment 101/1C/70)

found a clear response of the FAD fluorescence as shown in the third figure. Here, the addition of ATP in the absence of pyruvate to a sample of pyruvate dehydrogenase of heart is followed by a rapid, slight change of the NADH absorption. After a lag time of about 5 sec FAD fluorescence slowly disappears coinciding with the inhibition of the overall activity of pyruvate dehydrogenase which was followed during the time course in samples withdrawn from the cuvette and analyzed according to the method of O. Wieland and his collaborators. The

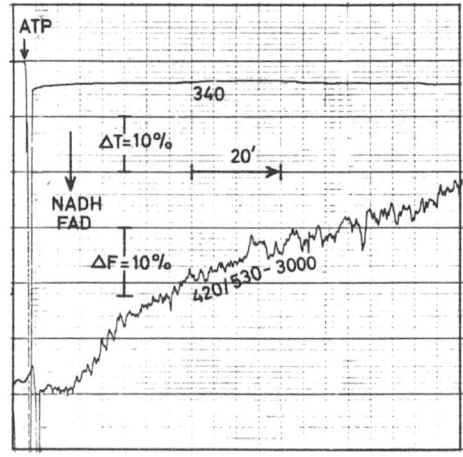

Fig. 3.   Pyruvate dehydrogenase of pig heart (specific activity 3.2 U/mg protein, 1.2 mg enzyme /1.5 ml buffer. 0.33 mM ATP (experiment 111/40/70)

fluorescence quenching on addition of ATP to pyruvate dehydrogenase
could be due to a conformation change of the binding sites for FAD
of the complex following the phosphorylation of pyruvate decarboxylase
units and subsequent dissociation of bound FAD. The latter is supported
by an experiment in which a mixture of pyruvate dehydrogenase and
α-ketoglutarate dehydrogenase as kindly supplied by O. Wieland and
his collaborators, was analyzed with the same technique. It could
clearly be shown as demonstrated in fig. 4, that the FAD is released
from the pyruvate dehydrogenase complex after ATP inhibition and
readily utilized by the α-ketoglutarate dehydrogenase complex. The

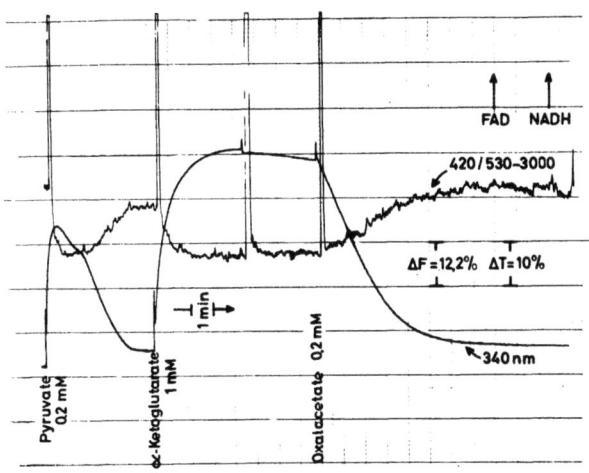

Fig. 4.   For explanation see text. 0.74 mM ATP was added 10 min.
before the reaction was started with pyruvate. PDH overall
activity was 0.25 U/mg. (experiment 121/I/70)

record shows the response of pyruvate dehydrogenase after partial
inhibition with ATP to the addition of pyruvate which does not release
anymore the full amplitude of the NADH cycle because of pyruvate
dehydrogenase inhibition. Then, on addition of α-ketoglutarate a full
reduction of NADH and FAD is initiated, which is later reversed by
addition of oxalacetate. It was found, that the total amount of FAD
being available in the system on the basis of dithionite reduction
does react after inhibition of pyruvate dehydrogenase now with
α-ketoglutarate dehydrogenase indicating a dissociation of FAD from
the pyruvate dehydrogenase complex after ATP inhibition. It would
be interesting in further studies to analyze the binding constant of
FAD to the lipoamide dehydrogenase moiety of the intact complex after
inhibition by ATP (B. Hess, unpublished experiments).

Enzyme-dependent Activation of Pyruvate Formate-lyase of *Escherichia coli*

Joachim Knappe, Hans-Peter Blaschkowski, and Rudolf Edenharder

Fachgruppe Biochemie der Universität Heidelberg, Heidelberg/Germany

The metabolic step by which pyruvate is converted to acetyl-CoA and formate according to eq. (1)

$$\text{pyruvate} + \text{CoA} \rightleftharpoons \text{acetyl-CoA} + \text{formate} \qquad (1)$$

is apt to attract some interest from at least two aspects despite the fact that its occurence is limited to certain bacteria. First, it is a unique type of reaction of an α-oxocarbonic acid, raising the question of its chemical mechanism. Second, in Escherichia coli it is subject to regulation, being operative only in the anaerobic state of the cell. It is established now, that a single enzyme species is responsible for this reaction, which is extremely sensitive towards inactivation by air (and other oxidants). This property immediately explains why the reaction does not proceed in the aerobic cell, but it is also responsible for the present low status of knowledge about the enzyme's structure and mode of action.

This report intends to briefly survey studies of our laboratory on this subject that have in particular been concerned with an enzymatic system by which the oxygen-inactivated formate-lyase is retransformed into the active state.

The Activation System

The first step towards the detection of this 'repair' process was the identification of S-adenosylmethionine as an indispensible cofactor of the formate-forming reaction when enzyme preparations were employed that were prepared from the cell extract without exclusion of air [1]. A rigorous protein fractionation subsequently led to not less than four proteins, 'enzymes I to IV', which only in combination yielded formate-lyase activity. The complex system could finally be disclosed to operate according to the following scheme [2]:

$$\text{Enzyme I} \xrightarrow{\text{Enzyme II(Fe), adenosylmethionine plus 'reducing system'}} \text{Lyase.}$$

Enzyme I, a species of molecular weight about 140000, was identified as the inactive form of the formate-lyase which is converted to the catalytically active form upon interaction with enzyme II, a species of molecular weight about 30000, which requires a preincubation with ferrous ion. Adenosylmethionine is involved in this process apparently as an allosteric effector. Enzymes III and IV were found to play the more indirect role of providing a reducing milieu (as evidenced by their replaceability by certain metal-thiol complexes). These proteins have recently been identified as flavodoxin and a thiamine diphosphate-dependent pyruvate: flavodoxin oxidoreductase [3].

How important it is not only to exclude oxygen from the activation reaction system but also to maintain the oxidation-reduction potential of the medium at a certain negative value, can be seen from the experiments in Fig. 1. Fe^{2+}/dimercaptopropanol was employed as the 'reducing system', the potentials being varied by previous addition of (traces of) oxygen. (The potentials of the complete reaction media were measured with a platinum electrode; only those experiments are plotted, where the potential did not change more than $\pm$ 10 mvolts during the whole activation period of 30 min.) Active lyase is therefore only formed at potentials more negative than about -390 mvolts, the maximum amount being observed at -440 mvolts. On the other hand, the lyase once formed and then complexed with its substrate pyruvate is much more stable. It breaks down, reforming the enzyme I-state, only at potentials more positive than about -170 mvolts. This large difference is of considerable practical value. For example, the lyase species can safely be trapped by adding pyruvate to the activation system.* In addition, the assay of its catalytic activity (via production of formate or ^{14}C-formate-pyruvate exchange) can be run at more positive potentials and by this means can be separated from the preceding activation reaction.

The list of 'reducing systems' found suitable to achieve the required potential which we have employed in our studies, comprises:
a) 3 mM Fe^{2+} plus 12 mM 2.3-dimercaptopropanol (this complex was previously suggested as a redoxbuffer for formate-lyase studies [5]);
b) 4 mM Co^{2+} plus 12 mM dithiothreitol; c) photoreduced ferredoxin or flavodoxin (employing chloroplast fragments and ascorbate/DCPI);

---

*The protecting effect of pyruvate was previously reported [4].

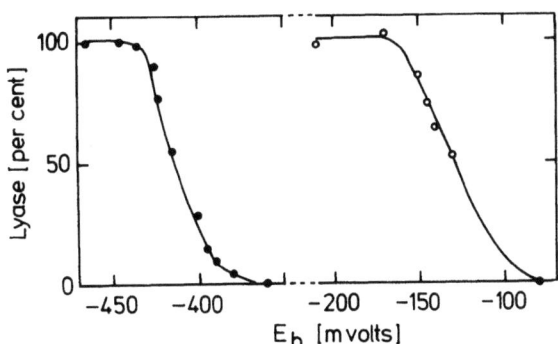

Fig. 1. Activation of lyase (•—•) and stability of the lyase-pyruvate complex (o—o) as a function of the oxidation-reduction potential of the buffer medium at pH 7.6

d) reduced flavodoxin (employing pyruvate:flavodoxin oxidoreductase and pyruvate/CoA). Since c) and d) contain further proteins they are obviously unsuitable for any closer examination of the activation process. A) and b) are the most convenient ones as they can directly be formed in the reaction tube from their components; b), however, forms precipitates, and a) has a more limited redox buffer capacity. On the basis of the amount of active lyase formed from enzymes I and II, the photosystem c) appeared to be definitely better than the metal complexes (Fig.2). It has recently been demonstrated that the difference stems from a higher efficiency of enzyme II; the amount of enzyme II required to completely transform enzyme I being least in this system.

Reducing System	Enzyme I (µg)	Enzyme II* (µg)	LYASE formed (units)
Co^{2+}- dithiothreitol	16	50	1.8
Fe^{2+}- dimercaptopropanol	16	50	1.2
Photoreduced Flavodoxin	16	50	8.6
	800	50	395

Fig. 2. Efficiency of various 'reducing systems' in the activation reaction. Time of activation, 60 min; units of lyase correspond to µmoles formate formed per hour at 28°C

Enzyme II has so far been purified about 200 fold but its state of purity is probably still very low. Its main feature is, that having been carried through several purification steps, it needs a preincubation with ferrous ion and thiols like dithiothreitol in order to be active in mediating the formate-lyase activation. It is a specific, not an oxygen-removing effect of ferrous ion (cobalteous ion, e.g.; competitively interferes), which suggests that Fe becomes bound to this protein as a prosthetic group. (It would redissociate only sluggishly, since the enzyme, once activated, can be incubated with chelating reagents and gelfiltrated without serious loss of activity; see Fig.3.)

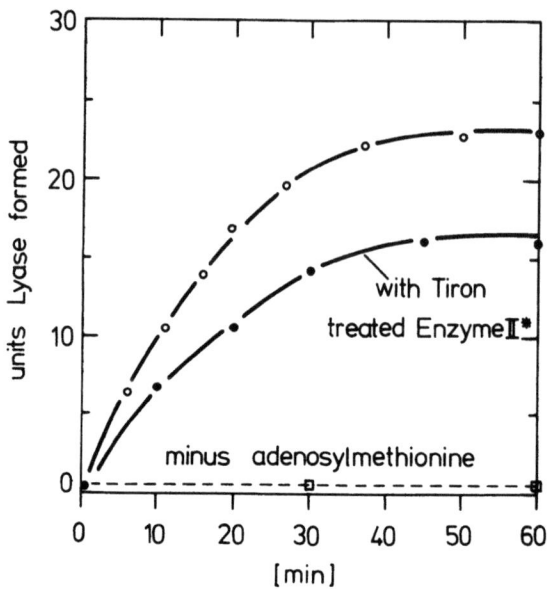

Fig. 3. Time course of activation reaction. Co^{2+}/dithiothreitol as 'reducing system'; 32 µg enzyme I, 217 µg enzyme II(Fe). -*Enzyme II(Fe) was incubated with 10 mM Tiron for 10 min, then gelfiltrated before use in this experiment

Consistent with the concept that lyase formation results from a protein-protein interaction, its rate is found to be a function of both the concentrations of enzyme I and enzyme II. The pH-optimum lies around pH 8. The K_M for adenosylmethionine is at about 10^{-5} M.

Gelfiltration on Sephadex G-100 is a convenient technique by which the lyase species can be separated from the components of the activation reaction. If the column is operated with an anaerobic buffer in which pyruvate is included, the yields are at about 80 per cent. As indicated in Fig.4, the lyase is eluted from the column at the same buffer volume as is found for the unmodified enzyme I in a separate experiment. That the activation is not accompanied by a significant change of the molecular dimensions of enzyme I had previously been suggested from the sedimentation velocities in the sucrose gradient [2]. Together with the fact that no enzyme II is detected in the lyase peak, it can thus be definitely ruled out that lyase is a complex of both proteins.

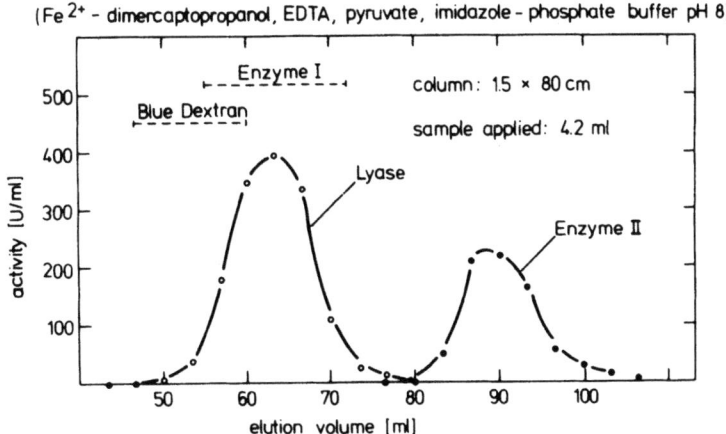

Fig. 4. Isolation of lyase by anaerobic chromatography on Sephadex G-100

Properties of Formate-Lyase

A double displacement reaction according to the following scheme

$$E + pyruvate \rightleftharpoons E \cdot "C_2" + formate \quad (1a)$$
$$E \cdot "C_2" + CoASH \rightleftharpoons E + acetyl\text{-}SCoA \quad (1b)$$

can be considered as a plausible minimal mechanism by which the pyruvate formate-lyase reaction occurs. It is suggested from the long known exchange of ^{14}C-formate with the carboxyl group of pyruvate which occurs in the absence of coenzyme A. Its rate is about 5 times higher than the overall reaction, which implies that the partial reaction where coenzyme A enters is the rate limiting step. The total reverse reaction is comparatively extremely slow (cf. [2]).

The inspection of enzyme I and of the lyase derived thereof has not resulted as yet in the detection of a prosthetic group or coenzyme. Enzyme I-preparations of a purity of about 10 per cent as judged from polyacrylamide gel electrophoresis (- this value would give an extrapolated specific activity of the homogeneous lyase of about 300 μmoles/min per mg) show normal protein spectra. Previous dialysis against 6 M urea (also of enzyme II-preparations) by which most conceivable coenzymes would be dissociated from the protein, does not prevent the transformation into lyase. Direct analysis for thiamine diphosphate were, as previously reported [2], surprisingly negative. We have also not detected 4'-phosphopantetheine as a prosthetic group linked covalently to the protein which could conceivably be involved in an acetyl-transfer step. (Note that $E \cdot "C_2"$ is not necessarily an acetyl derivative; it could also be an acetaldehyde derivative if the enzyme would start the catalytic turnover by acting as a reducing agent when splitting the C-C bond.) Furthermore, no radioactivity was found with the lyase peak (cf. Fig.4) when S-adenosyl-[Me-^{14}C]-methionine or an enzyme II, pre-incubated with ^{59}FeSO$_4$ had been employed in the activation reaction.

The present data thus suggest that lyase is a simple protein species, and that its formation from the enzyme-I state would comprise the reduction of a disulfide bridge of obviously unusual properties. - Preliminary experiments give no indication that the proposed reduction could possibly be achieved by chemical means.

It was recently found that the lyase can be inhibited by the pyruvate analogue, fluoropyruvate (Fig.5). This compound is an alkylating agent of sulfhydryl compounds (cf. [6]), though much weaker than

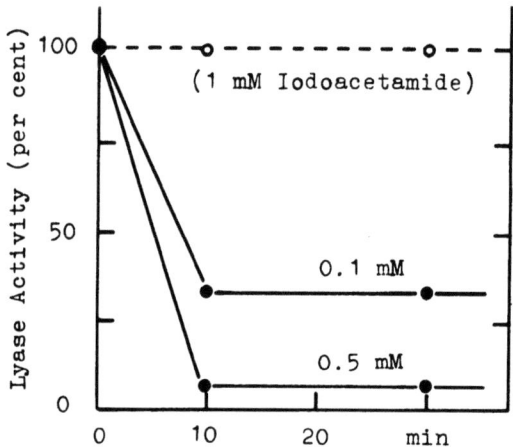

Fig. 5. Inhibition of lyase by fluoropyruvate. The reaction medium (pH 8) contained 30 mM pyruvate, 20 mM formate, Co^{2+}/dithiothreitol (4 mM/12 mM)

iodoacetamide; in fact, we observed a bimolecular rate constant about 100 times smaller than that of iodoacetamide. Therefore, when added to solutions of lyase which necessarily contain excessive sulfhydryl groups, the reagent is used up within a short time.

The extent to which the enzyme is inactivated upon addition of a constant amount of fluoropyruvate is lower if the concentration of pyruvate is increased, but becomes higher if there is formate present in addition. It is therefore most obvious that the inhibition occurs via binding of fluoropyruvate to the active site, and possibly reacting subsequently to form a covalent bond to the protein. (The completely inhibited enzyme is still inactive immediatly after gel filtration, but notably recovers activity spontaneously on standing at 25°C, pH 8, within a few hours; H. Klefenz, unpublished experiments.)

Metabolic Role of the Formate-Lyase Reaction

Henning [7] in 1963 found that of the two usual pathways of pyruvate conversion into acetate (acetylCoA) in Escherichia coli, the pyruvate dehydrogenase complex reaction is the one used by the aerobic cell whereas the formate-lyase reaction is only used by the anaerobic cell. This regulatory phenomenon, can impressively be demonstrated by the properties of two classes of mutants: Those with a defect in the dehydrogenase complex exhibit acetate requirement for aerobic, yet not for anaerobic growth. Conversely, mutants which we were able to isolate recently that neither produce active lyase nor enzyme I do not grow anaerobically unless acetate is added, yet they are acetate-independent under aerobic conditions (Fig. 6).

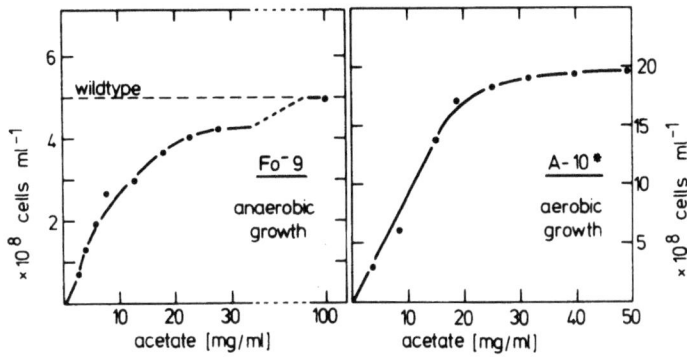

Fig. 6. Acetate auxotrophy of mutants defective in pyruvate formate-lyase (Fo⁻ 9) or in pyruvate dehydrogenase complex (A-10). Growth on glucose for 48 hours; inoculum, 6·10⁶ cells per ml. - *A-10, cf. Henning et al. [8]

The blocking of the dehydrogenase reaction at anaerobiosis has been explained by the inhibitory effect of NADH on the enzyme complex [9]. The cut-off of the formate-lyase reaction at aerobiosis, on the other hand, must quite obviously be due to the spontaneous reconversion of lyase to the inactive enzyme I-state at the more positive oxidation-reduction potential of the aerobic cell.

It would certainly be desirable to obtain direct evidence that the return of the ability to cleave pyruvate into formate and acetate, as observed when resting cells are transferred from aerobic to anaerobic conditions [7], is actually due to the enzyme II-mediated process studied in vitro. If the sole function of enzyme II is lyase activation, it should be possible to isolate mutants of the same phenotype as Fo$^-$9 (Fig. 6), but with a defect in the enzyme II protein. - It is to be noted in this context, that the synthesis of lyase protein per se is not impaired by oxygen to a significant extent; the amount of lyase and/or enzyme I is about the same in anaerobically and aerobically grown bacteria.

The thiamine diphosphate-dependent pyruvate:flavodoxin oxidoreductase reaction, which we had found in Escherichia coli as mentioned earlier, is in principle a third way from pyruvate to acetyl-CoA in this organism. However, because its constituent proteins are produced only in small amounts, or for any other reason, this system cannot contribute significantly to acetyl-CoA production in vivo, as is evident from the virtually absolute acetate-auxotrophy of the mutants discussed above. That it can serve as a reducing system in the in vitro activation of formate-lyase can operationally be understood from the very low oxidation-reduction potential of the flavodoxin species which is $E_o = -455$ mvolts at pH 7.7 for the $FlH\cdot/FlH_2$-transition [3]. A (indirect) participation in enzyme I $\rightarrow$ lyase transformation in vivo is not unlikely, as long as no other electron transferring system of Escherichia coli of similar reducing power becomes known.

These studies were supported by the Deutsche Forschungsgemeinschaft and by the Fonds der Chemischen Industrie.

References

1. KNAPPE, J., BOHNERT, E., and BRÜMMER, W., Biochim.Biophys.Acta, 107, 603 (1965).
2. KNAPPE, J., SCHACHT, J., MÖCKEL, W., HÖPNER, Th., VETTER Jr., H., and EDENHARDER, R., Europ.J.Biochem., 11, 316 (1969).
3. VETTER Jr., H., and KNAPPE, J., Hoppe-Seyler's Z.Physiol.Chem., 352, 433 (1971).
4. CHASE, T., and RABINOWITZ, J.C., J.Bacteriol., 96, 1065 (1968).
5. WOOD, N.P., Methods in Enzymology, 9, 718 (1966).
6. BERGMANN, E.D., and MIELEWITZ, A., J.Chem.Soc., 1963, 3736.
7. HENNING, U., Biochem.Z., 337, 490 (1963).
8. HENNING, U., HERZ, C., and SZOLYVAY, K., Z.Vererbungsl.,95, 236 (1964).
9. HANSEN, R.G., and HENNING, U., Biochim.Biophys.Acta, 122, 355 (1966).

Discussion:

Stadtman

In 1949-50 Chantrenne and Lipmann studied the formate-pyruvate exchange reaction and showed that CoA was essential for this reaction. I would assume from what you have just said that the exchange reaction itself is not dependent upon CoA, but that CoA would be involved in the reduced flavodoxin regenerating system.

Knappe

Right. The pyruvate-formate exchange reaction catalyzed by the lyase is definitely independent of CoA. Since the pyruvate: flavodoxin oxidoreduction involves CoA, one can readily observe a CoA-effect, when this reaction operates as the reducing system. The same holds for thiamine-diphosphate effects.

Helmreich

What is the role of adenosylmethionine?

Knappe

I don't know. As I mentioned we could not detect a transfer of the methyl group on to the enzyme I during lyase formation; but we did not check if possibly other parts become transferred. - I could tell by the way that thinking about adenosylmethionine to be sort of a regulatory molecule, we examined some time ago its concentration in aerobically and anaerobically growing bacteria. In fact we found - whatsoever it means - that the concentration is about two times higher in the anaerobic state.

Decker

Drs. Jungermann and Thauer in my laboratory have studied the lyase reaction in <u>Clostridium kluyveri.</u> In this organism the function of the enzyme is somewhat different from yours, because it appears that it is mostly connected with the production of formate for C_1-compound synthesis. With reference to the CoA-effect and the production of reduced flavodoxin in your system, we have found that ferredoxin in the clostridial system can be reduced by NADH, but this process is strictly dependent on the presence of acetyl-CoA. I wonder if you have tried to reduce the flavodoxin in your system with NADH in the presence of CoA or of acetyl-CoA. This would be the kind of flavodoxin or ferredoxin reducing system that we have found in the clostridial system. - Can you substitute your reducing system by NADH and a reductase which may be dependent on CoA?

Knappe

We have not looked if there is such an enzyme in $\underline{E.\ coli}$.

Mrs. Stadtman

I am a little confused about your enzyme II and the evidence for iron. You have a chelating agent taking away ferric ion and it looks like decreasing the activity. You really have evidence that you are putting iron on a protein?

Knappe

Well, the way how purified enzyme II-preparations have to be pretreated to get active strongly suggests that we are dealing with an incorporation of iron. However there is no definite proof as yet. We only suppose that enzyme II acts like a ferredoxin. I might say that if there is iron involved, it apparently sticks to the protein quite firmly, since iron chelating agents do not inhibit at the formation of lyase.

Mrs. Stadtman

The other question is on reducing systems. Can you use a hydrogen-platinum methylviologen type of system for reduction?

Knappe

We find that viologen dyes can replace the flavodoxin.

Segal

I understand that the oxidation is non-enzymatic as far as you can tell.

Knappe

Yes, I would think so. But I have not searched for an enzyme possibly involved in the inactivation of lyase.

Segal

There was a short report in Nature from a group in England of a similar phenomenon with a pyrophosphatase, that is an inactivation when the organism is grown aerobically and a reactivation under reducing conditions. Can you say whether it is similar to your system?

Knappe

I don't know the paper you are referring to. - Kaplan reported the oxygen sensitivity of a lactate dehydrogenase from $\underline{E.\ coli}$. We tried if our enzyme II system would facilitate the restoration of its activity, but did not detect any effect.

Sols

Only a small question - Has the formation of formatelyase in vivo, when shifting a population from aerobic to anaerobic conditions, been observed in the absence of protein synthesis?

Knappe

It readily occurs with resting cells.

Lactose Synthetase: Structure and Function

Robert L. Hill, Robert Barker, Kenneth W. Olsen, Joel H. Shaper, and Ian P. Trayer

Department of Biochemistry, Duke University Medical Center,
Durham, North Carolina/USA

In many respects, lactose is a unique disaccharide in living things. It is synthesized only in an adult, female mammal and has not been found as such in plants or lower animal forms. The ability of animals to synthesize lactose appears to be, therefore, a rather recent event in animal evolution and may have occurred some 200 million years ago as mammals evolved. Lactose synthesis is also a tissue specific reaction since it occurs only in the mammary gland, and then but for the short periods of lactation during adult life. Although it should be evident from these considerations that the synthesis of lactose must be strictly controlled, it is only recently that some insight has been obtained into its regulation. It is the purpose of this report to discuss recent developments in our knowledge of the regulation of lactose synthesis and to consider the properties of the components of the regulatory system.

General Properties of Lactose Synthetase

Lactose synthetase is composed of two protein components, a galactosyl transferase (1) and the well-known milk protein, α-lactalbumin (2). Together, the transferase and α-lactalbumin catalyze lactose synthesis as follows.

$$\text{UDP-galactose + glucose} \xrightarrow{Mn^{2+}} \text{lactose + UDP} \qquad (I)$$

The galactosyl transferase of lactose synthetase has been shown to be identical to the enzyme, UDP-galactose:N-acetylglucosamine galactosyl transferase (1), which catalyzes the following reaction.

$$\text{UDP-galactose + N-acetylglucosamine} \xrightarrow{Mn^{2+}} \text{N-acetyllactosamine + UDP} \qquad (II)$$

This enzyme is widely distributed among many tissues other than the mammary gland (3) and is known to occur in non-mammalian species (4). Its primary function does not appear to be the synthesis of N-acetyllactosamine (although this disaccharide is found in milk) but the incorporation of galactose into the oligosaccharide side chains of glycoproteins with terminal N-acetylglucosamine. Glucose is a much poorer acceptor for this transferase than N-acetylglucosamine, and its K_m is about 1 $\underline{M}$ as compared to a K_m of about 10^{-3} $\underline{M}$ for N-acetylglucosamine (5,6,7). In the presence of α-lactalbumin, however, the K_m for glucose is decreased about 1000-fold (5,8). Thus, lactose synthesis proceeds in the mammary gland by virtue of its unique ability to synthesize α-lactalbumin. α-Lactalbumin is only found in the mammary gland and its synthesis is under the same type of hormonal control as the other major milk proteins including casein (9).

On the basis of our knowledge of the galactosyl transferase and α-lactalbumin, it is possible to understand more clearly the unique aspects of lactose synthesis considered in the introduction. First, the ability to synthesize lactose must have occurred in mammalian evolution when α-lactalbumin evolved as a unique regulatory protein. Because the covalent structure of α-lactalbumin is strikingly similar to that of chicken egg-white lysozyme (10) it has been proposed that the genes for α-lactalbumin and lysozyme were derived from a common ancestor (10,11,12). Secondly, the fact that lactose is synthesized only in the mammary gland, and then only during lactation, is the consequence of the strict hormonal control of the mammary gland (9). During pregnancy, those hormones which prepare the gland to become a lactating, secretory organ, also regulate the expression of α-lactalbumin and the galactosyl transferase (13,14).

Structure of α-Lactalbumin

α-Lactalbumin has been studied in considerable detail since it was first shown by Ebner and coworkers (2) to be one of the two components of lactose synthetase. It was subsequently demonstrated by Brew et al. (10) to be not only similar in structure to egg-white lysozyme, but also a modifier of a galactosyl transferase in lactose synthetase (1). The complete covalent structure of bovine α-lactalbumin, as shown in Fig. 1, has been established (15,16,17). Knowledge of this structure has been particularly valuable in assessing several questions about α-lactalbumin, including its structural similarity with lysozyme, its evolutionary origins and the role of specific side chains in its action in lactose synthetase. Each of these aspects of α-lactalbumin will be considered briefly here.

Table I summarizes current information about the structural similarities between α-lactalbumin and lysozyme. The two proteins possess very similar covalent structures, and of the 123 residues in bovine α-lactalbumin and the 129 residues in hen egg-white lysozyme, 49 are identical when the two proteins are aligned as shown earlier. In addition, at least 26 residues appear to be conservative replacements. These similarities persist on comparison of human and guinea pig α-lactalbumin and human lysozyme. Clearly, the structural similarities are sufficient to support the view proposed earlier that the structural genes for α-lactalbumin and lysozyme are derived from a common ancestor (11,12). It has been suggested that because of their structural similarities, these two proteins also have similar three-dimensional structures (1, 20). Although this can only be answered by X-ray crystallographic analysis of

TABLE I

COMPARISON OF α-LACTALBUMIN AND LYSOZYME

Type of Comparison	Conclusion	Reference

1. Covalent Structure	a) Bovine α-lactalbumin vs. hen egg-white lysozyme
	    49 identical residues
	    26 conservative residues
	    4 identical disulfide bonds
	b) Similarities persist on comparison of molecules from other species
	    human lysozyme
	    guinea pig and human α-lactalbumin
	c) The genes for α-lactalbumin and lysozyme structures derived from common ancestor
		(15,16,17)
		
		(18)
		(19)
		
		(11,12)

2. Conformation
  a) Model Building
 	
  b) Optical Rotatory Dispersion
  c) Circular Dichroism
  d) NMR
  e) Low Angle X-Ray
	a) The sequence of α-lactalbumin can be accommodated to the conformation of lysozyme without major structural alterations
	b) Spectra of lysozyme and α-lactalbumin are very similar
	c) Same as b)
	d) Lysozyme and α-lactalbumin show similar spectral characteristics
	e) Lysozyme and α-lactalbumin differ significantly. Lysozyme and α-lactalbumin are very similar
		(20)
		
		(21)
		(21)
		
		(22)
		
		(24)

3. Immunological Cross-reactivity	a) Lysozyme and α-lactalbumin have little or no cross-reaction to a common antisera b) α-Lactalbumins from different species may have little or no cross-reactivity to a common antisera	(25) (25,26)

4. Chemical Modification
  a) Iodoacetate
 	
  b) 2-Hydroxy-5-nitro benzylbromide or N-bromosuccinimide
  c) Mercaptoethanol
 	
  d) Water soluble carbodiimides
  e) Tetranitromethane, iodine, tryosinase
	a) Histidine and methionine of α-lactalbumin react in accord with predicted conformation. Carboxymethylation does not completely abolish activity.
	b) Tryptophan in lysozyme and α-lactalbumin react differently. Extensive modification reduces activity markedly.
	c) Disulfide bonds in lysozyme and α-lactalbumin react at different rates
	d) Three carboxyl groups in lysozyme unreactive whereas all in α-lactalbumin react
	e) Loss of activity occurs and inactivation is closely correlated with extent of modification of tyrosyl residues
		(27)
		
		(28,29)
		
		(25)
		
		(30)
		
		(31)

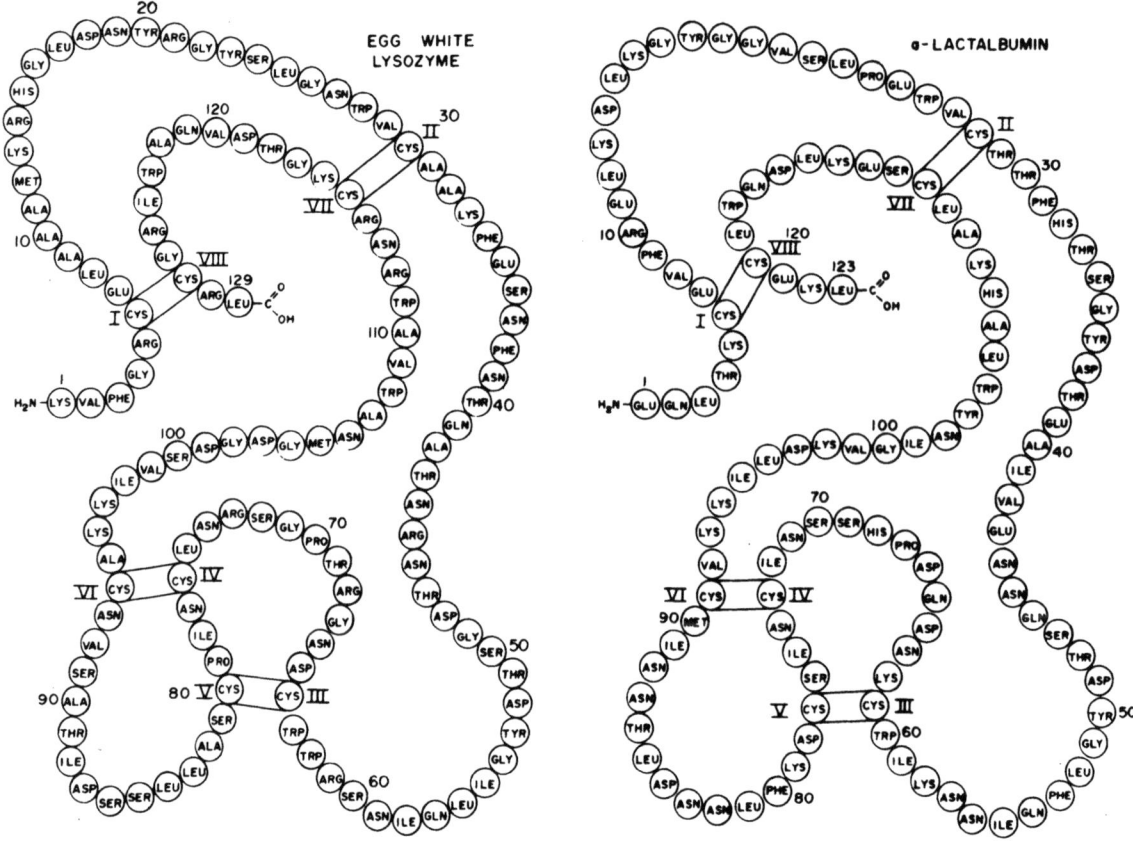

Fig. 1. Comparison of the covalent structures of bovine α-lactalbumin and hen's egg-white lysozyme. Taken from Ref. (17)

α-lactalbumin, several kinds of physical studies, as well as model building, support the view that α-lactalbumin and lysozyme have similar conformations. Of particular interest, however, is the recent report by Pessen et al. (24) that the two proteins have very similar structural features as judged by small-angle X-ray analysis. An earlier report (23) suggested considerable structural differences in the two molecules by this method of analysis. It is also of interest, that others have suggested that the two molecules are structurally different because neither show cross-reaction with antisera specific for one or the other protein. This does not seem a valid means for judging conformational differences, as pointed out earlier (26), since the antigenic determinants for a molecule are surface structures which may reflect little about conformation. Indeed, there may be little or no cross-reactivity between the α-lactalbumin from one species and an anti-α-lactalbumin antisera for α-lactalbumin from another species (32).

The reactivity of lysozyme and α-lactalbumin with specific reagents has been used to assess the similarities between the two molecules, but these studies have revealed little absolute information about this question. Although the reactivity of α-lactalbumin with iodoacetate appears to be in accord with the environment of histidine and methionine that is expected from the predicted conformation (20), its reaction with reagents which modify tryptophan, is not. Clearly, the microenvironments of specific side chains in the two molecules may differ considerably whether or not the molecules possess similar conformations. On the other hand, extensive modification of the side chains of histidine, tryptophan, tyrosine, aspartic acid and glutamic acid lead to extensive or complete loss of activity when tested with the galactosyl transferase. It is unclear whether inactivation is the result of modification of residues essential for binding, or if the modified derivatives are inactive as the result of denaturation of the molecule. Further interpretation of these kinds of studies must await additional knowledge of the nature of the interaction of α-lactalbumin and the transferase.

The Galactosyl Transferase

The galactosyl transferase of lactose synthetase appears to exist in the mammary gland bound to membranous structures of the Golgi apparatus (12,33,34). Although it can be solubilized by appropriate means, it exists in soluble form in the milk of most species, presumably as the result of the breakdown of secretory cells during lactation. Because bovine milk is easily obtained in unlimited quantities, it serves as the most convenient source of the galactosyl transferase. By means of conventional methods of enzyme purification including chromatographic procedures, the enzyme can be purified about 150-fold from raw, bovine, skim milk (35). Final purification, however, has been obtained only by affinity chromatography on columns of Sepharose to which α-lactalbumin has been bound covalently (35,36). It has been found that α-lactalbumin reacts with cyanogen bromide treated Sepharose (37) to give an insoluble product that retains its ability to stimulate lactose synthesis. In the presence of glucose (35) or N-acetylglucosamine (36), columns of the α-lactalbumin - Sepharose conjugate selectively bind the galactosyl transferase and allow inert protein to emerge unretarded. On removal of glucose or N-acetylglucosamine the galactosyl transferase can then be eluted in highly pure form and an essentially homogeneous preparation is obtained after rechromatography on the same column.

Recent studies by Barker et al. (38,39) have resulted in alternative methods for the purification of the galactosyl transferase by affinity chromatography. Hexanolamine derivatives of either UDP or N-acetylglucosamine as shown in Fig. 2, have been synthesized and reacted with CNBr-treated Sepharose. The hexanolamine moiety provides the amino group required for reaction with Sepharose and assures that the UDP or N-acetylglucosamine moieties protrude sufficiently from the Sepharose beads for binding with the transferase. Figs. 3 and 4 show the behavior of these conjugates in the purification of the galactosyl transferase.

Fig. 3 shows that the galactosyl transferase can be adsorbed quantitatively from whey on columns of UDP-Sepharose. In this experiment, whey was prepared from raw, skim, bovine milk by high speed centrifugation (35) and applied to the column.

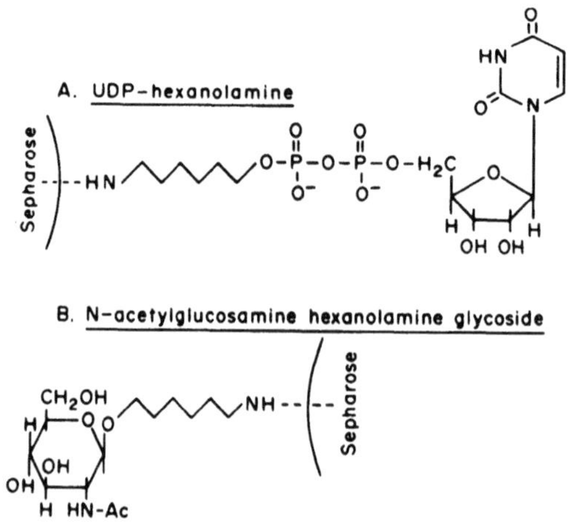

Fig. 2. The structures of the ligands that are bound to Sepharose to prepare affinity columns for the galactosyl transferase of lactose synthetase (38,39)

The column was then washed with buffer containing 0.005 M $MnCl_2$. Mn^{2+}, which is required for binding of UDP or UDP-gal to the transferase, is also essential for binding of the enzyme to the UDP-Sepharose. On elution with buffer containing EDTA and N-acetylglucosamine the enzyme emerges from the column. EDTA weakens the binding of the enzyme by chelating with Mn^{2+} and N-acetylglucosamine serves to stabilize the enzyme which becomes increasingly labile on purification (35). It is noteworthy that addition of urea (1.5 M) or borate (0.025 M, pH 8.5) accelerates the rate of elution of the enzyme. Urea may act by structurally altering the enzyme and thereby weaken its binding. Borate may act by forming complexes with the ribose moiety of the UDP, which would also be expected to weaken the binding of the transferase. The enzyme from whey is purified on the UDP-column about 150-fold and can be obtained in essentially homogeneous form by further affinity chromatography on columns of α-lactalbumin-Sepharose or N-acetylglucosamine-Sepharose.

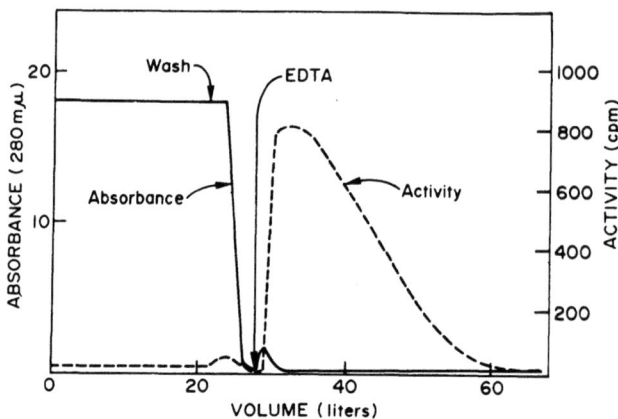

Fig. 3. Chromatography of whey from bovine milk on a column of UDP-Sepharose. Whey (22 liters) was applied to a column (4 x 80 cm) of UDP-Sepharose 4B equilibrated with 0.025 M cacodylate buffer, pH 7.4, containing 0.025 M $MnCl_2$ and 0.01 M β-mercaptoethanol. After the whey was applied, the column was washed with 6 liters of equilibration buffer. The transferase was eluted by washing the column with the same buffer without $MnCl_2$ and containing 0.025 M EDTA and 0.005 M N-acetylglucosamine

Fig. 4 shows the behavior of the galactosyl transferase on N-acetylglucosamine-Sepharose columns. In this experiment, the partially purified enzyme from UDP-columns (Fig. 3) was applied to the column in buffer containing UMP. Although the enzyme binds in the absence of UMP, its binding is considerably enhanced by either UDP or UMP, which are good inhibitors of the transferase. Borate and urea enhance elution of the active enzyme from the column, just as found with the UDP columns.

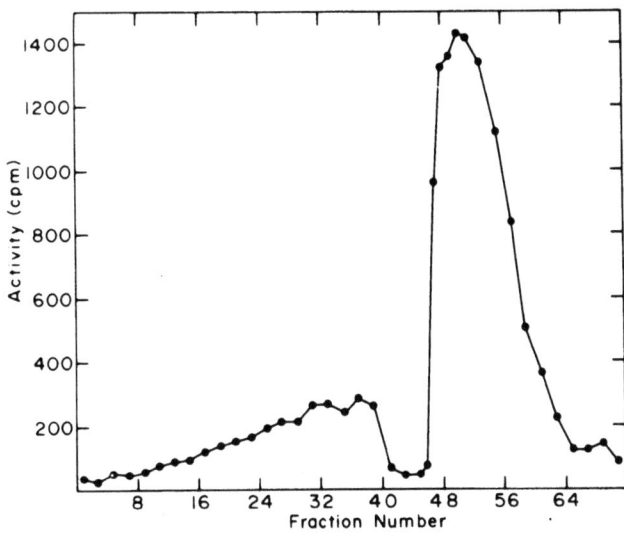

Fig. 4. Chromatography of the galactosyl transferase of lactose synthetase on columns of N-acetylglucosamine-Sepharose. Partially purified enzyme (Fig. 3) from the UDP-Sepharose column (100 ml) was applied to a 1 x 3 cm column of N-acetylglucosamine-Sepharose equilibrated with 0.025 M cacodylate buffer, pH 7.4, containing 0.025 M $MnCl_2$, 0.01 M β-mercaptoethanol and 0.5 mM UMP. The enzyme was applied in the same buffer. The column was then washed with 20 ml of the equilibration buffer and the enzyme then eluted with 100 ml of the 0.025 M cacodylate buffer, pH 7.4, containing 5 mM N-acetylglucosamine, 0.025 M EDTA, 0.01 M β-mercaptoethanol and 1.5 M urea

Essentially homogeneous transferase can be obtained solely by affinity chromatography of whey, first on UDP-Sepharose columns and then on N-acetylglucosamine or α-lactalbumin-Sepharose columns. Fig. 5 shows the SDS-gel electrophoretic patterns of the various fractions of the transferase throughout its purification. Final purification can be obtained on columns of either α-lactalbumin or N-acetylglucosamine.

The galactosyl transferase prepared by the method of Trayer and Hill (35) was found to have an apparent molecular weight of about 45,000 and to contain about 12% carbohydrate. It is clear from SDS-gel electrophoresis of these preparations as well as those prepared by affinity chromatography on UDP-, N-acetylglucosamine- or α-lactalbumin-Sepharose columns, that the enzyme contains two species of slightly different molecular weight. As judged by SDS-gel electrophoresis on 7.5% gels the species in major amount has an apparent molecular weight of about 50,000 and the species in minor amount of about 45,000. Both species contain carbohydrate as determined by staining of gels with the periodate-Schiff base reagent (40). The

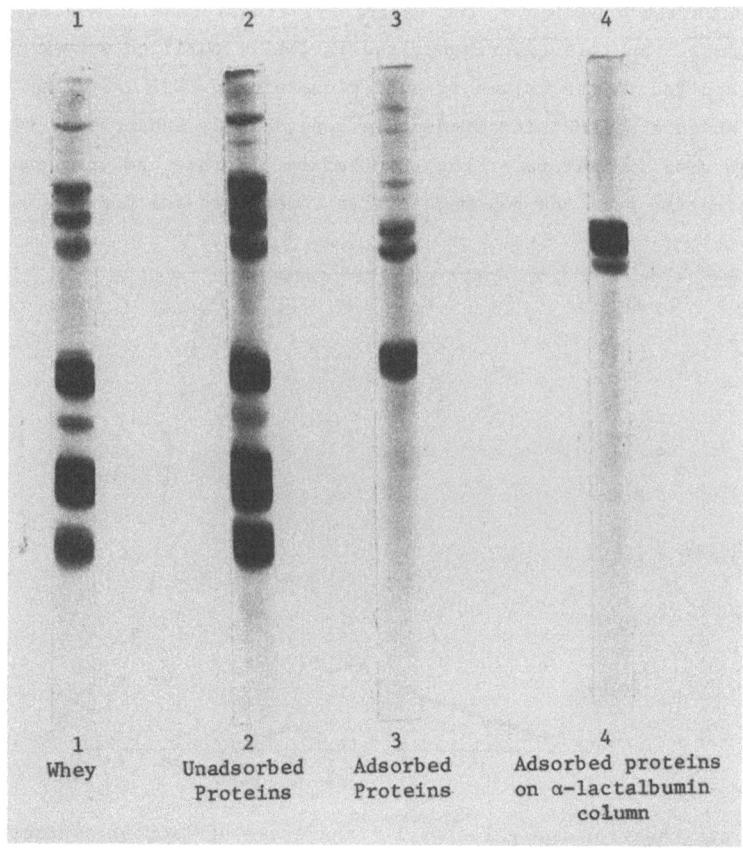

Fig. 5. SDS-gel electrophoretic patterns of fractions at different stages of the purification of galactosyl transferase from bovine milk. The pattern in 4 is obtained for the pure galactosyl transferase on columns of N-acetylglucosamine-Sepharose as well as α-lactalbumin Sepharose. The nature of the two bands is discussed in the text

amino acid compositions of the major and minor species are indistinguishable from one another as judged by analysis of hydrolysates of the appropriate bands excised from the gels by methods described earlier (40). The amino acid compositions of both bands are also indistinguishable from the composition of the enzyme of 45,000 molecular weight reported earlier (35). The amounts of the two species and the specific activity of the enzyme remains constant on repeated affinity chromatography on α-lactalbumin-Sepharose columns. For these reasons, it is believed that the galactosyl transferase is heterogeneous in size as isolated from milk and the two forms differ only in carbohydrate content. Whether this proves to be the case or not, it is noteworthy that the galactosyl transferase prepared by the method of

Trayer and Hill (25) appeared homogeneous on 5% SDS-gels as well as by sucrose density gradient centrifugation and sedimentation equilibrium in the ultracentrifuge. Reexamination of these preparations by SDS-gel electrophoresis on 7.5% gels, shows that the major and minor components are present. It appears that the 5% SDS-gels, which were used earlier (35), do not resolve the two components. In addition, the ultracentrifugal studies were not sufficiently precise to distinguish between the molecular weights of two proteins of such similar size. It should also be emphasized that the exact molecular weights of the two species cannot be determined exactly on SDS-gels and exact weights must be obtained by direct ultracentrifugal analysis of the purified components. Glycoproteins may give anomalous molecular weights on SDS-gels because the oligosaccharide side chains may interfere somewhat with micelle formation in SDS solutions (41).

Finally, it should be noted that the galactosyl transferase has a very high carbohydrate content. This is of interest since the transferase is one of the few membrane-bound proteins that has been obtained in highly pure form. Other membrane-bound proteins, such as those associated with tissue specific antigens (42), also have a high carbohydrate content.

Kinetics of the Galactosyl Transferase and Lactose Synthetase

Some insight into the mechanism of lactose synthetase has been obtained by kinetic analysis. To understand the lactose synthetase reaction it is convenient to consider first the nature of the galactosyl transferase reaction with N-acetylglucosamine as acceptor (Reaction II). Clearly, N-acetylglucosamine is the best acceptor for the transferase in the absence of α-lactalbumin (1). In addition, it has been found, in accord with the report of Klee and Klee (5), that glucose is a poor acceptor for the transferase and has a K_m about 1000 times higher than that of N-acetylglucosamine. The kinetics of the reaction with N-acetylglucosamine as acceptor, suggest that the enzyme catalyzes an ordered reaction, as reported recently by Morrison and Ebner (6) and shown schematically in Fig. 6. The purity of the transferase employed by these workers in their kinetic studies is unknown, but was

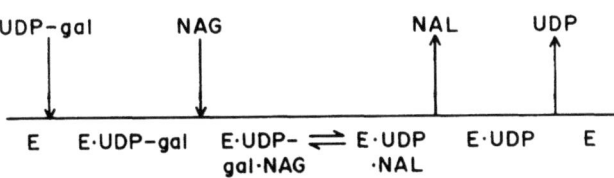

Fig. 6. The reaction scheme for the galactosyl transferase with N-acetylglucosamine as the galactosyl acceptor. Mn^{2+} is required for the enzyme but has been omitted here

apparently only partially pure. Nevertheless, analysis of the kinetics of the pure enzyme confirms that it acts by an ordered mechanism. Morrison and Ebner (6) were unable to distinguish unequivocally between ordered and partly random mechanisms from dead-end inhibition patterns, since no inhibitory analogue for N-acetylglucosamine was found. We have used UDP as an inhibitory analogue for UDP-galactose and glucose as an inhibitory analogue for N-acetylglucosamine. As shown in Figs. 7-10,

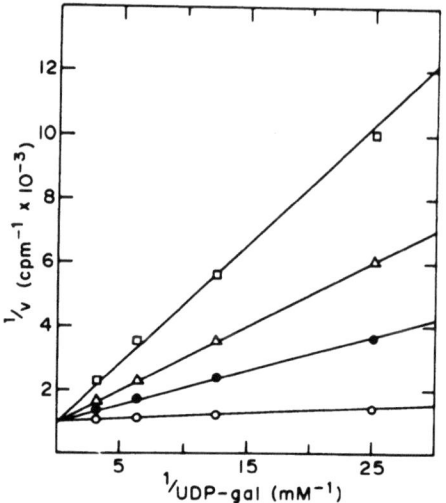

Fig. 7. The rate of N-acetyllactosamine synthesis as a function of UDP-galactose at fixed concentrations of UDP. The concentrations of UDP were o, 0; ●, 0.25 mM; Δ, 0.48 mM; and □, 0.95 mM. The data shown are for 5 mM N-acetylgalactosamine, but similar data are obtained at concentrations of 10 mM, 12.5 mM and 25 mM. Activity was measured by the radioactive method used earlier (35), under conditions which estimate initial velocity

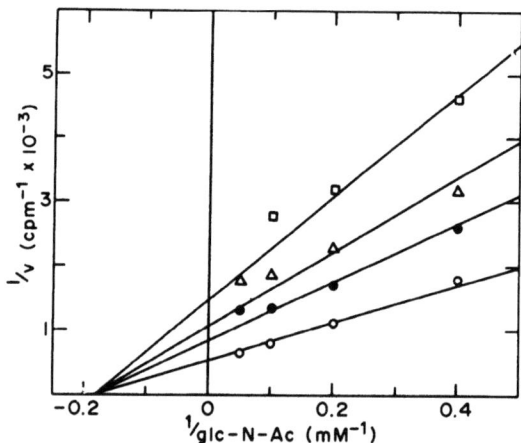

Fig. 8. The rate of N-acetyllactosamine synthesis as a function of N-acetylglucosamine concentration at fixed concentration of UDP. UDP concentrations are o, 0; ●, 0.24 mM; Δ, 0.48 mM; □, 0.95 mM. The data are for 0.16 mM UDP-galactose, but similar data are obtained at 0.63 mM and 2.52 mM concentrations. Initial velocities were measured as described in Fig. 7

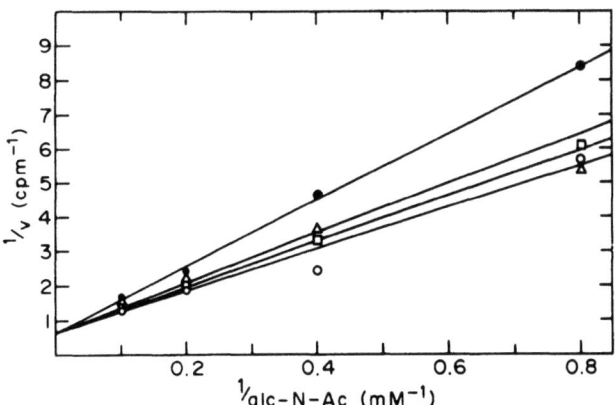

Fig. 9. The rate of N-acetyllactosamine synthesis as a function of the concentration of N-acetylglucosamine at different concentrations of D-glucose. The concentrations of glucose are o, 0; □, 0.2 M; Δ, 0.5 M; ●, 1 M. The initial UDP-galactose concentration was 0.32 mM. Similar data were obtained at 0.04 mM, 0.08 mM and 0.16 mM UDP-galactose. Initial velocities were measured by the radioactive assay (35) and N-acetyllactosamine synthesis was corrected for lactose synthesis

glucose is a competitive inhibitor with respect to N-acetylglucosamine and an uncompetitive inhibitor with respect to UDP-galactose. In these experiments, the lactose and N-acetyllactosamine formed in the reaction were separated by paper chromatography (n-butanol:pyridine:water; 6:4:3) in order to ascertain the exact

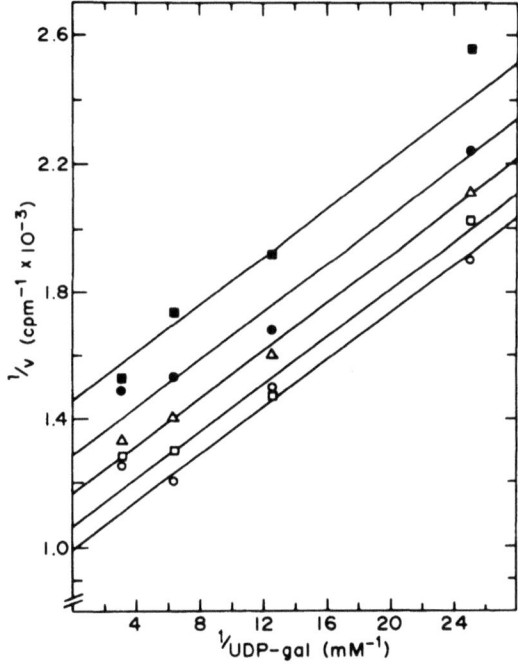

Fig. 10. The rate of N-acetyllactosamine synthesis as a function of the UDP-galactose concentration at fixed concentrations of glucose. The glucose concentrations were o, 0; □, 0.25 M; Δ, 0.5 M; ●, 0.7 M; ■, 1 M. The initial concentration of N-acetylglucosamine was 10 mM. Initial velocities were measured as described in Fig. 9

amounts of N-acetyllactosamine produced. UDP was found to act as a competitive inhibitor with respect to UDP-galactose and a noncompetitive inhibitor with respect to N-acetylglucosamine. As shown in Table II, this pattern of inhibition is

TABLE II

PREDICTED DEAD-END INHIBITION PATTERNS FOR GALACTOSYL TRANSFERASE

Variable Substrate	Mechanism	Inhibitor	
		Glucose	UDP
UDP-galactose	Ordered	Uncompetitive	Competitive
	Random	Noncompetitive	Competitive
N-Acetylglucos-amine	Ordered	Competitive	Noncompetitive
	Random	Competitive	Noncompetitive

consistent with only the ordered mechanism. The kinetic constants for these reactions are listed in Table III, and are in general agreement with those reported earlier (6).

TABLE III

KINETIC CONSTANTS FOR SUBSTRATES AND INHIBITORS OF GALACTOSYL TRANSFERASE

Substrate	Kinetic Constant	Apparent Value
UDP-galactose	K_A	0.029 mM
	K_{iA}	0.022 mM
N-acetylglucosamine	K_B	11.4 mM
UDP	K_{iP}	0.067 mM
Glucose	K_I	1.1 M

The initial rate equation employed is given below, where A refers to UDP-gal, B to N-acetylglucosamine and P to UDP. The kinetic constants have their usual meanings (43). V_1 was 2066 μmoles/min/mg enzyme.

$$\frac{E_0}{v} = \frac{1}{V_1}\left(1 + \frac{K_A}{A} + \frac{K_B}{B} + \frac{K_{iA}K_B}{AB} + \frac{K_A P}{K_{iP}A} + \frac{K_B K_{iA} P}{K_{iP}AB}\right)$$

Morrison and Ebner (7,8) have also examined the kinetics of the galactosyl transferase in the presence of α-lactalbumin with either glucose or N-acetylglucosamine as the galactosyl acceptor. They have concluded that the mechanism shown in Fig. 11 represents the action of α-lactalbumin in the transferase reactions. Although

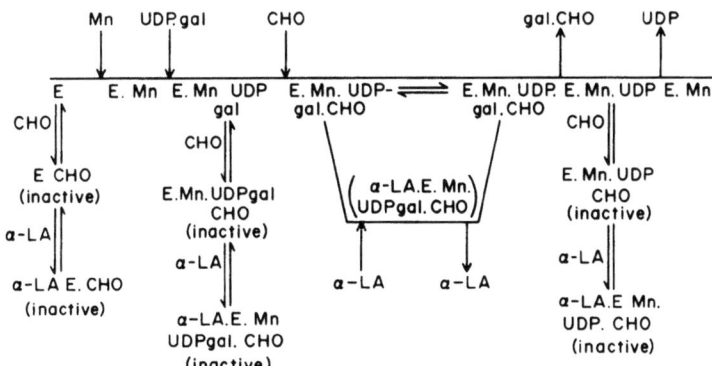

Fig. 11. A mechanism for the kinetics of the galactosyl transferase reaction in the presence of α-lactalbumin. From Ref. (8)

this scheme has been discussed thoroughly by these workers, it is important to note that it explains reasonably well the major effects of α-lactalbumin. In the absence of α-lactalbumin, reaction would proceed along the linear pathway (essentially identical to the scheme in Fig. 6) with both glucose and N-acetylglucosamine as acceptors, and the reaction flux along this path is a function of the concentrations of reactants and the K_m and V_{max} for each reactant. As noted above, the K_m for glucose is about 1000 times that of N-acetylglucosamine, whereas the V_{max} for each substrate is about the same (5,8). With N-acetylglucosamine as acceptor in the presence of α-lactalbumin, however, reaction can proceed by either the linear or branched pathway. Kinetic analyses reveal that α-lactalbumin lowers not only the K_m for this acceptor, but also the V_{max} for the reaction. Thus, α-lactalbumin can cause an apparent activation of the reaction at low N-acetylglucosamine concentrations, as noted earlier (1), since the K_m for N-acetylglucosamine is lowered and the reaction flux can proceed by both the linear and branched pathways. At higher α-lactalbumin concentrations, however, the reaction flux could proceed to a greater extent through the branched pathway. Because α-lactalbumin decreases V_{max} for the reaction with N-acetylglucosamine, it would cause an apparent inhibition of the reaction as its concentration increases, an observation first noted by Brew et al. (1). The dead-end complexes formed at higher α-lactalbumin concentrations would also decrease the apparent reaction rate.

With glucose as the acceptor in the absence of α-lactalbumin, the reaction flux proceeds also through the linear pathway but at a very low rate. In the presence of α-lactalbumin, however, the K_m for glucose is decreased markedly, and its V_{max} is essentially the same, thus reaction proceeds principally by the branched pathway.

The inhibition of lactose synthesis by high concentrations of α-lactalbumin (11), can be explained by formation of the dead-end complexes as indicated in Fig. 8.

The interaction of the galactosyl transferase with the Sepharose derivatives of either α-lactalbumin, UDP, or N-acetylglucosamine also lends some to support the mechanism shown in Fig. 11. Trayer et al. (35) noted earlier that glucose decreased the apparent K_m for α-lactalbumin in the lactose synthetase reaction. For these reasons, it was believed that the affinity of the transferase for α-lactalbumin-Sepharose conjugates should be enhanced by glucose as, indeed, proved to be the case. Andrews (36) has noted that N-acetylglucosamine also enhances the binding of the transferase to the α-lactalbumin conjugate. Barker et al. (38,39) have also shown that Mn^{2+} is required for binding of the transferase to UDP-Sepharose and binding is enhanced by N-acetylglucosamine. In addition, UDP (or UMP) enhances the binding of the transferase to N-acetylglucosamine-Sepharose conjugates. These are the effects expected from the mechanism in Fig. 11, and the proposed formation of the dead-end complexes. The Sepharose derivatives have not been used thoroughly, however, to evaluate all aspects of the mechanism and may on further study shed more light on the galactosyl transferase. It should be noted that an exact equation for the scheme shown in Fig. 11 has not been derived at present and more rigorous studies will be required before the scheme can be accepted unequivocally. Until this time it serves as a good model for the lactose synthetase system.

Concluding Remarks

The lactose synthetase system represents an interesting example of the regulation of synthesis of a tissue specific metabolite. It affords an opportunity to see how regulation is achieved at several different levels of organization of the animal organism. Evolution represents the most fundamental means for effecting the regulation of living things and their metabolic processes, and we can see how it has operated to limit lactose synthetase to a single class of animals, the mammals. Mammals alone have the ability to synthesize α-lactalbumin, and one of the unique genetic events of mammalian evolution was the duplication of a structural gene for a lysozyme-like molecule, which then led ultimately to a separate structural gene for α-lactalbumin. At a second level of animal organization, it is seen that lactose synthesis occurs specifically in the mammary gland, a uniquely mammalian organ. The mammary gland is specially organized to produce milk, including lactose, and consideration of the cellular organization of this gland offers an opportunity to see how lactose is formed and secreted. As noted by Brew (12,34), it is believed that α-lactalbumin is synthesized along with all other milk proteins in the rough endoplasmic reticulum of the cell and passes through the lumen of the endoplasmic reticulum until it reaches the Golgi region, where it can interact with the galactosyl transferase and specify lactose synthesis. The cells of the mammary gland are not secretory, however, until lactation is required, and the hormonal controls which regulate differentiation and development of mammary cells during pregnancy and lactation, also regulate synthesis of α-lactalbumin and the galactosyl transferase. The final regulation of lactose synthesis is at the molecular level, and is represented by the specific interaction of α-lactalbumin and the galactosyl transferase. Thus, lactose synthesis has a broad spectrum of regulatory mechanisms, including the evolutionary process, organ specificity, cellular structure, hormonal control, and enzymatic action.

References

1. BREW, K., VANAMAN, T.C., and HILL, R.L., Proc. Natl. Acad. Sci., U.S., 59, 491 (1967).
2. BROADBECK, U., and EBNER, K.E., J. Biol. Chem., 241, 762 (1966).
3. McGUIRE, E.J., JOURDIAN, G.W., CARLSON, D.M., and ROSEMAN, S., J. Biol. Chem., 240, PC4113 (1965).
4. Unpublished observations. Fluid in the embryonic chicken brain is a rich source of the transferase.
5. KLEE, W.A., and KLEE, C.B., Biochem. Biophys. Research Comm., 39, 833 (1970).
6. MORRISON, J.F., and EBNER, K.E., J. Biol. Chem., 246, 3977 (1971).
7. MORRISON, J.F., and EBNER, K.E., J. Biol. Chem., 246, 3985 (1971).
8. MORRISON, J.F., and EBNER, K.E., J. Biol. Chem., 246, 3992 (1971).
9. JUERGENS, E.G., STOCKDALE, T.G., TOPPER, Y.J., and ELIAS, J.J., Proc. Natl. Acad. Sci., U.S., 54, 629 (1965).
10. BREW, K., VANAMAN, T.C., and HILL, R.L., J. Biol. Chem., 242, 3747 (1967).
11. HILL, R.L., BREW, K., VANAMAN, T.C., TRAYER, I.P., and MATTOCK, P., Brookhaven Symp. in Biology, 21, 139 (1968).
12. BREW, K., "Essays in Biochemistry" edited by P. M. Campbell and F. Dickens, 6, 91 (1970). Academic Press.
13. TURKINGTON, R.W., BREW, K., VANAMAN, T.C., and HILL, R.L., J. Biol. Chem., 243, 3382 (1968).
14. TURKINGTON, R.W., and HILL, R.L., Science, 163, 1458 (1969).
15. BREW, K., and HILL, R.L., J. Biol. Chem., 245, 4559 (1970).
16. BREW, K., CASTELLINO, F.J., VANAMAN, T.C., and HILL, R.L., J. Biol. Chem., 245, 4570 (1970).
17. VANAMAN, T.C., BREW, K., and HILL, R.L., J. Biol. Chem., 245, 4583 (1970).
18. CANFIELD, R.E., KAMMERMAN, S., GOBEL, J.H., and MORGAN, T.J., Nature New Biology, 232, 16 (1971).
19. BREW, K., personal communication.
20. BROWNE, W.J., NORTH, A.C.T., PHILLIPS, D.C., BREW, K., VANAMAN, T.C., and HILL, R.L., J. Mol. Biol., 42, 65 (1969).
21. ROBBINS, F.M., and HOLMES, L.G., Biochem. Biophys. Acta, 221, 234 (1970).
22. COWBURN, D.A., BRADBURY, E.M., CRANE-ROBINSON, C., and GRATZER, W.B., Europ. J. Biochem., 14, 93 (1970).
23. KIRGBAUM, W.R., and KUGLER, F.R., Biochemistry, 9, 1216 (1970).
24. PESSEN, H., KUMOSINSKI, T.F., and TIMASHEFF, S.N., J. Agric. Food Chem., 19, 698 (1971).
25. ATASSI, M.Z., HABEEB, A.F.S.A., and RYDSTEDT, L., Biochem. Biophys. Acta, 200, 184 (1970).
26. ARNON, R., and MARON, E., J. Mol. Biol., 51, 703 (1970).
27. CASTELLINO, F.J., and HILL, R.L., J. Biol. Chem., 245, 417 (1970).
28. CASTELLINO, F.J., and HILL, R.L., Fed. Proc., 28, 405 (1969).
29. BARMAN, T.E., J. Biol. Biol., 52, 391 (1970).
30. LIN, T.Y., Biochemistry, 9, 984 (1970).
31. DENTON, W.L., and EBNER, K.E., J. Biol. Chem., 246, 4053 (1971).
32. LYSTER, R.L.J., JENNESS, R., PHILLIPS, N.I., and SLOAN, R.E., Comp. Biochem. Physiol., 17, 987 (1966).
33. SCHACHTER, H., JABBAL, I., HUDGIN, R.L., PINTERIC, L., McGUIRE, E.J., and ROSEMAN, S., J. Biol. Chem., 245, 1090 (1970).
34. BREW, K., Nature, 223, 671 (1969).

35. TRAYER, I.P., and HILL, R.L., J. Biol. Chem., 246, 6666 (1971).
36. ANDREWS, P., Fed. Eur. Biochem. Soc. Lett., 9, 297 (1970).
37. AXEN, R., PORATH, J., and ERNBACK, S., Nature, 214, 1302 (1967).
38. SHAPER, J.H., HILL, R.L., and BARKER, R., Fed. Proc., 30, 1265 (1971).
39. BARKER, R., OLSEN, K.W., SHAPER, J.H., and HILL, R.L., unpublished observations.
40. SCHWARTZ, M.L., PIZZO, S.V., HILL, R.L., and McKEE, P.A., J. Biol. Chem., 246, 5851 (1971).
41. REYNOLDS, J.A., and TANFORD, C., Proc. Natl. Acad. Sci., U.S., 66, 1002 (1970).
42. NATHENSON, S.G., SHIMADA, A., YAMANE, K., MURAMATSU, T., CULLEN, S., MANN, D.L., FAHEY, J.L., and GRAFF, R., Fed. Proc., 29, 2026 (1970).
43. CLELAND, W.W., Biochem. Biophys. Acta, 67, 104 (1963).

UDP-glucose: D-fructose-6-phosphate 2-glucosyltransferase: Metabolic Control of the Enzyme Activity

Maria A. R. De Fekete

Technische Hochschule Darmstadt, Fachbereich Biologie (10), Darmstadt/Germany

Sucrose has manifold physiological roles in plants : It is a major photosynthetic product, it can be a form of carbohydrate storage and is also a main form of carbohydrate transport. Two enzymes were described by Cardini and Leloir (1, 2) that catalyze, in vitro, the synthesis of sucrose:

$$\text{NDPglucose + Fructose} \xleftrightarrow{\text{Sucrose synthetase}} \text{Sucrose + NDP}$$

$$\text{UDPglucose + Fru-6-P} \xrightarrow{\text{Sucrose phosphate synthetase}} \text{Sucrose phosphate + UDP}$$

Information concerning the function of these two enzymes in vivo suggest that in many tissues sucrose synthetase catalyzes the formation of glucosyl nucleotides from the transported sucrose (3-8), while sucrose phosphate synthetase is responsible for sucrose synthesis (6-11). This last reaction can be made essentially irreversible by the action of sucrose phosphatase (12). The activity of sucrose phosphate synthetase of different plant tissues is too low to account for the rate of sucrose synthesis occurring in vivo (13). Therefore it could be inferred that the activity of this enzyme might be enhanced by some metabolite. A control at the site of sucrose phosphate synthetase should be expected, since, in general, regulation occurs at metabolic branchpoints (14) and fructose-6-phosphate is such a branchpoint, where competition by two or more enzymes occurs.

Our experiences on the influence of various metabolites upon the activity of sucrose phosphate synthetase of cotyledons from germinating Vicia faba seeds were at first quite conflicting. Some preparations were activated by citrate while others were inhibited. Further experiments showed that, depending on the state of the plant material used as a source of the enzyme, sucrose phosphate synthetase could be prepared either as an enzyme requiring citrate (tested with 5 mM Fru-6-P and 4 mM UDPG in glycylglycine buffer, pH 7,6), or one being also active in the absence of citrate. As shown in Fig. 1 the effect of citrate on these two preparations was quite different.

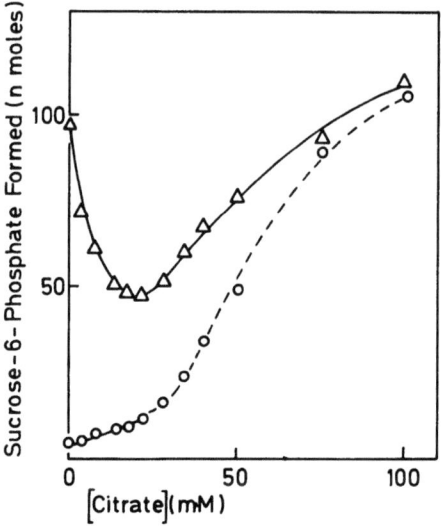

Fig. 1. Effect of citrate on the activity of sucrose phosphate synthetase. Circles represent the citrate-dependent enzyme, triangles the citrate independent enzyme. Protein 43 µg and 51 µg, respectively, per assay. Incubation time 30 min

For the citrate-dependent preparation a sigmoidal saturation curve with a 15-fold increase of activity at 150 mM citrate could be observed. The synthetase not requiring citrate was inhibited at lower citrate concentrations. A minimum of activity could be measured at 10 - 20 mM citrate. By increasing the citrate concentration reactivation occurred.

The citrate-independent enzyme could be rendered citrate dependent by repeated freezing and thawing and removal of the precipitate by centrifugation (15). The precipitate was able to activate the citrate-dependent enzyme if tested immediately after centrifugation, but lost its activating properties very rapidly. The search for the labile enzyme-bound activator was complicated by the fact that extracts of Vicia faba cotyledons contain other substances which activate the citrate dependent preparation. Thus besides citrate a number of carboxylic acids, EDTA and protamine are effective as activators of the citrate-dependent enzyme exhibiting sigmoidal saturation curves. Sucrose phosphate synthetase could also be prepared in effector-independent and effector-dependent forms from Solanum tuberosum tubers, from leaves of Spinacia oleracea and from Tetragonia expansa. Table I summarizes data concerning the activation of the effector-

TABLE I

Activation of sucrose phosphate synthesis by carboxylic acids

compound (50 mM)	Relative activity of the citrate-dependent sucrose phosphate synthetase prepared from		
	Vicia faba cotyledons	Potato tubers	Spinach leaves
None	1	1	1
Citrate	4,6	4,4	1,5
EDTA	3,7	3,2	
α-ketoglutarate	2,3	3,9	
Succinate	2,6	3,7	
Fumarate	2,4	4,4	
Malate	1,8	3,0	1,4
Pyruvate	1,8	1,7	0,9
Acetate	1,8	5,0	
Aspartate	2,0	2,7	

dependent sucrose phosphate synthetase form different sources. The enzyme from young spinach leaves was less effector-dependent than that prepared from potatoes or broad beans. The Vicia faba enzyme could be activated about 1,7 times by 20 mM 3-phosphoglycerate, ATP or UTP. Activation was not increased at higher effector concentrations and did not reach the maximum obtained with citrate. The citrate content is high in dormant legume seeds (16, 17). Therefore one might speculate that sucrose phosphate synthetase is exposed *in vivo* to citrate concentrations not far from those used in these experiments.

In contrast the citrate-independent enzyme from Vicia faba cotyledons was inhibited by nucleotides and inorganic phosphate. (Fig. 2) Complete inhibition could not be attained by any of these inhibitors. Also reactivation at higher effector concentrations, as produced at high citrate concentrations could not be observed.

Addition of citrate to the UTP-inhibited enzyme restored the activity. This suggested that both effectors share binding sites (15).

Enzyme preparations that where nearly inactive under normal assay conditions in the absence of citrate showed appreciable activity at saturating substrate concentrations. The saturation curves were sigmoidal in the absence of the natural activator or added citrate. In the presence of activators the curves were converted to hyperbolic forms. In the absence of citrate the K_m values of the dependent sucrose phosphate synthetase of Vicia were 4,3 and 9,8 mM for fructose-6-phosphate and UDPG, respectively. The presence of 0,1 M citrate lowered these values to 1,4 and 4,6 mM, respectively. These values are very similar to those obtained for the citrate-independent preparation (1,38 and 3,85 mM respectively).

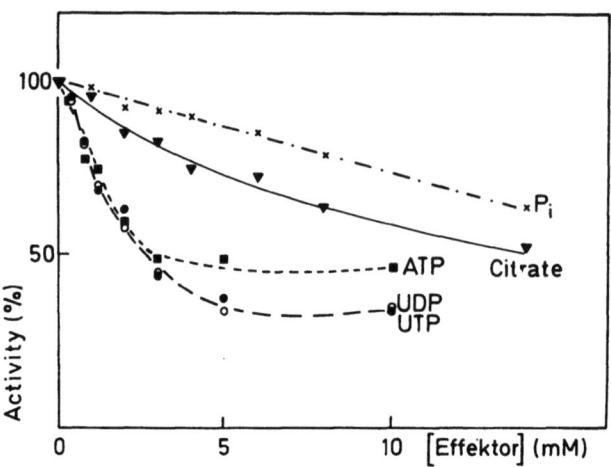

Fig. 2. Effect of ATP, UTP, UDP, citrate and inorganic phosphate on the activity of the citrate-independent sucrose phosphate synthetase. Protein per assay 28 µg. Incubation time 30 min

All these results would suggest that the natural activator is bound by two or more acidic (carboxyl- or phosphate)groups to the enzyme. It can be replaced by citrate. The conformation of the enzyme after binding of the natural activator is very similar to that after binding of citrate. Lower citrate concentrations or nucleotides displace the natural activator only partially from its binding sites resulting in a conformation with smaller activity.

The control of the activity of sucrose phosphate synthetase by the various effectors discussed seems of importance for the regulation of the rate of sucrose synthesis in vivo. The fact that citrate, on the orther hand, inhibits the activity of phosphofructokinase, which competes for fructose-6-phosphate (18), makes this regulation even more effective.

This investigation was supported by the Deutsche Forschungsgemeinschaft, Bad Godesberg, Germany. The author wishes to express her appreciation to Mrs. Ingrid Walter for technical assistance.

References

1. CARDINI, C.E., LELOIR, L.F., and CHIRIBOGA, J., J.Biol.Chem., 214, 149 (1955).
2. LELOIR, L.F., and CARDINI, C.E., J.Biol.Chem., 214, 157 (1955)
3. FEKETE, M.A.R. de, and CARDINI, C.E., Arch.Biochem.Biophys., 104, 173 (1964).
4. GRIMES, W.J., JONES, B.L., and ALBERSHEIM, P., J.Biol.Chem., 245, 188 (1970).
5. FEKETE, M.A.R. de, Planta (Berl.), 87, 311 (1969).
6. PRESSEY, R., Plant Physiol., 44, 759 (1969).
7. DELMER, D.P., and ALBERSHEIM, P., Plant Physiol., 45, 782 (1970).
8. SLABNIK, E., FRYDMAN, R.B., and CARDINI, C.E., Plant Physiol., 43, 1063 (1968).
9. FEKETE, M.A.R. de, Planta (Berl.), 87, 324 (1969).
10. MENDICINO, J., J.Biol.Chem., 235, 3347 (1960).
11. PREISS, J., and GREENBERG, E., Biochem.Biophys.Res.Commun., 36, 289 (1969).
12. HAWKER, J.S., and HATCH, M.D., Biochem.J., 99, 102 (1966).
13. HAWKER, J.S., Biochem. J., 105, 943 (1967).
14. ATKINSON, D.E., in "Metabolic Roles of Citrate" (edited by T.W. GOODWIN), Academic Press, London and New York 1968, p. 23.
15. FEKETE, M.A.R. de, Eur.J.Biochem., 19, 73 (1971)
16. TÄUFEL, K., and POHLOUDEK-FABINI, R., Biochem.Z., 326, 280 and 317 (1955).
17. MUNCH-PETERSEN, A., Acta Physiol.Scand., 8, 97 (1944).
18. KELLY, G.J., and TURNER, J.F., Biochem.J., 115, 481 (1969).

Discussion:

Gancedo

Is it possible to get lactose synthesis with the transferase and lysozyme instead of lactalbumin?

Hill

No.

Gancedo

Does lysozyme inhibit the reaction of the transferase with lactalbumin? Onemight assume that lysozyme could displace lactalbumin.

Hill

There is no effect of lysozyme.

Cori

I heard recently that the α-lactalbumin from the milk of an anteater has lysozyme activity. So this is a sort of an "ur"-protein, a primordial protein, from which evolution of α-lactalbumin and lysozyme could have originated.

Hill

Kangoroo milk is high in lysozyme as compared to bovine milk, but it also has α-lactalbumin. This α-lactalbumin has no lysozyme activity nor does the lysozyme have α-lactalbumin activity. Neither one shares the activity of the other.

Cori

It is known that in some animals, particularly in the seal there is no lactose present. Seal milk has a high fat content but no lactose. It would be interesting to know why the mammary gland of the seal does not synthesize lactose.

Shaltiel

Antibodies against α-lactalbumin do not crossreact with lysozyme. But when you open the molecule by reduction or carboxyamidomethylation you get crossreaction. This would indicate that antibodies recognize the similarity in amino acid sequence, but that folding generates structures which are unique and which are distinguished by antibodies.

Hill

We don't feel that this is a very good way of comparing the conformations of antigenic determinants. Crossreaction or the lack of it may only involve a group on the surface which is antigenic but not the conformation as a whole. I did not know about the unfolding experiments. This is rather interesting.

Frieden

You have tried to look at the interaction in solution between these two proteins and apparently you failed to see it. Yet, they must interact obviously and the question is whether you see an interaction on an affinity collumn. Can you give an estimate of the interaction constant?

Hill

Yes, the dissociation constant of the enzyme-α-lactalbumin complex is about 10^{-3}, that suggests a rather weak interaction. If you use a large amount of α-lactalbumin and a small amount of the enzyme, you can see the formation of the complex even in solution. I would agree that with the affinity column you are forming more stable complexes. When you use a small amount of transferase, a large amount of α-lactalbumin and add dimethylsuberimidate you find a one to one complex. I didn't have time to go into that.

Henning

I have a technical question. You seem to get agreement between M.W. determination by sedimentation equilibrium and dodecylsulfate gel electrophoresis. Now with glycoproteins this is very surprising, since usually the SDS does not work with glycoproteins. Did you use any special trick here?

Hill

As a matter of fact the values are a little bit high for glycoproteins on 7,5 % SDS gels. But go to 5 % or 4 % SDS, you get rather close agreement between sedimentation equilibrium and SDS gel values.

Lynen

Are the carbohydrates in the transferase essential or can they be removed by an enzymatic procedure.

Hill

We don't know.

Antonov

To what extent are the two species of transferases you have separated by gel electrophoreses, different?

Hill

We believe that they are identical except for their carbohydrates. Their amino acid composition is identical and the specific activity of a preparation containing both fractions does not change on affinity chromatography. The one experiment we have not done yet is to isolate the two species and determine independently their specific activities.

Sols

Does N-acetylglucosamine inhibit competitively lactose synthesis?

Hill

Not very much, but it may inhibit at very high concentrations.

Ziegler

You mentioned that mammary gland is the only tissue which can synthesize lactose. According to Richard Kuhn higher plants, for example Forsythia, can synthesize lactose. Do you have newer information on that?

Hill

No, but we would like to look at that in more detail.

Shaltiel:

Is your protamine effect and your citrate effect additive?

de Fekete

With the fully active enzyme both protamine and citrate have an effect and also UDP or UTP. You have competition between the different activators and inhibitors. The dependent enzyme has such high affinity for protamine that citrate is uneffective in the presence of protamine.

Decker

Could some of the effects of citrate or EDTA be due to an inhibition of the phosphatase rather than to activation of your enzyme. Have you tested phosphatase activity?

de Fekete

My enzyme was purified avout 15o fold and it was free of phosphatase.

Fischer

Is the natural activator a protein and secondly, is the concentration of the activator constant, while the interaction of the activator with the enzyme changes or _vice versa_.

de Fekete

In seeds in the absence of oxygen there is no activator present whereas in the presence of oxygen activator is present. There are also effects of light. To your first question: My fraction contained besides protein also lipoproteins and nucleic acids, so I do not know yet whether the activator is a protein.

Lynen

Acetyl CoA carboxylase is also activated by citrate. I wonder whether citrate causes aggregation of your enzyme also.

de Fekete:

We run some columns with Sephadex G 150 and found no differences in the elution patterns in presence or absence of citrate. I should like to add incidentally that my activator must be very small, not greater than about M.W. 10.000, because it does not influence greatly the citrate phosphate elution patterns.

Sols

I wonder if you could comment on the concentration of citrate in plants including potato tubers, and secondly whether the phosphofructokinase from these plants is markedly affected by changes in citrate concentrations.

de Fekete

The concentration of citrate is between 25 and 72 mM, so it is quite in the range where activation occurs. With potatoes we have a somewhat different kind of activation: Acetate is the most efficient activator and fumarate and succinate also activate. This correlates with the composition of acids in this plant tissue.

Holzer

I would like to come back to Dr. Fischer's question on the activator. As I understood you some material comes down on centrifugation after freezing and thawing, and in the supernatant the dependent enzyme remains without the independent enzyme. Can you convert the dependent to the independent enzyme for example by resuspending the dependent enzyme with the sediment?

de Fekete

Yes, I can reconvert, but only with fresh precipitate.

Mechanism of Interaction between Arginase and Ornithine Transcarbamoyl Transferase of *Saccharomyces cerevisiae*

J. M. Wiame, F. Messenguy, and M. Penninckx

Institute de Recherches, C.E.R.I.A., and Laboratoire de Microbiologie
de l'Université de Bruxelles, Bruxelles/Belgique

The study of arginine metabolism in Saccharomyces cerevisiae has revealed a type of regulatory mechanism which we shall call <u>epienzymatic regulation</u>. It implies a stoechiometric and reversible binding between enzymes, which is under the control of metabolites.

This regulation concerns <u>two concurrent pathways:</u> the biosynthesis and the degradation of L-arginine. The <u>biosynthesis</u> of arginine occurs when cells are growing with NH_4^+ as the only nitrogen nutrient. Two nitrogen atoms of arginine originate from glutamate via ornithine, one originates from glutamine via carbamylphosphate and the last comes from aspartate. With a slight simplification this is described by open arrows in fig. 1.

The synthesis of most enzymes involved in this pathway is repressed if arginine is added to the medium; this includes the ornithine carbamoyltransferase (OTCase) (1, 2) and the carbamylphosphate synthetase which is specialized in arginine biosynthesis

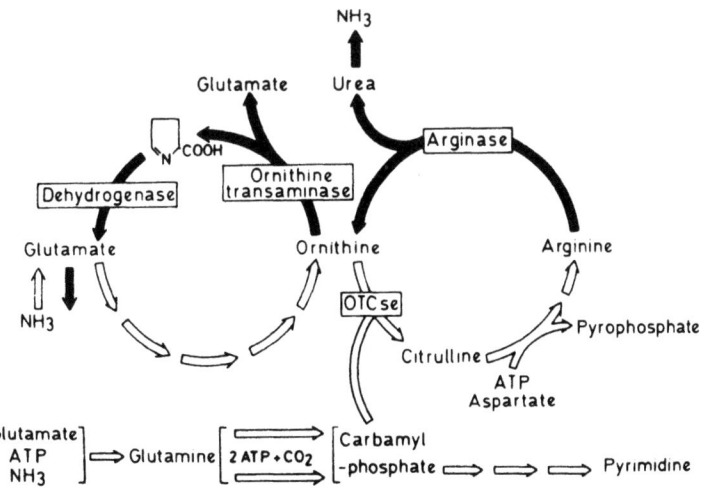

<u>Fig. 1.</u> Simplified scheme of arginine metabolism in <u>S.cerevisiae</u>.
⟹ anabolic reaction; ⟶ catabolic reaction. (5)

(another enzyme is involved in pyrimidine synthesis) (3). The metabolic flux from glutamate towards arginine is also controlled by a feedback inhibition exerted by arginine on the N-acetylglutamate kinase which starts the synthesis of ornithine (4). At the contrary to what one may wait for logically, there is no feedback inhibition on carbamylphosphate synthetase (3, 5). As a consequence, if ornithine is available, the synthesis of citrulline may remain active in spite of the presence of arginine, when the cells have not yet diluted enzymes by growth. This is precisely what occurs when arginine starts to be the nitrogen nutrient. In this case arginine is degradated into ornithine by arginase and will give one glutamate by transamination and another glutamate by the third enzyme of the <u>catabolic pathway</u>, the pyrroline carboxylate dehydrogenase (1). As it is usual for catabolic enzymes arginase and δ-L-ornithine transaminase are inducible and catabolically repressible (5, 6, 7).

In addition to these mechanisms all species of Saccharomyces have an arginase (epiarginase) which can bind and inhibit OTCase. As shown by molecular sieving a strong binding and maximal inhibition of OTCase is conditionned by the presence of arginine and ornithine at concentrations usually present in cells (8, 9). An example of inhibition is shown in fig. 2 when a crude extract rich in arginase (cells grown with arginine as sole nitrogen source) is added to a crude extract rich in OTCase and poor in arginase (cells grown with NH_4^+).

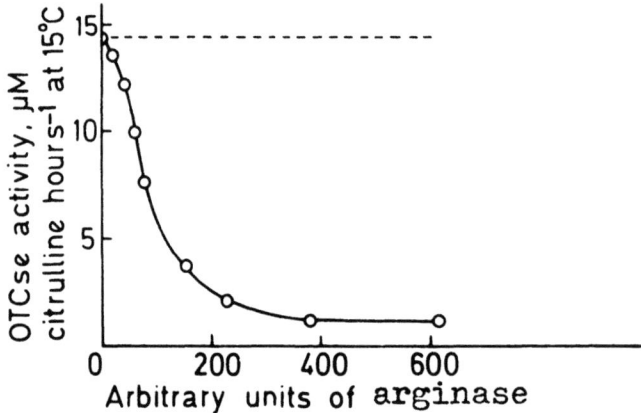

<u>Fig. 2.</u> Increasing amounts of crude arginase from cells grown on arginine as only nitrogen nutrient are added to crude OTCase (from cells grown in ammonia medium) corresponding to 2 mg of protein (5)

By this process, summarized in fig. 3, ornithine cannot be converted to an unnecessary arginine, an operation which would cost five energy rich bonds (see fig. 1) (8).

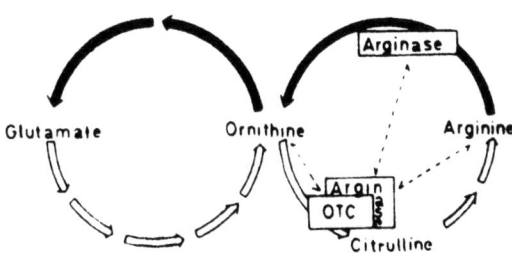

Fig. 3. Simplified scheme of arginine biosynthesis (⇒), catabolism (■⇒) and action of effectors (←----→) which contribute to OTCase inhibition (8)

The efficiency of this process can be shown by the early observation which raised this problem. When arginine is added to cells growing with NH_4^+ and if samples of cells are permeabilized in order to keep enzymes "in situ" the activity of OTCase drops dramatically. After one generation the activity is reduced to 10 % of what is expected by a dilution due to growth. The OTCase does not disappear, since if the activity is measured as usually, in extracts, it follows the usual evolution of a repressible enzyme, diluted by growth. This process of cancellation depends on protein synthesis as shown by action of cycloheximide (7).

It is interesting to mention here that regulatory mutations, argR, which lead to non repressibility of OTCase (2), simultaneously lead to absence of OTCase cancellation. This was taken as evidence that the process had a regulatory meaning (7). However the process was only understood by the finding that argR mutations have a double result: they give constitutivity for biosynthetic enzymes such as OTCase and unexpectedly suppress inducibility of the catabolic enzymes arginase and ornithine transaminase (10).
This opened a new problem which led to suggest the occurrence of regulatory genes responsible for the synthesis of an ambivalent repressor. Such a repressor would act on the expression of structural genes by an exclusion mechanism which balances anabolism and catabolism (1, 5), a mechanism with which we shall not be concerned here but which has some common biological finality with the epienzymatic regulation.

The Mechanism of Epiarginase and Ornithine Carbamoyl Transferase Interaction

Arginine, one of the two effectors of the interaction does not affect at all OTCase in the absence of epiarginase (11). Its action is exerted through epiarginase. Although a quantitative study of the catalytic and the regulatory functions has still to be done, one may think that the catalytic and the regulatory sites are identical. However the epiarginase may retain its regulatory function when its catalytic activity is lost by mutation (5, 8).

We have more informations on the way by which <u>ornithine</u> acts as an effector. As shown below, all data agree with the presence of a catalytic site distinct from a regulatory site.

The saturation function of OTCase by ornithine, at a constant carbamylphosphate concentration is diphasic (fig. 4 A and B). An hyperbolic relation is observed at low ornithine concentrations, while above 2 mM, ornithine partially inhibits the activity.

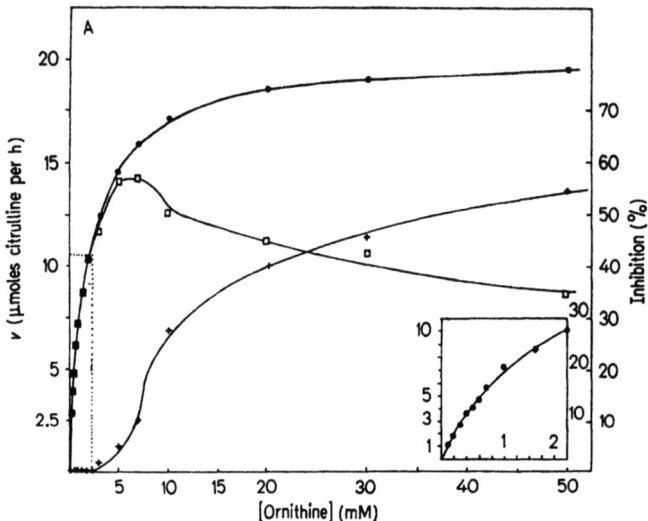

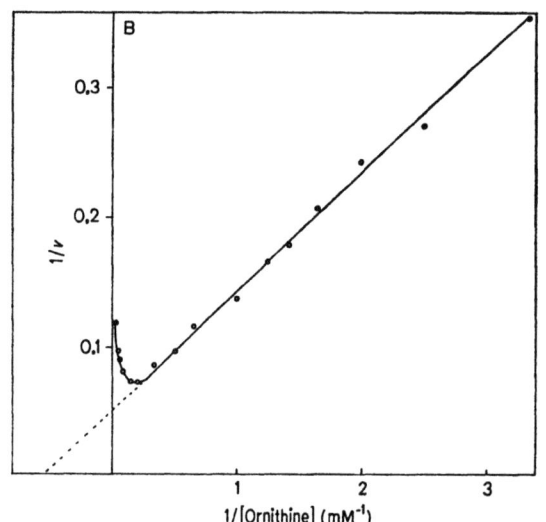

Fig. 4. Ornithine carbamoyltransferase activity as a function of ornithine concentration. Activity of par tially purified ornithine carbamoyltransferase at 15°C, as a function of ornithine concentration, carbamylphosphate is 10 mM. (A) , Experimental results; , Extrapolated hyperbola; +, inhibition function of ornithine carbamoyltransferase calculated from reciprocal plot. (B) Reciprocal plot of the ornithine carbamoyltransferase activity as a function of ornithine concentration (11)

The comparison of the experimental data with the extrapolated hyperbola expresses the inhibition which shows a cooperative effect.

The presence of epiarginase increases this inhibition (fig. 5). The simultaneous presence of epiarginase and arginine shows the strong inhibition promoted by arginine.

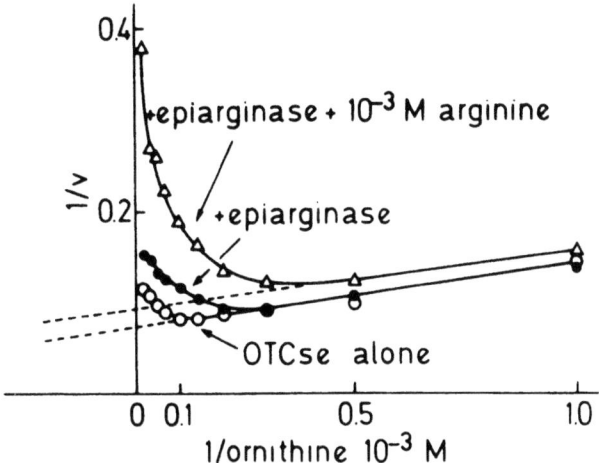

Fig. 5. Inhibition of OTCase by an epiarginase from mutant AG1 which has no arginase activity. This allows to study apparent affinity of OTCase at low concentration of ornithine in the presence of epiarginase and ornithine (5)

The experiment reported in fig. 5 was made with an epiarginase devoid of catalytic function. This allows to avoid the formation of ornithine during the experiment, which might modify ornithine concentration at low concentration. From this experiment it appears that, in the range of experimental uncertainties, at low ornithine concentration, the apparent affinity for ornithine is not affected by epiarginase + arginine, but that the final regulatory action is a strong amplification of a process which is already present in OTCase alone. For this reason excess substrate (ornithine) inhibition has been foreseen as a prerequisite for sensitivity of OTCase for epiarginase, and the two properties have been studied in parallel.

The Two Sites for Ornithine on Ornithine Carbamoyltransferase (11)

Desensitization for effectors was observed in many allosteric enzymes and was the main first indication of distinct regulatory and catalytic sites, as well as of the fact that regulatory interactions are indirect (12, 13, 14, 15). This includes the case in which substrate and effector are identical as in the inhibition of phosphofructokinase by ATP (16).

In the present case one of the most convincing evidences for distinct sites and of their indirect interaction is that one obtained by the coarse of <u>heating</u> which could hardly distinguish overlapping or close sites. When OTCase is heated at 58° at pH 6,5 in the presence of 50 mM ornithine, the relative importance of inhibition by ornithine decreases, the maximal capacity of the activity increases and is reached at increasing concentrations of ornithine (Fig. 6).

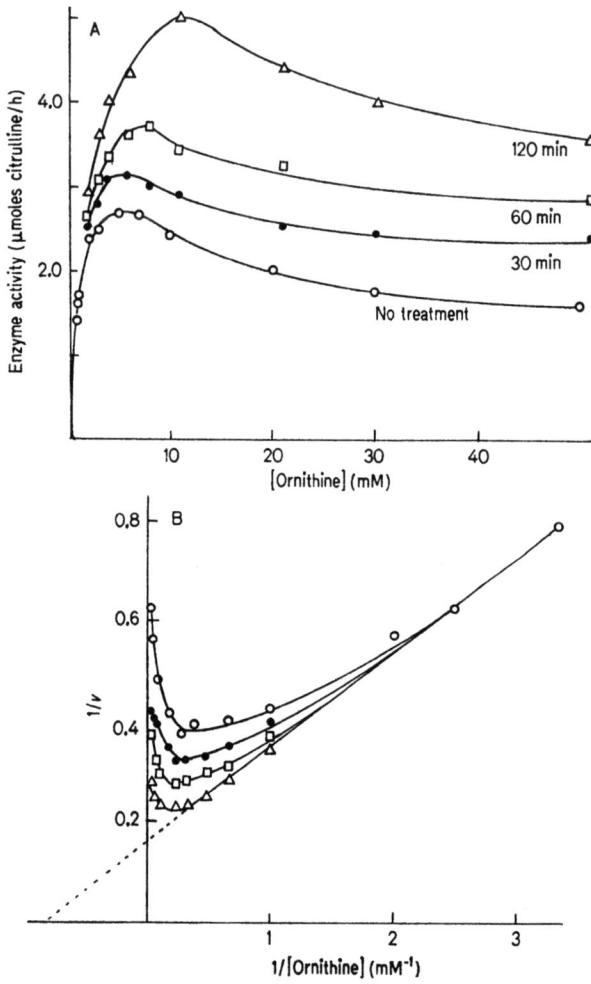

Fig. 6. Heat desensitization of ornithine carbamoyltransferase at 58°C. (A) 12,5 mg of a partially purified ornithine carbamoyltransferase extract are heated for different times at 58°C in 0.1 M Tris-HCl buffer pH 6.5 in the presence of 50 mM L-ornithine: o, no treatment; ●, 30 min heating; , 60 min heating; Δ, 120 min heating. The enzymes activity, v,- (μmoles citrulline per hour) is then measured at pH 8.0 and 15°C as usual (concentration of protein: 1.25 mg per ml) at different concentrations of ornithine. The concentration of carbamylphosphate is 10 mM. (B) reciprocal plot (11)

The saturation curve has a tendency to approach an hyperbolic function and it is interesting to note that the apparent K_m for ornithine at low concentration does not change by this treatment.

<u>Chemical modifications</u> could be crucial for future work such as mapping the sites for effectors and binding areas of the enzyme(s).

Beef liver OTCase is inactivated by acetylation with acetylimidazole, and ornithine protects against this action (17). The same occurs with Saccharomyces OTCase. In the presence of 50 mM ornithine, after acetylation by 0,2 M acetylimidazole for 60 min. at 15° in 0,02 M triethanolamine pH 8, a large part of the catalytic activity is retained but inhibition by ornithine disappears and the K_m for ornithine at low concentration is unchanged (fig. 7).

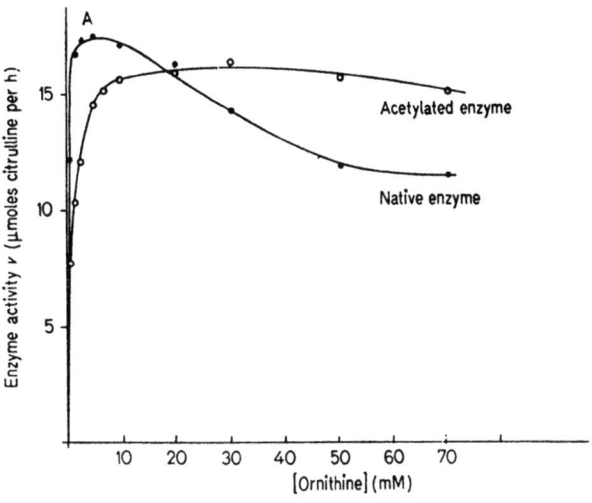

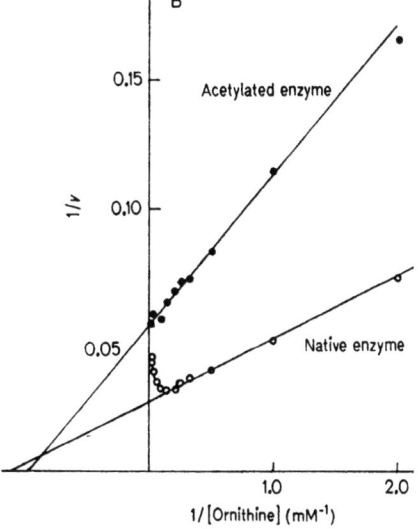

Fig. 7. Desensitization of ornithine carbamoyltransferase by acetylation. (A) 2.5 mg of protein of a partially purified ornithine carbamoyltransferase extract are incubated during 60 min at 15°C in the presence of 0.2 M acetylimidazole and 50 mM L-ornithine. After 60 min, aliquots are tested for activity as usual, with various ornithine concentrations and 10 mM carbamylphosphate. (B) Reciprocal plot (11)

The acetylated enzyme is no longer sensitive to epiarginase in the presence of 10^{-3} M L-arginine (fig. 8).

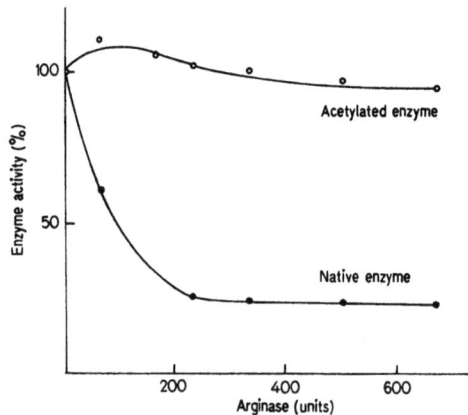

Fig. 8. Action of arginase on ornithine carbamoyltransferase in the presence of 15 mM L-ornithine and 1 mM arginine (11)

Most probably the regulatory site is not protected by ornithine because of its lower affinity for ornithine, but it contains an amino-acid residue such as tyrosyl common in both sites for recognition. It has been checked by molecular sieving (Sephadex-G 200) that after acetylation arginase does not bind OTCase anymore (11).

Mutation in the structural gene coding for OTCase may also affect its sensitivity. Starting from an OTCase negative mutant, a revertant (RF 90) has been obtained which has an OTCase distinct from wild type OTCase. This enzyme is no longer sensitive to inhibition by ornithine but unexpectedly it is activated by ornithine at high concentration (fig. 9). Again it has lost most of its sensitivity

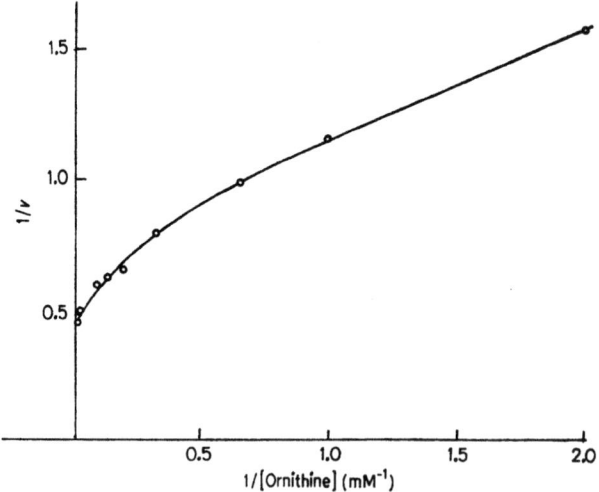

Fig. 9. Ornithine inhibition converted into activation in ornithine carbamoyltransferase from revertant strain RF90 (reciprocal plot)(11)

for arginase + arginine (fig. 10). This surprising behavior does not seem rare, it has been obtained in another revertant and the same type of inversion has been reported earlier in a mutated prephenate hydratase of Bacillus subtilis (20).

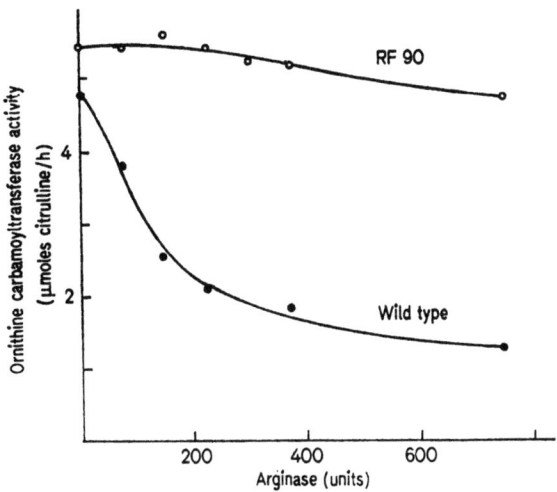

Fig. 10. Loss of sensitivity to arginase of PTCase from revertant strain RF90 (O) to be compared with the inhibition of wild strain ornithine carbamoyltransferase (●) (11)

A reversible desensitization of OTCase is observed under very simple conditions simultaneously with a non competitive inhibition by hydrogen ions. The saturation function by ornithine is hyperbolic at pH 7 and inhibition by excess of ornithine appears at higher pH (fig. 11).

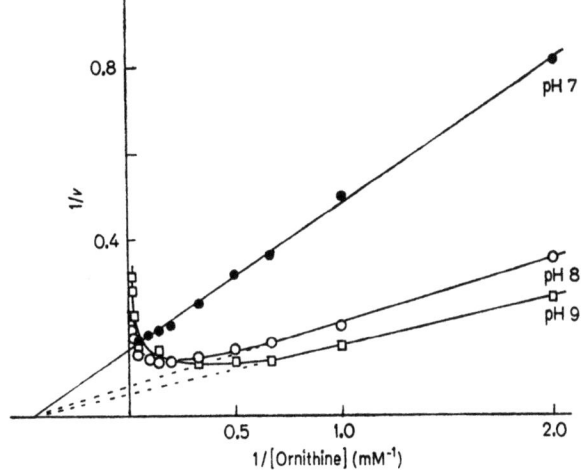

Fig. 11. Noncompetitive inhibition of ornithine carbamoyltransferase activity by H^+ (11)

On the contrary α, γ-<u>diaminobutyrate</u>, known to be a <u>competitive</u> inhibitor of Neurospora OTCase (18), when added to Saccharomyces OTCase in presence of 1 mM ornithine (which at that concentration does not show inhibition) does not provoke inhibiton (fig. 12).

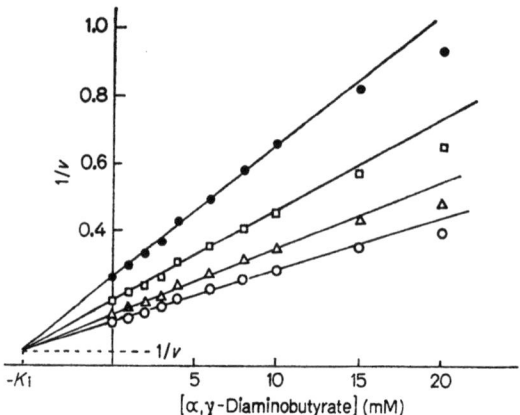

Fig. 12. Competetive inhibition of ornithine carbamoyltransferase by L-α,γ-diaminobutyrate. 0.66 mg of protein of a partially purified ornithine carbamoyltransferase extract are incubated at 15°C with o, 0.4 mM L-ornithine; □, 0.6 mM L-ornithine; Δ, 0.8 mM L-ornithine; ●, 1 mM L-ornithine, and with different amounts of L-α,γ-diaminobutyrate. The concentration of carbamylphosphate is 10 mM. v is μmoles citrulline formed per hour (11)

Apparently, analogues of ornithine which cannot compete at the catalytic site may <u>compete at the regulatory site</u> chasing ornithine but without the ability to perform its function. The result is a reduction of excess substrate inhibition without a modification of apparent affinity for ornithine at the catalytic site (fig. 13).

As mentioned before (5, 19) many yeasts outside the genus Saccharomyces do not possess the epienzymatic regulatory mechanism. A yeast like <u>Pichia fermentans</u> has an OTCase which is devoid of the inhibition by excess of ornithine and is insensitive to its own arginase as well as to the epiarginase of Saccharomyces. An interesting situation occurs in the yeast <u>Debaryomyces globosus</u>: its OTCase is sensitive to excess of ornithine but insensitive to its arginase; however, it is sensitive to epiarginase of Saccharomyces cerevisiae. This may be a way to explain the process by which epienzymatic mechanisms involving the specificity of two macromolecules could have been selected. In a first stage the selection of inhibition by excess of substrate, although partial, may give a first substantial advantage as realized in Debaryomyces globosus, and in a second independent selection, arginase may be modified into epiarginase, as in Saccharomyces.

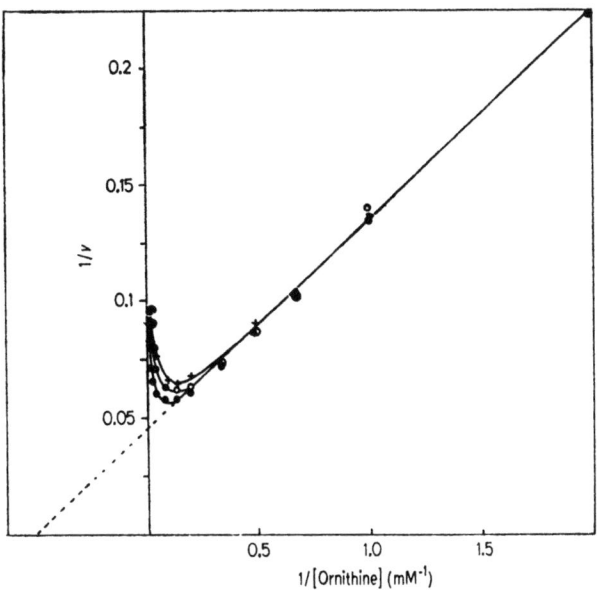

Fig. 13. Action of putrescine on the inhibition of ornithine carbamoyltransferase by ornithine (reciprocal plot). 0.66 mg of protein of a partially purified ornithine carbamoyltransferase extract are incubated at 15°C at different ornithine concentrations, +, without putrescine; O, with 5 mM putrescine; o, 50 mM putrescine. The concentration of carbamylphosphate is 10 mM. v is μmoles citrulline formed per hour (11)

However the second step is not a simple one. If one admits as a first model of regulation the one depicted in fig. 14, the modification of an arginase into an epiarginase implies the concourse of a binding area for OTCase and of a conformational transmission from the catalytic site of arginase unmasking the normally burried binding area.

Again one can separate these two events. As mentioned before an epiarginase has already an inhibitory action (weak binding in fig. 14 shown in fig. 5) and the transmission of arginine action may result from a next step. Thus, the whole system could result from three successive and independent events.

We acknowledge Academic press, FEBS Letters and Europ. J. Biochemistry for permission to reproduce figures.

This work was supported by a grant from the Fonds de la Recherche Fondamentale Collective. M. Penninckx is a boursier de l'Institut pour la Promotion de la Recherche Scientifique dans l'Industrie et l'Agriculture.

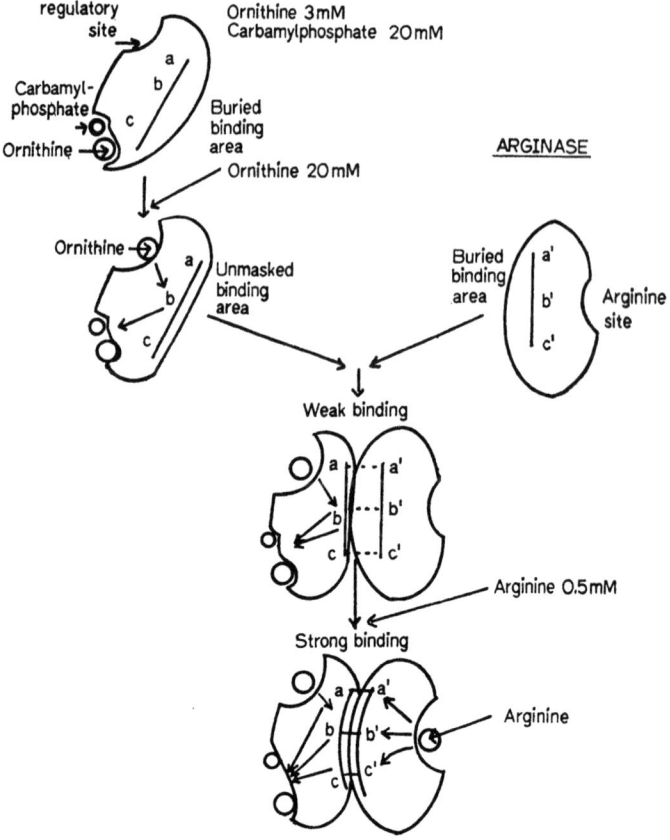

Fig. 14. Regulation of ornithine carbamoyltransferase (11)

References

1. WIAME, J.M., Report to the Xth Congress of Microbiology, Mexico 1970. (1971).

2. BECHET, J., GRENSON, M. and WIAME, J.M., Eur.J.Biochem., 12, 31 (1970).

3. LACROUTE, F., PIERARD, A., GRENSON, M. and WIAME, J.M., J.Gen.Microbiol., 40, 127 (1965).

4. DE DEKEN, R., Biochem.Biophys.Res.Commun., 18, 462 (1962).

5. WIAME, J.M., in "Current Topics in Cellular Regulation" (B.L. Horecker and E.R. Stadtman, eds.) Vol. IV, Academic Press, New York. (1971).

6. MIDDEHOVEN, W.J., Biochem.Biophys.Acta, 93, 650 (1969), ibidem, 156, (1967).

7. BECHET, J., and WIAME, J.M., Biochem.Biophys.Res.Commun., 21, 226 (1965).
8. MESSENGUY, F., and WIAME, J.M., FEBS Lett., 3, 47 (1969).
9. RAMOS, F., THURIAUX, P., WIAME, J.M., and BECHET, J., Eur.J.Biochem. 12, 40 (1970)
10. THURIAUX, P., RAMOS, F., WIAME, J.M., GRENSON, M., and BECHET, J., Arch.Int.Physiol.Biochim., 76, 955 (1968).
11. MESSENGUY, F., PENNINCKX, M., and WIAME, J.M., Eur.J.Biochem., 22, 277 (1971).
12. CHANGEUX, J.P., Cold Spring Harbor Symp.Quant.Biol., 26, 313 (1961).
13. GERHART, J.C., and PARDEE, A.B., J.Biol.Chem., 237, 891 (1962).
14. MARTIN, R.G., J.Biol.Chem., 237, 257 (1962).
15. MONOD, J., CHANGEUX, J.P., and JACOB, F., J.Mol.Biol., 6, 306 (1963).
16. SALAS, M.L., SALAS, J., and SOLS, A., Biochem.Biophys.Res.Commun., 31, 461 (1968).
17. GRILLO, M.A., and COGHE, M., Ital.J.Biochem., 18, 133 (1969).
18. HERMAN, R.L., LOU, H.F., and WHITE, C.W., Biochim.Biophys.Acta, 121, 79 (1966).
19. MESSENGUY, F., These de doctorat, univ. of Brussels. (1969).
20. COATS, J.H., and NESTER, E.W., J.Biol.Chem., 242, 4948 (1967)

Discussion:

Stadtman

In this model that you have proposed, Dr. Wiame, you have as receptor site on the arginase the catalytic site. This presents a problem, because arginine is converted to ornithine by hydrolysis. Possibly there is an allosteric site on the arginase.

Wiame

I can avoid this difficulty, because we have arginase devoid of hydrolytic activity, but which retained its regulatory function. In fact the experiment that I showed has been done with such an arginase to avoid transformation to ornithine.

Bücher

Dr. Wiame, am I correct that substrate inhibition by ornithine is not apparent at pH 7 ?

Wiame

Yes.

Bücher

You may then conclude that substrate inhibition is not due to the ornithinium, but is due to the ornithine free base.

Wiame

I do not know.

Dixon

It seems to me you could test the selective advantage of this mechanism: Could'nt you take yeast which has arginase inhibition and yeast which doesn't, and grow both in a medium with NH_3 as source and determine the generation time and then you grow them in the presence of arginine and see which one overgrows?

Wiame

In fact we are looking for yeast mutants which are very low in ornithine transcarbamylase in order to have an arginase which overcomes the synthetic capacity of the transcarbamylase. If we could show that then we could answer the question of the physiological relevance.

Stadtman

If the complex between arginase and the ornithine transcarbamylase has no arginase activity, and when there is an excess of arginase in relation to ornithine transcarbamylase, you could get around this problem.

Wiame

So far we have not yet a pure arginase free of ornithine transcarbamylase so when I mix arginase and ornithine transcarbamylase I cannot tell exactly about the molecular ratio of the two. But we have observed that the arginase activity does not decrease when we mix the two enzymes.

Katunuma

When you add an excess of arginine does the ornithine aminotransferase activity decrease?

Wiame

Yes, about tenfold.

Sols

If inhibition of ornithine transcarbamylase through arginase would be physiologically important, one would expect that there are more molecules of arginase in your cells than in derepressed cells. Is this the case?

Wiame

I think this is the case.

Sols

The second point is that the activity ratio in arginase induced cells and ornithine transcarbamylase derepressed cells must be high enough to prevent a futile cycle. Is this the case?

Wiame

Yes, this is the case also.

Gancedo

Concerning the rapid disappearance of certain enzymes when the organisms are shifted from one medium to another, I would like to add some comments on fructose-1,6-diphosphatase from yeast. When glucose is added to a derepressed yeast culture the rapid drop in enzymatic activity cannot be accounted for solely by repression and dilution. The inactivation appears to be irreversible, since reappearance of activity is dependent upon protein synthesis. We favour the hypothesis that inactivation results from proteolysis of fructose-1,6-diphosphatase. A similar phenomenon was observed with yeast malate dehydrogenase, by Mecke et al. in Freiburg.

The inactivation of fructose-1,6-diphosphatase does not occur in all yeasts. This raises the question of the possible selective advantage of this mechanism. We performed an "overgrowth" experiment similar to that proposed by Dr. Dixon in the case of the interaction of arginase and ornithin transcarbamylase. We took two yeasts, Saccharomyces cerevisiae and Torulopsis salmanticensis, with a similar generation time, one of which, S. cerevisiae, shows the inactivation phenomenon while T. salmanticensis does not. After derepression of fructose-1,6-diphosphatase, we placed the yeasts in a medium containing glucose and followed their growth. No significant difference in the growth pattern was observed.

Segal

How rapidly does fructose-1,6-diphosphatase disappear when you add cycloheximide?

Gancedo

With cycloheximide together with glucose you have the same rate of inactivation as with glucose alone.

Segal

When you add cycloheximide alone without glucose?

Gancedo

The enzyme is stable over many hours.

Segal

If proteolysis of fructose-1,6-diphosphatase would be involved in the inactivation you will have to assume that proteases are activated by glucose.

Gancedo

Not necessarily, fructose-1,6-diphosphatase could assume, in the presence of glucose, a conformation which is susceptible to proteolytic attack.

de Fekete

Dr. Gancedo, a couple of years ago Vogell et al. reported the activation of fructose-1,6-diphosphatase by phosphofructokinase; do you suppose that your effect could be due to phosphofructokinase and that as a consequence the altered phosphofructokinase no longer activates the phosphatase?

Gancedo

If I remember correctly phosphofructokinase only mimiced the action of another protein. I have tested that in the case of yeast, without success.

Henning

Dr. Wiame, looking at your first slide it seems to me, that the inhibition by arginase is cooperative. If that is so how do you reconcile that with a one to one binding ratio?

Wiame

There is cooperativity in the presence of arginase. Moreover the inhibition of ornithine transcarbamylase at subrate excess is also cooperative. We would like to speculate that ornithine transcarbamylase and arginase are oligomers.

Enzymatic Modification of Basic Chromosomal Proteins in Developing Trout Testis

G. H. Dixon, G. S. Bailey, E. P. M. Candido, A. J. Louie, M. M. Sanders, and M. T. Sung

Department of Biochemistry, University of British Columbia, Vancouver/Canada

(Abstract)

During the differentiation of trout testis cells, there are major changes in the complement of basic chromosomal proteins bound to DNA. At a specific stage in development, new, sperm-specific, highly arginine-rich proteins-the protamines-are synthesized on small cytoplasmic polysomes, phosphorylated on their seryl residues by a cyclic AMP-dependent protein kinase and rapidly transported into the nuclei of spermatid cells. After binding to chromatin, the protamines progressively replace the histones on DNA, transforming nucleohistone into the very much more compact nucleoprotamine. During this replacement the O-phosphoryl groups of protamine are removed and, in the mature sperm, de-phospho-protamines become the sole basic proteins in complex with DNA.

Since there are large changes in testis cell volumes during spermatogenesis, cells at various stages of development can be separated readily using the "staput" technique in which cells are sedimented at one gravity in a 1-3% bovine serum albumin gradients. Such separations show the following:-

(1) Protamine is synthesized and phosphorylated most rapidly in middle spermatid cells sedimenting at 1.5 mm/hr.

(2) Following protamine synthesis and phosphorylation, the middle spermatids are converted to smaller cells, late spermatids, (1.0 mm/hr) which have lost their histones almost completely.

(3) Finally, there is a dephosphorylation of the phospho-protamine species correlated with the contraction of the nucleus and the further decrease in cell size (0.6 mm/hr) characteristic of the mature sperm.

Trout testis histones are extensively modified by acetylation and phosphorylation and this process takes place during the earlier stages of spermatogenesis in cells sedimenting at 3.5 and 2.8 mm/hr. The following sites of modification, all in basic regions, have been established:-

Histone I(f1): One site of phosphorylation in the sequence: Ala-Ala-Lys-Lys-Ser (PO$_4$)-Pro-Lys located in the C-terminal 2/3 of the molecule but <u>no</u> acetylation.

Histone IIb$_2$(f2a2): One site of phsophorylation (Ser 1) and one site of acetylation (Lys 5) in the N-terminal region: Ac-Ser^1 (PO$_4$)-Gly-Arg-Gly-Lys^5 (ε-N-Ac)-Thr-Gly-Gly-Lys-Ala^{10}-Arg

Histone IV(f2a1): One site of phosphorylation (Ser 1) and four sites of acetylation (Lys 5,8,12 and 16) in the N-terminal region: Ac-Ser^1 (PO$_4$)-Gly-Arg-Gly-Lys^5 (ε-N-Ac)-Gly-Gly-Lys (ε-N-Ac)-Gly-Leu^{10}-Gly-Lys (ε-N-Ac)-Gly-Gly-Ala^{15}-Lys (ε-N-Ac)-Arg-His-Arg-Lys (ε-N-Me) Histone IIb$_2$(f2b) and III(f3) are also phosphorylated and acetylated but the precise sites of modification are not yet established.

Both histone phosphorylation and acetylation occurs at a much higher rate in the larger, earlier cells (2.8 and 3.5 mm/hr), which are also rapidly synthesizing DNA and histone, than in the later, smaller cells (1.5 and 1.0 mm/hr) in which DNA and histone synthesis has ceased. Thus, histone phosphorylation appears to occur (as in the case of protamine) in the period following synthesis. It is suggested, therefore, that modification by acetylation and phosphorylation of the highly basic regions of the histones is an essential prerequisite for the correct binding of these regions to DNA. For example, a molecular model of the N-terminal 18 residues of histone IV in the α-helical conformation fits precisely into the major groove of DNA and the four lysyl residues (5,8,12,16) bind to four adjacent DNA phosphates in one strand of the double helix. However, the question arises of how this region of histone IV could assume an α-helical conformation with the strong electrostatic repulsion between the positively charged lysyl ε-amino groups. The acetylation of lysines 5,8,12 and 16 could provide a mechanism by which α-helix formation might be favoured. Once the **α-helical** N-terminal region bound to the major groove then deacetylation could occur and the histone would be firmly locked into place by the strong interaction between the positive ε-amino groups and the negative DNA phosphates.

(Supported by the Medical Research Council and National Cancer Institute of Canada).

Discussion:

Fischer

Rabinowitz and Lipmann have shown that some of the phosphates of phosvitin do exchange in reversible reaction. They have assumed that phosvitins have high energy phosphates. Hence I ask: Do you have in protamine phosphates of different reactivity? Secondly: Do you have ever seen reversibility of the protein kinase reaction, that is ATP synthesis?

Dixon

The second question first: I was not able to detect the reversal. As to the phosphates of different reactivities: we do not have a direct way to look at that, we only have a clue that might indicate that there are differences.

Shaltiel

I wonder, whether β-elimination of serinphosphates might have physiological meaning. This would be a way of chopping down your peptides. I wonder, whether you have any indication of degradation of this type.

Dixon

Phosphorylated protamine is very labile, because it fragments.

O. Wieland

Have you tried to look at the possible effect of cyclic AMP on the acetylating enzyme?

Dixon

Yes, we did this experiment with high hopes a long time ago, it doesn't work.

Wallenfels

Do you have any idea where exactly methylation of lysine occurs?

Dixon

In histone 4 on lysine 2o.

Th. Wieland

What techniques did you use for separation and staining of phosphoproteins?

Dixon

You may have noticed that one can separate phosphoprotamines very readily on albumin lactate starch, but you can't separate them as well on polyacrylamide, because the bands are too close together. An other advantage of this gel is that you can use different kinds of staining methods: You can stain with amido black and then instead of washing the stain out you can develop like you develop a photographic print by putting it briefly into N sulfuric acid. The unbound amido black then goes in the micelles and the background disappears and you get immediately a very black staining of protein. This method is two orders of magnitude more sensitive than the normal method. This enables us to see the proteins in a small number of cells.

Heilmeyer

From your slides it seemed to me that the histones disappeared at different rates during the development.

Dixon

This is a very interesting story in itself. In fact you can get intermediate stages. If you take whole chromatin and you shear it in a waring blendor you get two fractions: A fraction which contains all the protein. The nuclear histone can be solubilized by this treatment. Now if you extract the acid soluble proteins from the very compact fraction you find all the protamine, very large amount of histone 1, and the very lysine rich histone. Histone 1 seems to be the last one to be displaced. Now this is very curious, because if you take chromatin and you treat it with salt, histone 1 is the first to dissociate and the arginine rich ones are the most difficult to displace. Biologically it is completely the reverse: The arginine rich histones are displaced first biologically, and the lysine rich ones last. This tells us immediately that the displacement reaction cannot be merely ion exchange or something, it must be some kind of a dynamic process.

Decker

I have a question concerning the acetylase reaction. Do you know where the acetyl-CoA comes from?

Dixon

In our experiments it was added.

Decker

You added free acetate?

Dixon

We have added free acetate but in _in vitro_ experiments we also added acetyl-CoA.

Decker

I wonder whether you have an activation enzyme in the nucleus or if acetyl-CoA penetrates into the nucleus.

Dixon

I would think that protamine cannot penetrate, but acetyl-CoA may.

Hill

Can you correlate changes in the histone-protamine pattern as a function of time with cellular development?

Dixon

Of course we are looking through a narrow window, we are just looking at the acid extract of the proteins, but we have just begun a program of looking at changes with time.

Shaltiel

Do you have some idea regarding the specificity of the kinase? Would it work on copolymers of arginine and serine?

Dixon

Tom Langan has found that there is a phosphorylation site in histone 1 which is at serine 38 near the N terminal. He has a kinase from rat liver which phosphorylates at that position. With histone 1 from testis we get phosphorylation of a completely different sequence near the C terminal. So we used the rat liver kinase of Langan and found that the specificity depended on the source of the histone. Testis kinase phosphorylates serine 38 of rat liver histone and both the rat liver and the trout testis kinase phosphorylate the trout testis histone 1 at the C terminal side. So the different histones have different phosphorylation sites. We did a general survey of histone 1 in a whole series of animals and we found that there were very large differences in the patterns of phosphopeptides as you went from fish to amphibians to birds to mammals. There is a good deal of specificity in the histone 1 in terms of what's available for phosphorylation.

Krebs

Would you care to comment on the phosphorylation by the kinase of free protamines and histones as compared to protamines and histones bound to chromatin?

Dixon

When we take the kinase reaction *in vitro* to its maximum we can only label about 10 % of the serines of protamine. On starchgel we find only monophosphoprotamine. Tryptic digests of the monophosphoprotamine show that there are several sites, but *in vitro* the kinase reaction does not go beyond the first stage. *In vivo* however, we see the whole series of mono-di-tri-tetraphosphoprotamines. Either there is an other kinase we have not yet found or there is, *in vivo*, regulation which we do not know.

Active Site-directed Side Chain Modification

Hugo Fasold, Franz W. Hulla, and Akitsugu Kenmoku

Institut für Biochemie, Universität Frankfurt, Frankfurt (Main)/Germany

Two families of reagents have served for active site mapping of proteins. Many experiments make use of reagents suited to modify a certain kind of amino acid side chains, if possible without side reactions. Acylation of ε-amino-groups of lysines by acid anhydrides or acid chlorides may be named as examples, or the reaction of tetranitromethane with tyrosines. It is usually not possible, however, to guide the reaction to an unequivocal modification of only one kind of functional group, and a careful evaluation of the number of side chains derivatized will be necessary. Thus, iodoacetic acid derivatives, although entering reaction with SH groups much faster, will also alkylate imidazoles and amino groups, and the reaction conditions for one given side chain may be changed considerably by its environment on the protein molecule. Usually, experiments of this kind are evaluated by comparing the amount of reagent introduced into the enzyme monomer with the decrease of catalytic activity. A major problem is posed by the difficulty of excluding an unspecific denaturation or conformational change by physical or chemical means.

The second line of investigations uses substrate analogues, designed to form covalent bonds with the protein. Again, stoichiometric incorporation with inhibition or activation, though often more easily established, should be checked with caution. Moreover, the preferential modification of only one side chain in the enzyme monomer should be demonstrated. Tryptic or other peptide maps of the protein are frequently used for this purpose. Again, a specific inhibition without interfering denaturation of the enzyme may be most difficult to verify.

An example for both methods is given with preliminary results on the inhibition of rabbit muscle actin polymerization by diazonium-1H-tetrazole, and the covalent activation of rabbit muscle phosphorylase b by an AMP analogue.

Materials and Methods

Actin from rabbit muscle was prepared as described previously (1). During experiments verifying the results reported, actin prepared by the procedure of Spudich and Watt (2) was also used. Viscosity changes during polymerization were followed with Ostwald-type viscosimeters (Schott, Mainz, G 20) at $25°$ C. The average flow time of water was 60 sec. The concentration of actin was 9 mg in a total volume of 3 ml of 0.2 mM ATP-0.2 mM ascorbate- 1 mM Tris-HCl- buffer pH 7.5. Characterization of chymotryptic cysteine peptides after labelling with 3,2'-dicarboxy-4'-iodoacetamidoazobenzene was carried out as described in a previous publication (1). Myosin was prepared from rabbit muscle, and its ATPase activity was measured as described previously (1).

Phosphorylase b was prepared from rabbit muscle according to Fischer and Krebs (3), and recrystallized three times. Enzymatic activity was determined by measuring the reaction velocity in direction of glycogen synthesis essentially as described by Cori et al (4). Inorganic phosphate relased during the reaction was determined in an Auto-Analyzer system (Technicon). The enzyme was inactivated in aliquots of the incubation mixture by addition of 1/10 of their volume of 60% v/v $HClO_4$.

^{14}C-Amino-1H-tetrazole was synthesized by heating a solution of 0.01 mCi of barium cyanamide (Farbwerke Hoechst), 11 mMoles of cyanamide (E. Merck Co., Darmstadt), and 10.5 mMoles of sodium azide in 5 ml of 6.5 N HCl to $80°$ C for four hours. After cooling, the compound crystallized out in 80 % yield. The specific activity was a 1,2 uCi/mMole. ^{32}P-labelled 6-(purine-5'-ribotide)-5-(2-nitrobenzoic acid)-thioether and 6-(purine-5'-riboside triphosphate)-(2,4-dinitrophenyl)-thioether were prepared as described previously (5,6). Cross-linking reagents on the basis of p-diamino-azobenzene linked to two oligoproline chains were synthesized according to Wetz et al (7).

Protein content determinations were carried out by the biuret method (8). Diazonium-1H-tetrazole (DHT) was prepared just prior to use by dissolving 100 mg of the amine in 1 N HCl (2,5 ml), and the addition of 69 mg $NaNO_2$. The pH was then adjusted to 5.0, and the total volume brought to 5.0 ml. s_{20w} values were determined in a Beckman Model E ultracentrifuge with automatic scanning equipment, and Schlieren optics. Radioactivity was measured in a Packard scintillation counter, model 2704, with 3% gain and a discriminator setting at 50 - 700 for ^{32}P, 12% gain and a discriminator setting at 50 - 900 for ^{14}C. Samples were dissolved in 1 ml of "Digestin" (E. Merck Co., Darmstadt) prior to the addition of the scintillator fluid (4 % Scintimix, E. Merck Co., Darmstadt) in toluene.

The labelled protein was prepared for fingerprinting by carboxymethylation. To this purpose, cyanate-free urea was added to a thoroughly dialyzed solution (10 ml) of actin to 8 M concentration. 0,3 ml of 5 % EDTA solution and 3.0 ml of 1.4 M Tris-HCl buffer, pH 8.5, were also added, as well as 1 ul of mercaptoethanol. After 2 hours standing in a closed vessel, the solution was poured into a solution of 20 mg of iodoacetic acid in 0.1 ml of dilute NaOH solution at pH 7.0. After 30 minutes, the protein was dialyzed against water for 48 hours.

For peptic fingerprints, dialyzed protein solutions were adjusted to pH 1.9 with HCl, and 0.5 % (w/weight of protein) of bovine pepsin were added. 12 hours later the pH was readjusted, and the same amount of pepsin was added. Tryptic and chymotryptic digestions were carried out as usual (9). Lyophilized peptide mixtures were applied to 62 x 35 cm paper sheets and subjected to electrophoresis at 100 V/cm in pyridine-acetate buffer, pH 6.5. Chromatography was carried out in n-butanol-pyridine-acetic acid water (30:20:6:24). For autoradiography, fingerprints were placed between two sheets of x-ray film, pressed between 2 mm steel plates. Exposure time was 2-8 weeks. CD spectra were taken in a Cary 61 automatically recording apparatus.

Results and Discussion

As shown in previous experiments, N-ethylmaleinimide (NEM) will become attached to a single SH group on the actin monomer. Since DHT is known to react readily with sulfhydryls, the labelling of actin by this reagent was compared to the reaction of NEM-pretreated actin. The complete alkylation of the superficial SH group by NEM could be verified by the reaction of the modified actin with 3,2'-dicarboxy-

4'-iodoacetamidobenzene in 6 M urea, and the consecutive chymotryptic fingerprint of the azodye carrying peptides.

When G-actin and NEM-treated G-actin were treated with various amounts of ^{14}C-DHT up to 15-fold molar excess at pH 7.4, the evaluation of polymerization inhibition and the amounts of the reagent covalently bound to the protein resulted in the curves shown in Fig. 1.

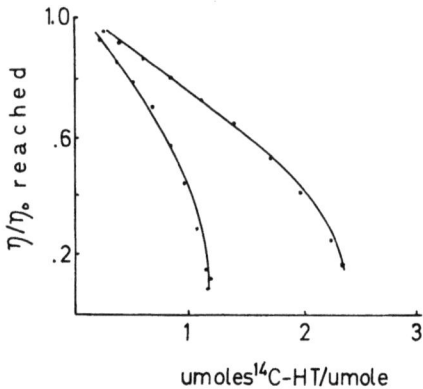

Fig. 1. Polymerization inhibition of actin (right) and NEM-treated actin (left) by ^{14}C-DHT. Protein concentration was 12 mg/ml of a 0.2 mM ATP-0.2 mM ascorbate-1mM Tris-HCl buffer, pH 7.4. The pH value was carefully maintained by addition of small amounts of 0.2 N NaOH during the addition of DHT solution prepared as described in Methods. Polymerization was induced by addition of 1/20 of the total volume of 2M KCl-2 mM MgCl$_2$ solution in the viscosimeter immediately after the labelling reaction and after 24 hours dialysis against the ATP-ascorbate-Tris buffer. Radioactivity was determined in lyophilized samples of the protein after five days' dialysis against the buffer

Since the amount of ^{14}C-tetrazole bound to actin during the inhibition of polymerization experiments suggested the labelling of a side chain essential for this reaction, it immediately became necessary to demonstrate, that one amino acid had indeed become preferentially modified. To this end, completely inhibited actin from the two series described in Fig. 1 was carboxymethylated, digested with trypsin, and subjected to fingerprinting as described in Methods. The autoradiographs of the ^{14}C-HT-actin showed two distinctly labelled peptides, while two very weak additional spots appeared. The two weak spots and only one strongly labelled peptide also appeared on the print of NEM-pretreated HT-actin. We therefore concluded that the diazonium reagent had become attached to the superficial SH group, when protection by NEM was omitted, and to a second side chain, which should be involved in the polymerization reaction, before reacting with probably two other functional groups on the protein at a slower rate.

The nature of the amino acid side chain thus preferentially labelled by DHT could not be unequivocally determined from difference spectra of the modified protein (Fig. 2). Comparison with the spectra of model compounds obtained by reaction of DHT with a tenfold molar

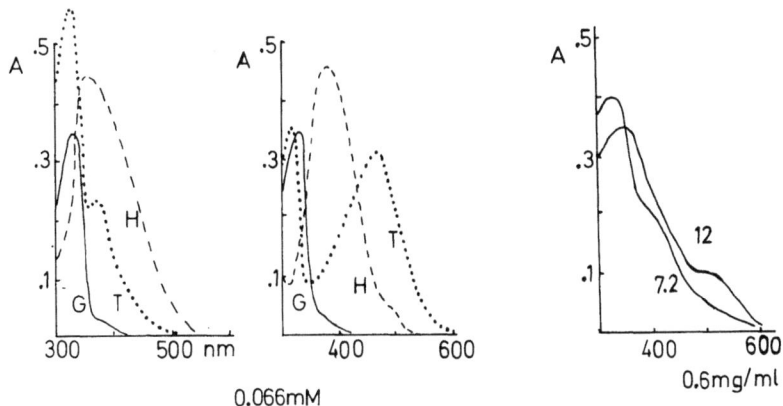

Fig. 2. Spectra of the mono-azo derivatives of glutathione (G), N-acetyl-histidine (H), and N-acetyl tyrosine (T) at pH 7 (left) and pH 12 (right). The compounds were obtained by reacting ^{14}C-labelled DHT with a large excess of the amino acid derivatives. Paper chromatography in sec butanol-formic acid-water (70:15:15) confirmed the presence of only the mono-azo derivatives in the case of N-acetyl histidine and N-acetyl tyrosine, and elution and counting of the spots served to determine their concentration in the solution. Below: Difference spectra of labelled NEM-pretreated actin at pH 7.2 and pH 12. The excess of DHT had been dialyzed out from the solutions for 24 hours with frequent changes. Blank cuvettes contained actin at the same concentrations and pH values.

excess of glutathione, N-acetyl histidine and N-acetyl tyrosine provides evidence, that only a small amount of histidine or tyrosine side chains had been modified, probably to less than 0.4 residues per monomer. Even after pretreatment with NEM, however (Fig. 2), the spectrum of the labelled actin indicated the modification of approximately one SH group per molecule. Final evidence for the nature of this functional group can only be furnished after the isolation and investigation of the strongly labelled tryptic peptide from NEM-pretreated material.

It was attempted to establish the native conformation of modified actin by difference and CD spectra. When the protein becomes denatured by the action of alkali or excessive labelling with

DHT or tetranitromethane to at least four modified side chains in the molecule, the difference spectrum to native actin shows a new broad peak at 240 nm. HT-actin does not give this peak in a difference spectrum to native actin. It appears, however, when one of the treatments mentioned above is applied. The CD spectrum of the modified actin is almost identical with that of native actin, upon denaturation by urea or alkali, both spectra are changed in the same manner. The best evidence for native properties of HT-actin could be furnished by an intact binding capacity for myosin. Though the modified protein can still activate myosin ATPase, and change its susceptibility to Mg^{++}, it will do so only to an extent of about 30% of maximum activity. This, however, could be due to conformational restriction of the modified protein, keeping it in the typical G-form even after the addition of salt. We are currently investigating the binding properties of native and modified actin for Subfragment 1 of myosin in the analytical ultracentrifuge (10).

As in most experiments with side-chain specific reagents, the selective blocking of one active site, keeping the native conformation of the molecule intact, is most difficult to establish. Obstacles of this sort are usually much easier overcome when working with quasisubstrates, expecially when one can bring about an activation of the enzyme studied by the modification. Phosphorylase b from rabbit muscle, which is activated by AMP, furnished a good example of this kind (5).

The enzyme was incubated with [1] for 0.5 to 2.0 hours at $20°$ C

in 0.01 M sodium glycerophosphate-HCl, 0.001 M EDTA buffer, pH 8.0 at a concentration of 30 mg/ml. The reagent was applied in hundredfold excess. The reaction mixture, now brought to pH 6.5, was then passed over a Sephadex G 50 column to remove excess reagent. The specific enzymatic activity and radioactivity were determined in aliquots, this was repeated after the labelled enzyme had been passed over another or several more Sephadex or charcoal columns, in order to establish the covalent bond between the nucleotide and the enzyme.

Activation of phosphorylase b in this manner is characterized in Fig. 3.

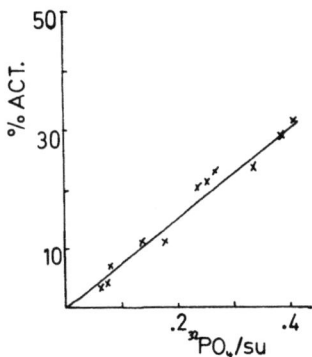

Fig. 3. Activation of phosphorylase b by covalently bound nucleotide after reaction with [1]. The ordinate denotes enzymatic activity as fraction of maximal activity after saturation with AMP

The specific radioactivity of the labelled enzyme remained unchanged after dialysis for 2 days at pH 2.5 in the cold. In control experiments samples from the same enzyme preparation were treated in the same manner, but AMP was added instead of the reagent [1]. They gave the expected specific enzymatic activity. When AMP was added to samples of the enzyme partially activated by treatment with [1] at saturation concentrations, the enzymatic activity was restored to the level of the control samples.

Again, the specific labelling of a single side chain of the enzyme had to be proved. Therefore, peptic fingerprints were carried out as described in Methods. With samples activated up to 25 % by the covalently bound nucleotide, only a single radioactive spot was found in autoradiographs (Fig. 4). Its position was distinctly different from that of the reagent or other analogues. With samples activated to a higher degree, additional radioactive spots appeared, indicating some slower side reactions of the reagent with surface SH groups. The s_{20w} value of the labelled phosphorylase was 8.3; the protein sedimented as a single symmetrical peak.

Since the activation of an allosteric enzyme should be a sufficiently specific interaction with low molecular weight effectors,

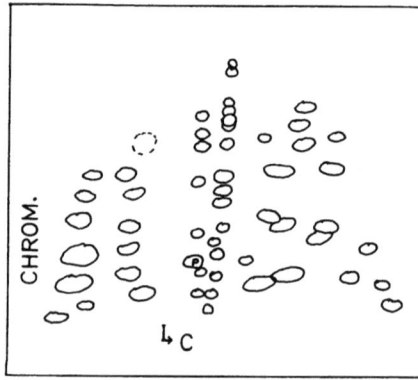

Fig. 4. Peptic fingerprint and autoradiograph of ^{32}P-nucleotide-labelled phosphorylase b (20% activation). The radioactively labelled peptide is denoted by P. Broken circles indicate the position of the reagent, the corresponding riboside, and inosinic acid

the possibility of an unspecific labelling of the peptide observed in autoradiographs, as in Fig. 4, also of a partial denaturation of the enzyme, seems remote. From control experiments with other amino acids (5) it became evident that the reagent will modify aliphatic SH groups by forming a new, and more stable thioether between the nucleotide moiety and a cysteine side chain. No other side chains were so far observed to react with the compound.

The synthesis of protein-reactive nucleotides has now been carried further to obtain an analogue of ATP [2]:

[2]

It was used in labelling experiments with actin. The F-form of the rabbit muscle protein will depolymerize readily under the influence of this nucleotide, and repolymerize readily to about 60 % of the amount of actin invested (6). After addition of ATP and lowering of the salt content, however, the F-actin formed could be only partially depolymerized again, and retained the nucleotide moiety. In this manner, we hope to label the ATP binding site of the protein.

The reaction of proteins with bifunctional reagents may be cited as a special case of side chain modification, successfully used with hemoglobin, chymotrypsinogen, and other enzymes for the estimation of distances within a tertiary or quaternary structure (11). The principle should also be applicable to cross-linking protein oligomers, as with F-actin, or associated proteins, as in actomyosin, if the cross-linking reagents were of sufficient length, and not too hydrophobic in character. Compounds of this sort have now been synthesized, and an example is given [3] :

$$JCH_2CO-N'[Pro]_{9,10}-CO-NH\text{-}\bigcirc\text{-}N=N\text{-}\bigcirc\text{-}NH-CO-[Pro]_{9,10}\text{-}N\text{-}COCH_2J$$ [3]

In protein aggregates, these reagents may serve to investigate active sites. Thus, it was possible to apply [3] to the cross-linking of F-actin, as each monomer carries one SH group on its surface (1). Artificial dimers obtained by depolymerization of this product and isolation by chromatography on Sephadex G 150 are now being characterized as to the kinetics of their polymerization, accessibility of side chains for DHT, and ATP binding properties.

References

1. FASOLD, H. and LUSTY, C.J., Biochem., 8, 2933 (1969).
2. SPUDICH, J.A., and WATT, S., J.Biol.Chem. 246, 4866 (1971)
3. FISCHER, E.H., and KREBS, E.G., J.Biol.Chem., 231, 65 (1958)
4. CORI, C.F., CORI, G.T. and GREEN, A.A., J.Biol.Chem. 151, 39 (1943)
5. HULLA, F.W. and FASOLD, H., Biochem., in the press
6. FAUST, U., Diplomarbeit 1970, Fachbereich Chemie, University of Frankfurt/Main
7. WETZ, K., MEYER, Chr. and FASOLD, H., in preparation.
8. WEICHSELBAUM, T.E., Am.J.Clin.Pathol., 7, 40 (1946)
9. FASOLD, H., Biochem.Zschr., 342, 294 (1965)
10. LOWEY, S., SLAYTER, H.S., WEEDS, A.G. and BAKER, H., J.Mol.Biol. 42, 1 (1969).
11. FASOLD, H., KLAPPENBERGER, J., REMOLD, H. and MEYER, Chr., Angew. Chem., in the press.

Discussion:

Helmreich

When you carry the reaction with your AMP-thioethernitrobenzoate derivative so far as possible - there is one binding site per M.W. of 100 000 - can you still activate with AMP?

Fasold

We did not go beyond 40 % activation, because side reactions are setting in and the enzyme begins to precipitate. However, when we go to 30 or 40 % activation and then add AMP we further activate and now get 100% activation. We can say therefore that the phosphorylase protein has not changed its properties with respect to AMP activation.

Fischer

You have shown that your AMP derivative activates phosphorylase b like AMP. AMP is of interest for us also because in the presence of AMP you cannot split the phosphate from phosphorylase a with phosphorylase phosphatase. Do you know whether your derivative would like AMP inhibit the phosphatase action on phosphorylase a?

Fasold

We have not yet done that, but it seems a very good experiment.

Pette

Is the polymerisation of actin a random or a vectorial process?

Fasold

Kinetic studies on the polymerization of actin suggest, that the contact sites of the G-actinmonomers are not identical and that each monomeric unit has a head and a tail.

Metabolic Interconversions of Enzymes: Relation to the Hysteretic Response

Carl Frieden

Department of Biological Chemistry, Washington University School of Medicine, St. Louis, Missouri/USA

I would like to thank the Organizers of this Symposium for inviting me to present some ideas on the concept of hysteresis. At first, it seemed to me that such a concept might be somewhat removed from the regulation of metabolic control by the enzymatically catalyzed interconversion of enzyme forms. However, I later realized that under some conditions one can apply the predictions of the hysteresis concept to the kind of interconvertible enzyme systems which are the topic of this Symposium. I appreciate the efforts of the Organizers for their roundabout way of bringing this fact to my attention.

The Hysteretic Enzyme Concept: Development of the hysteretic concept (1) was an outgrowth of the recognition that models for describing the kinetic properties of regulatory enzyme systems were based on ligand binding assumptions rather than kinetic characteristics (2,3). Thus, it was implicitly assumed that any transformation between different enzyme forms occurred rapidly relative to enzymatic activity. However, it is now clear that such transformation may in fact be quite slow relative to the rate of the experimentally measured enzymatic activity. When this happens and when the forms of the enzyme display different kinetic characteristics, then we achieve the time dependent response which characterizes the hysteretic enzyme.

I found it interesting that, in this Symposium which has been organized in terms of the enzymatic interconversion of one form of an enzyme or protein to another, some of the enzymes discussed could be considered as hysteretic enzymes as well. Phosphorylase *a* is certainly an example. Addition of glycogen to this enzyme results in a time dependent increase in activity. For the frog phosphorylase *a*, the half-time for the activation process at 10°, for example, is about 5 minutes (4) and activation

---

This investigation is supported in part by the United States Public Health Service (Grant # AM 13332) and the National Science Foundation (GB 26583-X).

apparently occurs concurrently with depolymerization of the enzyme to one-half the molecular weight. Glutamine synthetase has been shown by Kingdon et al. (5) to be catalytically inactive at pH 7.1 but to regain activity under conditions of the assay due to a time dependent metal and glutamate ion activation.

These are not isolated examples, because examination of the literature (1) reveals a relatively large number of enzymes which display the characteristics of a hysteretic enzyme. Furthermore, many of these enzymes, such as phosphorylase a or glutamine synthetase exhibit complex kinetic behavior and function at key points in metabolic regulation.

It seems to me that the experimental determination of rate constants for conformational changes of the type discussed here is an essential part of the relation of enzyme structure to enzyme activity, aside from the questions of metabolic significance. One can ask a number of pertinent questions: For example, is the conformational change a reflection of a distinct step in the catalytic mechanism; is such a change generated through a subunit containing enzyme in a concerted or a sequential manner; why are some conformational changes slow while others are quite rapid; is a ligand necessary or do the different conformational forms exist in the absence of the ligand; does specific protein-protein interaction influence the rate or extent of such changes and so on. So far we have very few answers to these important questions concerning protein structure and function.

Originally, hysteresis was discussed in terms of three types of mechanisms: those in which an isomerization step involving the interconversion between two kinetically different forms was slow; those in which a change in protein-protein interactions (which indeed may reflect an isomerization) was slow; and those in which the displacement of one ligand by another was slow either as a result of a sluggish off rate or a slow isomerization step preceding the dissociation of the ligand from the enzyme.

For a number of cases one general equation appears to be quite useful. That is:

$$v_t = v_f + (v_o - v_f)e^{-k't} \qquad [1]$$

where v_t is the velocity at time t, v_f is the final velocity, v_o the beginning velocity and k' represents some collection of rate constants and ligand concentrations. In the simplest case, one in which we assume only a single substrate, and for which we can write a mechanism of the type

$$E + S \xrightleftharpoons{L_1} ES \xrightleftharpoons[k_{-3}]{k_3} E'S \xrightleftharpoons{L_2} E' + S$$

$$\phantom{E + S \xrightleftharpoons{L_1} } k_5 \downarrow \phantom{\xrightleftharpoons{k_3}} k_6 \downarrow \phantom{E'S \xrightleftharpoons{L_2} E' + S}$$

$$E \xrightleftharpoons[k_{-2}]{k_2} E'$$

$$+ \qquad +$$
$$P \qquad P$$

                                                                                                      I

the apparent rate constant has the form:

$$k' = \frac{k_3 S + k_2 L_1}{L_1 + S} + \frac{k_{-3} S + k_{-2} L_2}{L_2 + S} \qquad [2]$$

L_1 and L_2 are dissociation constants for the binding steps and the rate constants k_2, k_{-2}, k_3 and k_{-3} reflect the conformational changes of either the enzyme or the enzyme substrate complex. A simple example shows the consequences of slow rate constants. If $L_2 \ll L_1$, i.e. if the affinity of S for E' is greater than for E, and the transformation takes sufficient time, the velocity of the reaction at a given substrate concentration will slowly increase from its initial value after addition of the substrate to the E form of the enzyme, as the substrate level becomes relatively higher with respect to the dissociation constant of the induced E' form. The apparent rate constant, k', can be evaluated fairly easily from this type of experimental data.

    A few comments at this point as to the time dependence of various steps in an enzyme catalyzed reaction may be appropriate in relation to the meaning of the v_t, v_o, and v_f terms of the general equation. The most rapid steps in an enzyme catalyzed reaction is the binding of the enzyme and substrate molecule. Theoretical estimates as well as experimental determinations of this step result in values on the order of $10^7 - 10^9$ sec^{-1} moles^{-1} and most such interactions, at reasonable substrate and enzyme levels are over well within a millisecond. Although quite variable of course, the turnover number of a large variety of enzyme is such to suggest that per active site, one molecule of substrate is converted to product at times on the order of 10-500 msec. Thus, conformational changes or ligand release times which are much longer than a second are those which might be slow relative to the initial velocities, v_o and v_f. In the past we have made the assumption that the substrate concentration does not change during the conversion of one form of the enzyme to another while the reaction is going on, which makes the interpretation of v_t, the velocity as a function of time, considerably easier. This could be valid under one of two conditions: either the substrate is at a saturating level or it is being produced approximately as rapidly as it is being used (as in a metabolic pathway, for example). On the other hand, these conditions may not always be obtainable experimentally and the substrate and product concentrations may change during the transformation. Clearly this will give rise to some difficulty in the interpretation of the kinetic data since there are other valid kinetic reasons which result in velocity changes throughout the course of the reaction. Indeed, abortive complex formation (i.e., the combination of an enzyme-coenzyme complex with the product of the reaction) could itself give rise to an apparent hysteretic effect, as is apparently the case for lactic dehydrogenase (6). One of the possible approaches to a proper interpretation of such kinetic data is a careful analysis of the time course of the reaction at a variety of substrate concentrations and different enzyme levels. With the availability of computer programs to simulate kinetic data by iterative methods (that is, by solving the differential equation directly) such approaches may become much more useful than in the past. The variation of the enzyme concentration in particular is important because of the wide number of protein-protein interaction cases in which such interaction affects the kinetic parameters (7). The stopped-flow equipment now generally available and the more sophisticated data analysis computer programs should also be quite useful in analyzing for these effects.

One must superimpose on the time dependent hysteretic change the fact that v_o, which reflects the kinetic properties of the enzyme prior to the transformation and v_f, which reflects those after the transformation impart to the overall reaction their own parameters. Thus other enzyme forms may exist which could give rise to complex kinetic behavior as a function of substrate concentrations even if one could prevent the hysteretic transformation. Furthermore, a more realistic situation would be if the substrate concentration were changed rapidly from one level to another, rather than from 0 to some concentration.

The simple kinetic picture may then become more complex, because velocity changes will reflect not only the substrate dependence of the reaction velocity but also the substrate dependence of the hysteretic response.

Because many hysteretic enzymes are affected by non-substrate ligands, we should also derive the pertinent expression for enzyme systems involving substrate and modifier ligands as well as for cases where there are several binding sites for a particular ligand. Such equations have been derived and can quickly get out of hand with respect to their complexity even with a relatively large number of simplyifying assumptions. Specific equations will not be presented here, but it is of interest to note that in many cases the same basic equation for the time dependence of the velocity holds:

$$v_t = v_f + (v_o - v_f)e^{-k't}$$

Of course, it is the expression for the apparent rate constant which becomes quite complex (1).

Protein-protein interactions, including in their simplest form, polymerization or depolymerization reactions may also give rise to a hysteretic response, but may reflect intramolecular conformational changes. In a few cases, the second order rate constant for the association reaction of proteins has been determined and found to be on the order of $10^5 - 10^6$ sec^{-1} moles^{-1} (8-10). On the other hand, the rates for ligand-induced polymerization reactions are frequently slower than reflected by such a rate constant. These and ligand-induced depolymerization reactions then are almost certainly controlled by isomerization steps. An example of such a case is the depolymerization induced in the bovine liver glutamate dehydrogenase by the addition of DPNH and GTP shown in Fig. 1. The apparent rate constant for the depolymerization as such is 60 sec^{-1}. Under these conditions (1.5 mg/

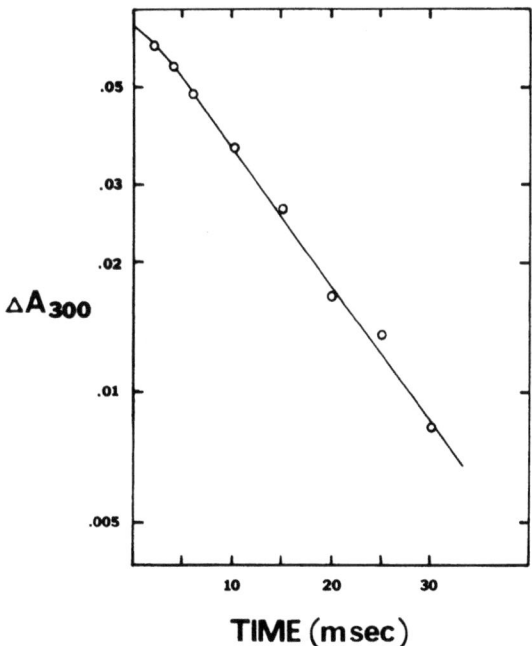

Fig. 1. Depolymerization of bovine liver glutamate dehydrogenase induced by adding GTP to the enzyme-DPNH complex. The turbidity change was measured at 300 nm using a Durrum stopped flow spectrophotometer (2 cm light path). Final concentrations were: enzyme, 1.5 mg/ml; DPNH, 120 µM; GTP, 500 µM. Reaction carried out in 0.1 M Tris-acetate buffer containing 1 mM phosphate and .1 mM EDTA at 10°, pH 7.45

ml; 120 µM DPNH), the turnover number for the enzyme, in the absence of GTP, is about 5-10 sec^{-1}. In this case the conformational change is faster than the overall reaction rate and a hysteretic response would not be expected (nor is one seen experimentally).

Hysteretic responses may be seen using glutamate dehydrogenase, but the mechanism is displacement of one ligand by another. Fig. 2 shows the activity response observed when ADP (an activator) displaces the inhibitor GTP from the enzyme DPNH-GTP complex. The apparent rate constant for this process is about .3 sec^{-1} and considerably slower than the overall activity of the enzyme (20 sec^{-1}) under conditions of ADP activation.

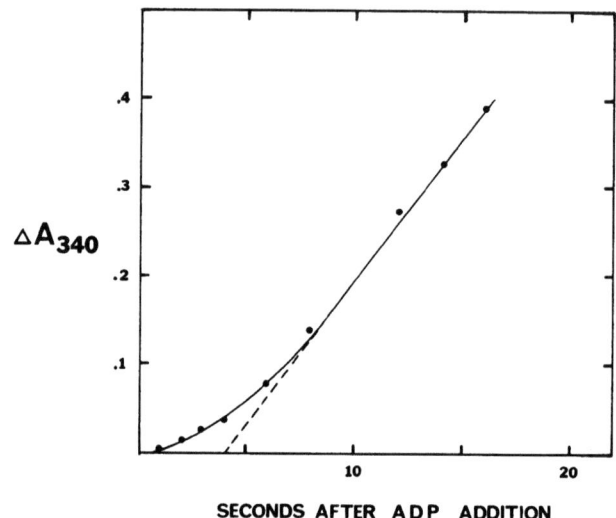

Fig. 2. Stopped flow experiment in which ADP and α-ketoglutarate in one syringe were mixed with glutamate dehydrogenase, DPNH, GTP and NH$_4$Cl in the other syringe. Final concentrations after mixing were: enzyme, 1 mg/ml; DPNH, 150 µM; GTP, 50 µM; NH$_4$Cl, 500 µM; α-ketoglutarate, 500 µM; ADP, 5 mM. Reaction conditions same as given in legend to Fig. 1

Displacement reactions of this type are in fact an important mechanism for inducing an hysteretic response. A few calculations show what I mean. For example, suppose that the ligand-enzyme dissociation constant is on the order of 10^{-7} M, a low, but not unreasonable value. If the "on" rate (diffusion controlled) is assumed to be 10^8 sec^{-1} moles^{-1}, then by definition the "off" rate of the ligand is 10 sec^{-1}, corresponding to a half time of about 70 msec. This value represents a relatively rapid reaction although it can be slower than the turnover number of the enzyme. Clearly, it would be 10 times longer if the dissociation constant were 10^{-8} M, 100 times if 10^{-9} M and so on. However, experimental measurements of displacement reactions have indicated that in fact the so-called "off" rate maybe considerably slower than that obtained by the simple calculation given here and ligands with reasonable dissociation constants may be displaced relatively slowly. In fact, the steady state rate of several dehydrogenases appears to be controlled by the off rate of the coenzyme product. It is not yet clear whether such cases reflect a slow isomerization step prior to the release of the ligand or the off rate as such but this makes little difference as far as the mechanism is concerned. In recent studies of the rates of nucleotide induced depolymerization of glutamate dehydrogenase, we came to the conclusion that there is an isomerization of the enzyme-reduced coenzyme complex preceding the reduced coenzyme being released from the enzyme (11). The isomerization step could therefore control the supposed off rate of the coenzyme. Of course, this type of mechanism may be as applicable to the off rate of an allosteric modifier as well as to the substrate or product of the enzymatic reaction.

Off rates of this type may be regulatory in various systems. One wonders, for example, whether the off rate of cyclic AMP (or its displacement by other nucleotides) from the regulatory subunit of a protein kinase could give rise to a hysteretic type response which may be important in the control of the enzymatic activity of this enzyme.

Such a mechanism is not limited to enzyme-related cases. For example, Riggs et al. (12) have shown that dissociation of the lac repressor-operator complex is quite slow, with a half life of 19 minutes in the absence of inducer. This rate is markedly enhanced by the inducer isopropylthiogalactoside and increases proportionally to the inducer concentration between 10^{-7} and 10^{-5}. IPTG apparently induces a conformational change in the repressor while still bound to the DNA which allows dissociation. On the other hand, o-nitrophenyl-β-D-ducoside, an inhibitor of induction, decreases the rate of the operator-repressor dissociation.

The Interconverting Enzyme System: The Organizers of this Symposium have now induced me (a slow conformational change) to add another mechanism which can be treated in a manner similar to that of the hysteretic concept. This mechanism involves the enzymatically catalyzed conversion of one form of an enzyme to another form. In many ways and under certain conditions this mechanism should be equivalent to a unimolecular isomerization of one enzyme form to another. Thus we can start with the following system:

$$E_1 + E_2 \rightleftharpoons E_1 E_2 \longrightarrow E_1' + E_2$$

and $\qquad\qquad\qquad\qquad\qquad\qquad\qquad\qquad\qquad\qquad\qquad$ II

$$E_1' + E_3 \rightleftharpoons E_1' E_3 \longrightarrow E_1 + E_3$$

where E_2 and E_3 are enzymes which interconvert the forms of the enzyme in question, E_1. E_2 and E_3 may be for example, a kinase and a phosphatase. Under certain conditions, the rate of formation of E_1' can be represented as proportional to E_1 and E_2 and the rate of formation of E_1 as proportional to E_1' and E_3. If the rate of the reaction catalyzed by E_2 and E_3 are considered, for the time being, as constant, then

$$E_1 \underset{k_{-2}}{\overset{k_2}{\rightleftharpoons}} E_1'$$

Similarly

$$E_1 S \underset{k_{-3}}{\overset{k_3}{\rightleftharpoons}} E_1'S$$

If one now introduces the fact that E_1S and $E_1'S$ can react to give product, we can expand the mechanism to

$$E_1 + S \rightleftharpoons E_1 S \underset{k_{-3}}{\overset{k_3}{\rightleftharpoons}} E_1'S \rightleftharpoons E_1' + S$$

$$E_1 \underset{k_{-2}}{\overset{k_2}{\rightleftharpoons}} E_1'$$

$$+$$

$$P$$

and this is formally identical to Mechanism I discussed earlier in the paper. If some factor initiates a change in the steps involving the interconversion of E_1 and E_1' or E_1S and $E_1'S$, then we should be able to obtain a hysteretic response for this system.

There are a number of factors which could effect the interconversion of the two forms although some are similar to the isomerization case discussed earlier, including substrate or modifier dependent susceptibility of the interconversion of the two enzyme forms by E_2 and E_3. Another factor, however, would be a change in the activity of E_2 or E_3 or both. Because of the number of ways in which the activity of the enzymes catalyzing the interconversions could be influenced, this characteristic gives this type of system a great deal more flexibility than was obtainable in the simple isomerization case first described. Although I have greatly oversimplified the system, the results should be similar, in terms of some type of hysteretic response, even if the situation is a good deal more complex. In any event, if the hysteretic concept as originally defined has any in vivo meaning, then certainly the metabolic interconversion system must also.

Metabolic Role for Hysteretic Enzyme System: Regardless of the mechanism for inducing the hysteretic response, the most important question is whether such slow responses serve an important role in metabolic regulation. An idea of the time dependence of various aspects of enzyme related rates was discussed earlier with the conclusion that conformational changes or ligand release times which are much longer than a second or so may fall into the hysteretic enzyme category. It is of interest that many of the enzymes which exhibit such responses are regulatory enzymes which occupy key points in metabolic pathways. One can build, therefore, a teleological argument for the fact that the hysteretic response is important in regulation. However, this may not be a very satisfactory approach. Furthermore, pertinent data on this point may be difficult to obtain. For example, such data would necessarily correlate known hysteretic responses to in vivo changes in metabolite concentration. It is difficult to predict what the magnitude of the concentration change would have to be since this will depend upon the kinetic characteristics (like degree of cooperativity) for the particular system. Whether such changes occur is not really a question since an organism at equilibrium cannot survive. Obviously, however, how fast they occur is of considerable importance. Again this type of data may be very difficult to obtain, but it is exactly this point which incorporates the in vivo rationale for the hysteretic response. Thus, one idea is that hysteretic enzymes serve to buffer the cell against large changes in metabolite concentration which would otherwise produce undesirable affects in terms of allosteric control. Another is that, under certain conditions, the hysteretic response could regulate the flux of metabolite into two or more biosynthetic pathways which intersect at a common point. We examined, under some very restrictive sets of conditions how hysteretic enzymes at branch points in metabolic pathways may be able to control the rate at which substrate may get shunted from one pathway to another (1). For a simple branch point, this situation is relatively easy to describe. For a complex point of regulation where several metabolic paths may intersect and where two or more of the enzymes may show an interacting hysteretic response, an analytical description may take some time. In this regard, the kinetic behavior of a full blown allosteric hysteretic enzyme has not been examined for any system, but it is easy to predict that it would indeed be quite complex.

References

1. FRIEDEN, C., *J. Biol. Chem.*, **245**, 5788 (1970).

2. MONOD, J., WYMAN, J. and CHANGEUX, J.-P., *J. Mol. Biol.*, **12**, 88 (1965).

3. KOSHLAND, D. E., NEMETHY, G. and FILMER, D., *Biochemistry*, **5**, 365 (1966).

4. METZGER, B. E., GLASER, L. and HELMREICH, E., *Biochemistry*, **6**, 2021 (1968).

5. KINGDON, H. S., HUBBARD, J. S. and STADTMAN, E. R., *Biochemistry*, **7**, 2136 (1968).

6. GUTFREUND, H., CANTWELL, R., McMURRAY, C. H., CRIDDLE, R. S. and HATHAWAY, G., *Biochem. J.*, **106**, 683 (1968).

7. FRIEDEN, C., *Ann. Rev. Biochem.*, **40**, 653 (1971).

8. FINLAYSON, B., LYMN, R. W., TAYLOR, E. W., *Biochemistry*, **8**, 811 (1969).

9. NOBLE, R. W., REICHLIN, M. and GIBSON, Q. H., *J. Biol. Chem.*, **244**, 2403 (1969).

10. KELLETT, G. W. and GUTFREUND, H., *Nature*, **227**, 921 (1970).

11. HUANG, C. Y. and FRIEDEN, C., submitted for publication.

12. RIGGS, A. D., NEWBY, R. F. and BOURGEOIS, S., *J. Mol. Biol.*, **51**, 303 (1970).

Discussion:

Buc

If one would measure the relaxation spectrum of an hysteretic enzyme one would have a good check of Dr. Frieden's kinetic scheme.

Frieden

I agree, that the relaxation times should correlate with the constants determined by the kinetic analysis.

Buc

In interconverting enzyme systems, one has to consider in addition to the rate constant for the isomerization, which is concentration independent, the rate constants for the interconversion of the various enzyme forms which are dependent on enzyme concentration.

Frieden

That is correct and therefore the scheme for this particular case is more complicated. But at the same time it has the advantage of adding flexibility as well.

Helmreich

If one takes into account the large rate acceleration of the glycolytic system as a whole in response to muscle contraction which occurs without large changes in the levels of metabolites one would have to conclude that glycogen phosphorylase is a hysteretic enzyme, whose activation damps out metabolite fluctuations.

Frieden

Yes, that is what I had in mind. However, it is very difficult to take _in vivo_ observations and apply them directly to an _in vitro_ model.

Stadtman

Would you agree, that a hysteretic response would also result in a sigmoidal response to metabolite concentration.

Frieden

This should entirely depend on the kinetic behaviour of the enzyme before and after the hysteretic transformation. If the allosteric transition of an enzyme occurs rapidly in relation to the rate of the reaction, one would obtain a sigmoidal response to metabolite levels, after the transition from one state to the other. However, the kinetic response at intermediate substrate concentrations could be much more complicated than a sigmoidal response. The kinetic response may represent the weighted averages of two sigmoidal responses.

Stadtman

This would be an additional reason for the biological importance of such hysteretic responses, if one wants to attach any significance to sigmoidal response.

Segal

Would you say, that one biological significance of hysteretic responses is that they damp overshoots in metabolite concentrations.

Frieden

Yes, we postulated initially that it would serve as a buffering system.

Segal

With regard to Dr. Stadtman's question and to Gregorio Weber's idea, that isomerization leads to sigmoidicity of response, we should keep in mind that if the isomerization is very rapid, the situation is different from a hysteretic response.

Frieden

Yes, that is right. I should have made it clear, that the point of interest is that most regulatory enzymes can undergo a variety of changes. Not only are there T and R states, but intermediate states as well and each of these different conformational states has different relaxation times, different responses to substrate concentration, etc..

Fasold

We have been working with an enzyme, histidine ammonia lyase, which initiates the breakdown of histidine. We have been worried about this enzyme, because it always showed an increase in rate no matter what you did with it. If one would extrapolate this behaviour to the situation in the living cell one would have to assume that the enzyme really never reaches its full activity, but always is waiting for the histidine concentration in the cell to pile up before it gets slowly activated.

Krebs

If I interpret your comments concerning the protein kinase correctly, then you mean, that the leaving rate of cyclic AMP from the binding protein is extremly slow and that therefore a ternary complex between cyclic AMP, the regulatory subunit and the catalytic subunit exists.

Frieden

Kinetically at least, that seems to be the case.

Helmreich:

Is phosphofructokinase a hysteretic enzyme?

Frieden

Oh, sure at least with respect to pH changes it is a hysteretic enzyme.

O. Wieland

Dr. Frieden, would you think that oscillations can be explained by a hysteretic response?

Frieden

Well, I was afraid, that someone would ask me that question. Actually, I never have been able to think it out in my mind what will happen past the first cycle of oscillation. Therefore I really don't know, what the relationship of oscillations to hysteresis may be.

Pette

To the question of Dr. Helmreich, concerning phosphofructokinase I would like to mention the work of Dr. Hofer, who studied the association-dissociation of the rabbit muscle enzyme. He found that the associated form of that enzyme differs considerably from the dissociated enzyme. At high concentrations, the enzyme is present as a tetramer and in this form it is inhibited to a much lesser extent by ATP and it has high affinity for fructose-6-P. The cellular concentration of the enzyme is so high that it should exist in the tetrameric state. The polymerization of the enzyme is probably under metabolite control.

Frieden

These are very interesting experiments. We did different experiments with phosphofructokinase but our results agree with these conclusions.

Cori

After listening to all this, I really would like to have a definition of hysteresis. Is it just the slowness of the reaction, why you call it a hysteretic reaction?

Frieden

Hysteresis as I defined it and as it was used earlier by Gregorio

Weber means really as I interpret the word " to lag behind ". And this is exactly what I had in mind.

Palm

By the definition given for this type of mechanism one would assume that hysteresis would occur in many systems. Now, take glutamic dehydrogenase: GTP is released slowly from the enzyme. Hysteresis only refers to the slow release of the ligand but it does not say how the reaction is influenced by the slow release of the ligand. Although hysteresis may easily be deduced from the kinetic treatment of the dissociation of the ligand it is not at all clear, whether it has biological significance.

Frieden

The function of glutamic dehydrogenase depends on the free nucleotide concentrations. At least in the case of the liver enzyme the direction of the enzymatic reaction, whether it uses NADH or NAD or NADP as cofactor, depends on concentration. Therefore it could be important, which one of the nucleotides happens to be bound to the enzyme, and thus buffers the change in pyridine nucleotide concentrations.
This buffering effect could be important.

Why Are Enzymes Interconvertible?

Richard H. Haschke, Ludwig Heilmeyer, Jr., and Ernst Helmreich

Physiologisch-chemisches Institut, Universität Würzburg, Würzburg/Germany

A logical progression from an enzyme with no regulatory capacity (i.e. active in the presence of substrate) to an enzyme with the capacity for activity modulation is not hard to visualize. There is a definite selective advantage gained by an organism that does not waste energy in producing unnecessary products. At present, in evolutionary time, the wide occurrence of allosteric control provides ample evidence of this fact (1). We want to consider, however, a possibly higher level in the sophistication of enzyme regulation; that is covalent interconversion of enzymes. If indeed this represents a more sophisticated type of metabolic control, examples should provide evidence of an advantage to be gained by such modification.

All known enzymes subjected to allosteric and covalent modification are oligomeric proteins. Since some of the enzymes with the highest turnover numbers are found to be single chain enzymes such as catalase (2), multi-chain enzymes are not exclusively selected for higher catalytic efficiency but for a broader and more effective response to regulative modifiers. From what we know today it seems that the monomeric subunits of oligomeric enzymes show only low catalytic activity (3-5). An explanation for this phenomenon may lie in natural selection for a polypeptide chain that is not necessarily a fully active catalyst per se, but one that has the proper requisites to become active in regulated catalysis as an oligomeric enzyme.

The possibility that the monomer is not capable of binding the substrates is excluded in at least the example of glycogen phosphorylase in which glycogen, glucose-1-P and 5'-AMP (an allosteric effector) still bind to the inactive predominantly monomeric apophosphorylase b with the same association constants as to the active dimeric enzyme (5,6). In the case of another

allosteric protein (glutamine synthetase) the subunit binds all
of the necessary effectors and substrates, yet no appreciable
catalysis occurs (7). A possible explanation for these observations
is that the subunits possess all of the requisite chemical functions,
but are in a catalytically inactive conformation. Data in support
of this argument have been recently provided by Dr. Feldmann from
this laboratory.

Phospho-Dephospho Hybrids of Phosphorylase

Phosphorylase on CNBr-activated Sepharose-4B matrix	Activity (arbitrary units)	
	+ AMP (1 mM)	- AMP
Dimer b	49	0
Monomer b (imidazole-citrate, washed) residual activity	05	0
Dimer hybrid phos b - phos a (PCMB monomer a, treated with ME)	32	14
Conversion of hybrid to phos a dimer (kinase)	48	38
Dimer a	78	68
Monomer a (PCMB, washed with ME) residual activity	06	06
Dimer hybrid phos a - phos b (PCMB monomer b, treated with ME)	45	18
Dimer hybrid assayed +5 mM G-6-P	41	10
Conversion of hybrid to phos b dimer (phosphatase)	73	02

Phosphorylase b was dissociated into monomers in 0.8 M imidazole
citrate buffer pH 6.2 (8). p-Chloromercuribenzoate was used to
dissociate phosphorylase into monomers (9). The PCMB was
removed with 0.1 M 2-mercaptoethanol. The experiments were
carried out at 20°C in 50 mM glycero-P-buffer, pH 6.8. For
further details, see text. The differences in activity of dimer b
and dimer a are due to different amounts of enzyme used in the
two experiments.

Stable phospho-dephospho hybrids of glycogen phosphorylase may be formed if phosphorylase b monomers, covalently linked to Sepharose, are associated with monomeric subunits of phosphorylase a formed by treatment with p-Chloromercuribenzoate (9). Hybridization is achieved in three steps: 1) Coupling of phosphorylase b to CNBr activated Sepharose, 2) dissociation of the insoluble enzyme into subunits by imidazole-citrate (8) and 3) reassociation of the monomeric subunits of phosphorylase b and a. This procedure produces stable phospho-dephospho hybrids. As is seen in the table, the newly formed species is differentiated from a mixture of phosphorylase b and a by its glucose-6-P sensitivity. Therefore, there is an induction effect in which activity appears on combination of two subunits with little catalytic capability. This experiment shows that the subunits of phosphorylase cannot express their full catalytic potential because they lack the proper conformation.

Functional organelles and interconvertible multienzyme complexes

If one agrees that oligomeric enzymes, as compared with single chain enzymes, have a vastly increased regulatory potential, one now might ask if the arrangement of several of these enzymes in the form of a functional organelle such as is found with glycogen phosphorylase in muscle, confers added regulatory advantages. The protein-glycogen complex isolated from rabbit muscle contains both the anabolic and catabolic enzymes of glycogen metabolism. Moreover the regulation of several of these enzymes is modified in the complex as compared to a mixture of the soluble component proteins (10,11,12). Phosphorylase kinase is activated by calcium ions at the same concentration required for initiating muscle contraction, thereby linking the two functions (10,11).

Simultaneously with the kinase activation and resultant conversion of phosphorylase b to a, phosphorylase phosphatase activity is reversibly inhibited, thus preventing wasteful ATPase activity (12). The reversible inhibition of the phosphatase is lost when the protein-glycogen complex is dissociated. The phosphorylase a produced in this complex also shows properties not found in the solubilized enzyme in that its activity is modulated by the allosteric effector glucose-6-P (11). The behavior of phosphorylase b in the glycogen complex is also different and it may explain why phosphorylase b is inactive in a resting muscle. It has been pointed out that the activity of purified phosphorylase b under simulated rest conditions is several orders of magnitude too high mainly as a result of the relatively large amount of 5'-AMP present in the resting muscle cell (13,14). When AMP activation of phosphorylase b is measured in the complex, it is found that the enzyme is activated to a much lesser extent even though the nucleotide is still bound to phosphorylase b (15). Thus, in this complex heterologous protein-protein interactions play an important regulatory role and add another facet to the modulation of enzyme activity. It appears, however, that nature has paid a price for gaining a rapid on-off switching mechanism of glycogenolysis which is synchronized with muscle contraction. If the total activity of glycogen phosphorylase is measured in the intact stimulated muscle by the rate of lactic acid formation (assuming phosphorylase to be the rate limiting reaction), one finds that at most only 5-10% of the maximum potential phosphorylase activity is used. This is based on the specific activity of isolated purified phosphorylase (16). Since natural selection is usually for efficient systems, it is hard to imagine why a protein is synthesized in such excess that only one tenth is needed. Since phosphorylase is ca. 10 mg/ml of intracellular water, this amounts to ca. 10% of the soluble intracellular protein. A more plausible explanation is that the specific activity of the enzyme is much lower when it is integrated in the protein-glycogen organelle. Thus in this case the organization of an enzyme, in a multienzyme complex, resulted actually in a decrease in catalytic efficiency.

There appears to be an example, however, in which the formation of the multienzyme pyruvate dehydrogenase complex affords an increase in the catalytic rate, as compared to the individual enzyme in solution. This complex is subject to covalent control by a phosphorylation-dephosphorylation of the pyruvate decarboxylase moiety (17). The response time is, however, much slower than that of the phosphorylase system. The modification occurs over time periods of hours and correlates with nutritional changes of the cell. The resulting modified enzyme is also much more stable (18). This suggests that the biological function of interconversion of this multienzyme complex differs from that of an oligomeric enzyme composed of homologous subunits. In a multienzyme complex, protein interactions may be much more restrained than in oligomeric enzymes. Accordingly, reactions of enzymes in this mammalian multienzyme complex seem not to be allosterically facilitated. Thus, in a multienzyme system which is structurally integrated in mitochondria as in the case of the pyruvate dehydrogenase complex, interconversion might serve a different purpose. It might be compared with induction and repression of enzyme synthesis de novo. Both respond in mammalian tissues to nutritional and hormonal stimuli.

Stabilization of enzyme conformations by covalent modification

The chemical modification freezes an enzyme in a conformation required by cellular function. This may be an active or inactive conformation, or one which responds differently to allosteric regulators. Chemical modification overrides allosteric control and stabilizes certain conformations. Several well documented examples are found here. The first, E. coli glutamine synthetase has a different sensitivity to allosteric modifiers after adenylation. The γ-glutamyl transferase activity is more strongly depressed by cumulative feedback inhibitors after covalent modification (19). The second example is glycogen synthetase, which upon phosphorylation by the general protein kinase is converted from the I form (does not require glucose-6-P for activity) to the D form (requires glucose-6-P for activity) (20). The effect is then to freeze the glycogen synthetase in either an active or inactive conformation. Another case is the formation of phospho-dephospho hybrids of glycogen phosphorylase. In the fully phosphorylated species (phosphorylase a), the activity is not affected by glucose-6-P. However, the hybrid, which is composed of a phosphorylated and non-phosphorylated subunit, is sensitive to glucose-6-P (21).

The importance of protein-protein interactions in enzyme interconversions

Any discussion of covalent modification as a mechanism for regulation should include the control of the modifying enzymes. A possibility would be a rapid turnover of the modifying enzyme with the actual amount of active catalyst being determined by the relative rates of synthesis and degradation. The quantity of protein substrate converted would then be directly proportional to the amount of modifying enzyme. This type of regulation is wasteful because of the large energy requirement for enzyme synthesis. A more efficient control can be achieved by a reversible protein-protein interaction to modulate the activity of the modifying enzyme. In the case of the protein kinase a regulatory subunit complexes with the catalytic subunit in the absence of cyclic AMP and results in an inactive enzyme. Cyclic AMP activates the kinase by binding to the regulatory subunit and dissociating the complex (22).

Another interesting example of protein-protein interaction is the effect of 5'-AMP on the activity of phosphorylase phosphatase. The AMP binds to phosphorylase a and produces a substrate that is only poorly attacked by the phosphatase (16). When the two enzymes are contained in the protein-glycogen complex, in contrast to solution, the nucleotide has no inhibitory effect. This is presumably due to protein-protein interactions that prevent the AMP from producing a conformational change as occurs with the purified enzymes. The interactions between the proteins are relatively weak, because a ten fold dilution of the complex is sufficient to dissociate the components and to allow AMP to regain its potent inhibitory action (12).

A now classical example of a protein-protein interaction which modifies the specificity of an enzyme is the lactose synthetase reaction. α-Lactalbumin, the specifier protein which is only produced in the lactating mammary gland, binds to a galactosyltransferase and alters its acceptor specificity. The enzyme-lactalbumin complex now transfers the galactosyl residue from UDP-galactose to glucose rather than to N-acetyl-glucosamine and thus functions as a lactose synthetase. The considerable sequence homology between the specifier (α-lactalbumin) and lysozyme has interesting evolutionary implications in that both enzymes catalyze the same general type of glycosyl transfer reaction (23,24). These examples all involve regulation of enzyme activity without covalent modification, although, in the first two cases the enzyme subjected to control by protein interaction was itself a modifying enzyme involved in the covalent modification of another enzyme.

Covalent modification allowing for synchronous regulation of different pathways

There is now good evidence mainly from the work of E.G. Krebs that the interconvertible enzyme systems which catalyze two independent pathways, glycogen synthesis and glycogen breakdown, are joined by a modifier enzyme common to both pathways. The common link is the general protein kinase. When activated by the cyclic AMP signal, the kinase phosphorylates both phosphorylase kinase and glycogen synthetase, activating the former and inactivating the latter (20). This concerted effect initiates glycogen breakdown while inhibiting glycogen synthesis.

The cyclic AMP activated protein kinase appears especially suited for the function of synchronizing several pathways in that the enzyme is relatively unspecific towards the substrate. The phosphorylation of serine residues in a variety of protein substrates readily occurs in vitro (25). In terms of biological function, however, the protein kinase appears rather specific, since in muscle, the enzyme phosphorylates glycogen synthetase I and phosphorylase kinase, but not phosphorylase b (26). Therefore, the specificity of this enzyme in vivo seems to be determined by the recognition of a certain structure of the protein substrate. The synchronization of several pathways in a single cell by this protein kinase can be extended to parallel regulation of metabolic pathways in different cells. When epinephrine is discharged, the resulting second messenger (cyclic AMP) can activate the protein kinases in a wide variety of organs of differentiated cells (27). This results in the coordinated acceleration of catabolism in several organs.

Identity of interconverting enzymes

As nature evolved more complex control systems as interconversion cascades, additional enzymes had to be synthesized. An unnecessary expenditure of energy could be avoided by employing the same modifying enzyme for more than one function in an interconversion cascade, as has been discussed for the protein kinase. Another possibility is to utilize energy rich bonds in the chemical modification reaction, which would allow one enzyme to catalyze both the forward and reverse reaction. In the case of E. coli glutamine synthetase, enzyme modification causes inactivation through the formation of an adenylyl-tyrosine bond with a high transfer potential. The adenylylating enzyme, after association with a "specifier" protein, reactivates glutamine synthetase by now transferring AMP from the tyrosyl residue to P_i rather than to water. The result is a recovery of a high energy bond as ADP (19).

Reversal of the phosphorylation of yeast phosphorylase has recently been shown by Fischer et al. (28) to involve the same phosphorylase b kinase, which activates the enzyme by phosphorylation and conversion to the a form. The same enzyme seems also to catalyze the dephosphorylation reaction (conversion to the b form). This is quite different from the case of rabbit muscle phosphorylase a in which the seryl-phosphate ester is a low energy bond and not reversible by the kinase. The phosphorylated residue in yeast phosphorylase has not yet been identified (28).

Interconversion cascade as a means of signal amplification

Recently, A. Sols has summarized the cellular concentrations of enzymes participating in interconversion cascades (29). The concentrations are large and differ by no more than ca. one order of magnitude. Moreover these enzymes seem to have relatively low turnover numbers. Therefore, a catalytic rate amplification is difficult to visualize. An amplification of signal response is more likely. The cellular concentration of chemical signals such as Ca^{++} and cyclic AMP are in the order of 10^{-6} M (30). Usually the concentration change which effectively turns on the interconversion cascade is not greater than one order of magnitude. Specific binding proteins for each signal with dissociation constants in the same range would be expected to be found (see G. Zubay, this volume).

An example for signal amplification is the response of the glycogen phosphorylase interconversion cascade to cyclic AMP. Concentrations of cyclic AMP in the micro molar range produce an effect resulting finally in the formation of tenth mM concentrations of active phosphorylase a (16). This amplification actually results in a greater amount of active catalyst. One may easily visualize, that in a cascade system where one catalyst activates another catalyst, an amplification effect is more readily achieved than merely by binding micro molar amounts of cyclic AMP directly to the last anzyme to be activated in an interconversion cascade (in this case phosphorylase b). The same applies to the response of the phosphorylase interconversion cascade in muscle to calcium. The work of S. Ebashi, E.G. Krebs and E.H. Fischer and their associates on the reversible activation of the isolated phosphorylase b kinase (31, 32) and on the regulatory properties of the glycogen organelle from rabbit skeletal muscle suggests that the interconverting enzymes of glycogen phosphorylase in the living muscle can respond without delay to the 10^{-6} M concentrations of calcium (11). This signal triggers muscle contraction and phosphorylase *b* kinase activation. As is the case with cyclic AMP, this small amount of calcium is capable of quickly producing a *ca.* hundred fold larger amount of phosphorylase *a*.

Conclusions

My remarks were by no means intended to give a comprehensive or even representative summary of this symposium. Obviously some of the most interesting aspects I have not even touched. For example, how important chemical modification actually is for turning enzymatic activity on or off. Here the I strain mice come to mind (see C.F. Cori, this volume).

Moreover, I only remind you of the exciting aspects of regulation involving protein interaction and chemical modification of histones, enzymes and proteins involved in transcription and translation of genetic information (see G. Dixon *et al.*; O. Hayashi *et al.*; Zillig, Rabussay *et al.*, this volume). Nor did I even mention interconversion reactions involving limited proteolysis. (see Chapeville *et al.*, Katunuma *et al.*, this volume). I have obviously only taken some examples from a rather specific area of this ever broadening field and have restricted my remarks mainly to the properties of the phosphorylase system. I have continued more or less where Dr. Cori started out at the beginning of our meeting. I have done so however, not only because phosphorylase is an "historical" enzyme and the oldest example of an interconvertible enzyme and also not entirely because I am more familiar with the work on phosphorylase, more important was that I felt that one could from the properties of this system arrive at certain generalizations about interconvertible enzymes. I have therefore given a highly speculative and personal account, which needs to be criticized and clarified. Perhaps some points will be taken up in the following panel discussion. One thing however seems clear to me: The thread which may guide us through the labyrinth of metabolic regulation with its almost baroque diversity of regulatory variations of ever increasing complexity will be an understanding of homologous and heterologous protein interactions about which we still know rather little.

Here one is reminded of a sentence in Jaques Monod's recent book: Le hasard et la necessité "What concerns regulation by means of an allosteric protein, the important point is that everything is possible".

References

1. STADTMAN, E.R., Advances in Enzymology, 28, 41 (1966).
2. NICHOLLS, P., and SCHONBAUM, G.R., The Enzymes, 8, 147 (1963).
3. CHAN, W.C., Biochem. Biophys. Res. Commun., 41, 1198 (1970).
4. GRAVES, D.J., TU, Jan-I, ANDERSON, R.A., MARTENSEN, T.M., and WHITE, B.J., this volume.
5. FELDMANN, K., unpublished results.
6. KASTENSCHMIDT, L.L., KASTENSCHMIDT, J., and HELMREICH, E., Biochemistry, 7, 3590 (1968).
7. HOLZER, H., and DUNTZE, W., Annual Reviews of Biochemistry, 40, 345 (1971).
8. HEDRICK, J.L., SHALTIEL, S., and FISCHER, E.H., Biochem., 8, 2422 (1969).
9. MADSEN, N.B., and GURD, F.R.N., J.Biol.Chem., 223, 1075 (1956).
10. MEYER, F., HEILMEYER, Jr., L.M.G., HASCHKE, R., and FISCHER, E.H., J. Biol. Chem., 245, 6642 (1970).
11. HEILMEYER, Jr., L.M.G., MEYER, F., HASCHKE, R., and FISCHER, E.H., J. Biol. Chem., 245, 6649 (1970).
12. HASCHKE, R., HEILMEYER, Jr., L.M.G., MEYER, F., and FISCHER, E.H., J. Biol. Chem., 245, 6657 (1970).
13. HELMREICH, E., and CORI, C.F., Advances in Enzyme Regulation, 3, 91, (1965).
14. MORGAN, H.E., PARMEGGIANI, A., J. Biol. Chem., 239, 2440 (1964).
15. HASCHKE, R., unpublished results.
16. FISCHER, E.H., HEILMEYER, Jr., L.M.G., and HASCHKE, R., Current Topics of Cellular Regulation, 4, 211 (1971).
17. LINN, T.C., PETTIT, F.H., HUCHO, F., and REED, L.J., Proc.Natl. Acad.Sci., 64, 227 (1969)
18. SIESS, E., WITTMAN, J., and WIELAND, O., Hoppe-Seyler's Z. Physiol. Chem., 352, 447 (1971).
19. SHAPIRO, B.M., and STADTMAN, E.R., Ann. Rev. Microbiol., 24, 501 (1970).
20. SODERLING, T.R., HICKENBOTTOM, J.P., REIMANN, E.M., HUNKELER, F.L., WALSH, D.A., and KREBS, E.G., J. Biol. Chem., 245, 6317 (1970).
21. HURD, S.S., TELLER, D., and FISCHER, E.H., Biochem.,Biophys. Res. Commun., 24, 79 (1966).
22. TAO, M., SALAS, M.L., and LIPMANN, F., Proc.Natl.Acad.Sci., 67, 408 (1970).
23. BREW, K., VANAMAN, T.C., and HILL, R.L., Proc. Natl. Acad. Sci., 50, 491 (1968).

24. CASTELLINO, F.J., and HILL, R.L., J. Biol. Chem., 245, 417 (1970).
25. REIMANN, E.M., WALSH, D.A., and KREBS, E.G., J. Biol. Chem., 246, 1986 (1971).
26. KREBS, E.G., personal communication.
27. KUO, J.F., and GREENGARD, P., Proc. Natl. Acad. Sci., 64, 1349 (1970).
28. FISCHER, E.H., COHEN, P., FOSSET, M., MUIR, L.W., and SAARI, J.C., this volume.
29. SOLS, A., and GANCEDO, C., Biological Control Mechanisms, E.Kun and S. Grisolia, ed., in press.
30. POSNER, J.B., STERN, R., and KREBS, E.G., J. Biol. Chem., 240, 982 (1965).
31. OZAWA, E., HOSOI, K., and EBASHI, S., J. Biochem., 61, 531 (1967).
32. BROSTROM, C.O., HUNKELER, F.L., and KREBS, E.G., J. Biol. Chem., 246, 1961 (1971).

General Discussion

Stadtman (Moderator)

I think, Dr. Helmreich, you have covered every important aspect of the problem. It seems to me, that the questions, which you have asked are very pertinent to the discussion this morning and they will form the basis of most of what follows. I would like to say at the outset that we have a small group of people sitting at this table, whose function it is to lead the discussion, to raise additional questions along the lines of those which have been presented by Dr. Helmreich.

I would urge however, that all of the members of the audience be prepared to enter into the discussion, this panel is only to serve as a catalyst and we hope that we can arouse a general discussion through participation of others.

It is hard to decide where to begin. So many points have been raised by Dr. Helmreich which are worthy of discussions. One point we might get some further opinion on is whether or not there is a fundamental difference in the regulatory control of complex systems such as the phosphorylase-glycogen synthetase organelle and the multienzyme-systems, of the type Dr. Reed and Dr. Wieland are working with. Dr. Helmreich seems to think, that there is. I am not sure that one can generalize on the basis of only two different systems and I wonder if the multienzyme-complex of the type that Reed has demonstrated is fundamentally very different from enzyme aggregates such as the glycogen-synthetase-phosphorylase organelle in which glycogen synthetase and glycogen phosphorylase make up a complex exhibiting unique characteristics. While it is true that glycogen is an essential component of the enzyme aggregate its function may be similar to that of membranes which serve as a solid surface upon which metabolically related enzymes can be organized to facilitate metabolism. In any case are there members of the panel, who would like to comment on this subject.

Fischer

I think that Ludwig Heilmeyer has some data on that point. The question was whether the enzymes in the glycogen organelles respond differently to allosteric effectors than when they are present in the isolated, solubilized state. The same question was raised by Dr. Cori in his introductory remarks, and it is, indeed, difficult to answer. For example, what is the state of activity of phosphorylase in the muscle? Is there enough AMP to activate phosphorylase b and how important is this form of activation as compared to phosphorylation of the enzyme? As was pointed out, no more than 5 percent of the phosphorylase needs to be activated to account for all the lactic acid produced under conditions of maximal muscle contraction. As also pointed out by Dr. Cori, if one considers the intracellular concentrations of all substrates, intermediates and allosteric effectors (and their fluctuation), one should have enough active phosphorylase to account for most of the lactic acid produced during stimulation. Yet, in resting muscle, practically no lactic acid is formed. How is this possible?

Perhaps Dr. Heilmeyer could show some of his results which indicate that AMP does not activate phosphorylase when this enzyme is bound to muscle glycogen particles.

Heilmeyer

With respect to the interaction of glycogen phosphorylase with AMP there are two observations made in intact muscle, which are not in agreement with the behaviour of the purified enzyme. First it is well established that the dephosphorylation of crystallized phosphorylase *a* to the *b* - form by phosphatase action is completely blocked by AMP in an *in vitro* - system. (Nolan, F.C., Novoa, W.B., Krebs, E.G. and Fischer E.H., Biochem. 3, 542, 1966). The concentration of 0.5 mM AMP, found in resting muscle is ca. 100 fold higher than that exerting half maximum inhibition *in vitro*. Therefore, one should expect this AMP inhibition also to occur in the living cell, which is certainly not the case. A second point, which Dr. Fischer raised, is that phosphorylase *b* activity in the living cell must be somehow supressed and that the enzyme cannot be activated by AMP. Dr. Haschke and I studied therefore a glycogen organelle, isolated from rabbit muscle, which contains all the enzymes involved in glycogen metabolism and in interconversion of phosphorylase.

In Table 1 the action of the phosphatase on phosphorylase *a* in the glycogen organelle is shown. The substrate of the phosphatase is ^{32}P labeled phosphorylase *a* and the release of radioactive phosphate is measured. In that system only a very small inhibition of the phosphatase reaction by 1 mM AMP is observed. If the complex is dissociated simply by 20 fold dilution, the AMP sensitivity of the phosphatase reaction appears. Since AMP inhibition results from a binding of this nucleotide to the substrate phosphorylase *a*, rather than to the phosphatase itself, these experiments indicate that probably heterologous protein interactions render phosphorylase insensitive to this effector. Upon dilution also the phosphatase changes its behaviour, as is shown in Fig. 1. The affinity of the phosphatase for phosphorylase a is much higher in the isolated state than in the glycogen organelle.

Table 1

EFFECT OF NUCLEOTIDES AND Ca^{++} ON PHOSPHORYLASE PHOSPHATASE ACTIVITY IN A DILUTED AND UNDILUTED 30 PELLET SUSPENSION

EFFECTORS	% INHIBITION			
	30 PELLET SUSPENSION UNDILUTED		30 PELLET SUSPENSION DILUTED 1 : 20	
	SUBSTRATE PHOSPHORYLASE a	PHOSPHOPEPTIDE	SUBSTRATE PHOSPHORYLASE a	PHOSPHOPEPTIDE
NONE	0	0	0	0
ATP-Mg-Ca	73	51	85	11
ATP-Mg	12	19	74	11
AMP-Mg-Ca	5	--	64	--
IMP-Mg-Ca	7	0	62	0

The 30 pellet material, the protein-glycogen complex, was suspended in 50 mM glycerophosphate, 1 mM EDTA, pH 6.8, yielding a protein concentration of ca. 100 mg/ml. Phosphatase activity was assayed by following the release of radioactivity from ^{32}P labeled phosphorylase a or from a phosphopeptide obtained by CNBr cleavage of phosphorylase a. Each sample contained 5 mM Mg^{2+} and 0.3 M free Ca^{2+} when added. Nucleotide concentration was 1 mM and inhibition was measured 30 sec after addition of the nucleotides.

AMP activation of phosphorylase b was measured with AMP-1-N oxide, because this derivative is not attacked by AMP deaminase, which is present as a contaminant in the protein-glycogen complex. As you can see in Fig. 2, AMP-1-N oxide activation of phosphorylase b bound to the glycogen particle amounts to only ca. 1/4 of that of the isolated enzyme. Fig. 3 shows that this lack of response of phosphorylase b is not due to an inability of the complex bound enzyme to bind AMP. Use was made of Dr. Fasold's thioether-nitro-benzoic acid derivative of AMP, which by covalent linkage of the nucleotide moiety

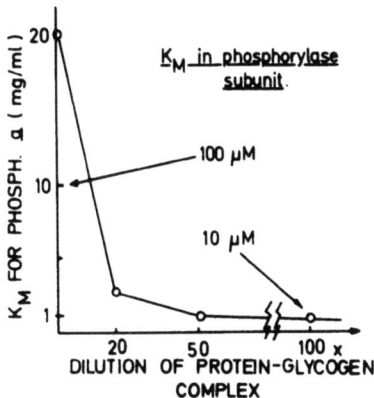

Fig. 1. The protein concentration of the undiluted protein-glycogen suspension was ca. 100 mg/ml. Phosphatase activity was measured as a function of substrate concentration at each dilution as described in Table 1

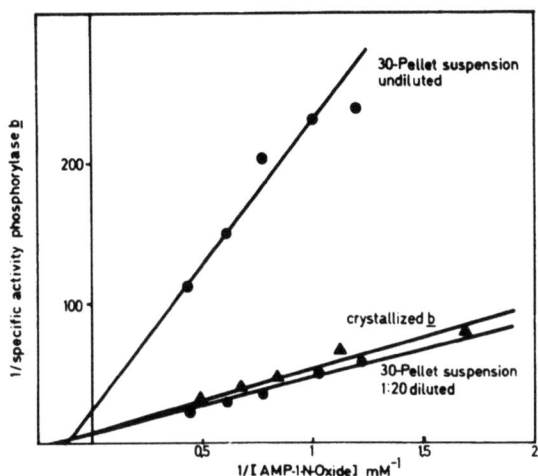

Fig. 2. The undiluted 30 pellet suspension contained ca. 10 mg phosphorylase b/ml. Activity was measured as a function of AMP-1-N oxide concentration by following the production of glucose from glycogen in presence of 20 mM arsenate. To the diluted suspension glycogen was readded to the same concentration of 10 mg/ml as in the concentrated material

of this derivative to phosphorylase b activates the enzyme. (This volume) This AMP derivative reacts with phosphorylase integrated in the glycogen organelle with the same rate as with the purified enzyme. In Fig. 4 we demonstrate differences in the reactivity of cysteinyl sidechains in isolated and complex bound phosphorylase.

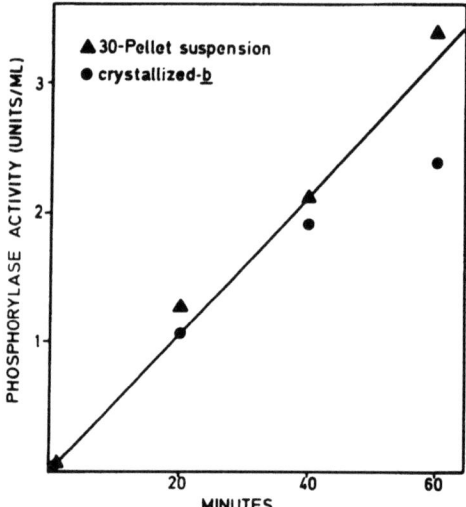

Fig. 3. Crystallized phosphorylase b, freed of AMP, was used in the same concentration of 15 mg/ml as in the 30 pellet suspension. The concentration of the modifying agent

(5-P-Ribose-AMP-6-S-⟨O⟩(COOH)(NO$_2$)) was in both experiments 5 mM. The reaction was stopped by 1000 fold dilution, and activity of phosphorylase was determined without addition of AMP

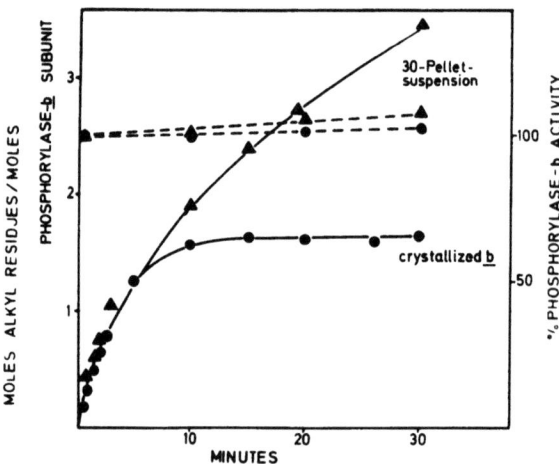

Fig. 4. Carboxyamidomethylation of phosphorylase b in the presence of 10 mM ^{14}C-iodacetamide was followed by measuring TCA precipitable radioactivity. Aliquots were removed and enzyme activity was measured after 1000 fold dilution. The alkylation reaction was stopped by addition of 50 mM cysteine and phosphorylase was isolated from the 30 pellet suspension through chromatography on DEAE cellulose prior to the determination of covalent bound radioactivity

It is known that, without loss of activity or change in quarternary structure in crystalline phosphorylase b two cysteinyl residues per monomer of 100,000 daltons can be carboxyamidomethylated by iodoacetamide. If a third cysteinyl residue is substituted, the enzyme becomes inactive. In contrast to the isolated enzyme, phosphorylase in the glycogen organelle tolerates modification of three SH-groups without loss of activity. This suggests that phosphorylase has a different conformation in the glycogen organelle than in the isolated state.

Helmreich

One might conclude from these experiments, that all what one can measure with an allosteric enzyme in the isolated state is its regulatory potential, but one does not know, how much of that regulatory potential is utilized in the living cell. The glycogen organelle is a very suitable system, to find out how much of the regulatory potential of the isolated enzymes is actually expressed in the living cell.

Madsen

I just want to comment on the point that phosphorylase b in solution is much too active whereas this enzyme is inactive in resting muscle. But perhaps what one has forgotten is the effect of pH. Usually the activity of the isolated phosphorylase b in the presence of activators and inhibitors is measured at pH 6.8, if the pH decreases the enzyme is much less active. For example, at pH 6.2 phosphorylase b in the presence of all the modifiers has virtually no activity. We have observed (1) a very steep rise in activity as the pH is increased from 6.4 to 6.7 in the presence of allosteric modifiers and low substrate concentration. The activity reaches a maximum at pH 7, remaining constant to more alkaline pH's, with a value which is 1.5 % of V_{max}. In addition to the alkaline shift, the steep sigmoidal shape of the activity vs. pH is in marked contrast to the normal bell-shaped curve but is very similar to that found by Trivedi and Danforth (2) for muscle phosphofructokinase. Therefore I would like to suggest that a decrease in pH might account for the inactivity of phosphorylase b in the glycogen organelle, or for that matter in the resting muscle. pH changes in the course of muscle contraction are known to occur. Next, I would like to comment on Ernst Helmreich's experiments with the phosphorylase monomers bound to activated sepharose. You may recall that Dr. Buc (3) found that phosphorylase b is active in the absence of AMP when the concentrations of P_i are very high. We have extended the work of Buc and have shown that other anions besides P_i at very high concentrations also activate phosphorylase b in the absence of AMP (4). Their activity follows the Hoffmeister series. One might say, that these anions force the enzyme into an active conformation by ionic interactions. Have you studied the effects of anions on the activity of your monomers attached to sepharose?

References

(1) Madsen, N.B., in "Symposia of the 1st Meeting of the Pan American Association of Biochemical Societies", Caracas, 1971; Academic Press, New York, in press.
(2) Trivedi, B. and Danforth, W.H., J.Biol.Chem. 241 : 4110 (1966).
(3) Buc, H., Biochem.Biophys.Res.Commun. 28 : 59 (1967).
(4) Engers, H.D. and Madsen, N.B., Biochem.Biophys.Res.Commun. 33 : 49 (1968).

Helmreich

No, we have not tried that yet. These are rather recent experiments. But, if I might comment on that point: If one accepts my argument, namely that the monomer of an oligomeric enzyme has all the requisites for activity, but it lacks the proper conformation, it would seem reasonable to expect that a variety of factors including salting out effects by anions or hydrophobic interactions as in the case of formamide as shown by Donald Graves might induce the conformation of the monomer required for activity.

Graves

We all agree, that interactions of monomers in an oligomeric enzyme are important for regulation. But we have to wait and see whether the monomers are active or not. Probably, as more monomers of oligomeric enzymes are studied for activity, some will be found to be active and some which are inactive. On your slide, the monomers showed certainly activity, 16 %; although we have to be careful to express this activity quantitatively, because there may be changes in the affinity for substrates and activators when the monomers are formed and bound to a matrix. Activity measurements have to be done under saturation conditions. My other question is directed to Dr. Heilmeyer. It concerns the AMP-N_1-oxide. Did AMP-N_1-oxide inhibit phosphorylase phosphatase in the glycogen organelle?

Heilmeyer

We have studied the AMP-N_1-oxide effect during flash activation in the glycogen organelle. That is in the presence of Ca^{++}, ATP-Mg^{++}: Phosphorylase b is rapidly phosphorylated under these conditions also in the presence of AMP-N_1-oxide, but there is no inhibition of the phosphatase. If we dissociate the phosphatase from the glycogen complex by dilution, the phosphatase becomes sensitive to inhibition by the AMP-N_1-oxide.

Buc

I really do not want to discuss phosphorylase, but it seems to me that there remains one important problem. It might be difficult to fully activate phosphorylase b by allosteric modification in the living cell, because the allosteric constant is very large and in the order of 1000. But we should keep in mind, that in order to stabilize the active conformation, allosterically, even in the case of such a big allosteric equilibrium constant we only need about 4 kcal per mole. This is really not much energy. It can easily be furnished by subunit interactions of one way or the other. Therefore, since conformational changes are energetically so cheap, the main question that arises is: Why was nature not content to use that trick, that means stabilization of conformation by noncovalent interaction, but instead evolved stabilization of conformation by covalent modification. This still puzzles me. Another point relates to the coupling of different metabolic pathways. Some examples, discussed at this symposium suggest, that coupling of different pathways is effectively achieved by heterologous association of two different enzymes. This involves direct protein interactions and does not rely on the metabolite pool. Another point, which I would like to make, relates to Dr. Dixon's experiment but it also applies to phage infection, sporulation a.s.o.

Apparently for development and differentiation, we need a time ordered sequence of reactions. This is made very clear by the modification of histones in Dr. Dixon's experiments, where you have a series of changes which indicate time ordered transformations each of which becomes irreversible by covalent modification. This seems to me are examples of control of events which cannot be achieved by allosteric regulation. Here, we have indeed gained something new, by employing covalent modification.

Stadtman

I would like to come back to the question of aggregation and point out, what has been alluded to already that aggregation of enzymes in the form of the multienzyme organelle, which Dr. Fischer and Dr. Krebs have looked at effects the catalytic efficiency and the regulatory potential of the enzymes. You may achieve the same through interaction of one enzyme molecule with another enzyme molecule of the same kind. This is the case with high concentrations of glutamic dehydrogenase, for example. I think, that one of the real problems enzymologists must face, is to measure catalytic activities at high enzyme concentrations since these are conditions, which most closely resemble the state of the enzyme in the living cell.

For example, glutamine synthetase accounts for about 0.5 % of the total protein in the E. coli cell. In vitro studies have shown that at such high protein concentrations the enzyme undergoes paracrystalline aggregation in response to relaxation and subsequent tightening as provoked by removal and addition of divalent cations. This raises the possibility that a similar phenomenon may occur intracellularly and brings up the question of whether paracrystalline aggregation affects the catalytic efficiency of the enzyme. The more general question of whether intracellular aggregation of enzyme molecules is physiologically important has been previously discussed by Dr. Frieden.

Frieden

In the case of glutamic dehydrogenase it became quite clear that inhibition by GDP changed considerably with increasing enzyme concentration, although the specific activity of the enzyme in the absence of modifiers remained rather constant over a wide range of concentrations.

Dr. Pette showed me some data yesterday on phosphofructokinase, which show that the specific activity of this enzyme is also very much dependent on protein concentration although this may result from changes in the effect of the allosteric modifiers. There are really quite a number of enzymes, many of which may be allosteric enzymes which polymerize at high concentrations and whose properties may be quite different at these concentrations. Another point which I would like to add: A number of people have looked for protein-protein-interactions in solution and have been unable to see interactions between enzymes which are in the same metabolic pathway and which are close to each other. If one attributes importance to protein-protein-interactions, one should see such interactions between enzymes, but these have virtually never been seen. The only systems which readily undergo reversible interactions in solution are the self polymerizing systems. Perhaps the sepharose technique might be a general technique for looking for this type of protein-protein-interactions.

Stadtman

With regard to the reasons why some enzymes such as those concerned with pyruvate oxidation and fatty acid synthesis are organized in multienzyme complexes, I wonder if this form of organization is not related to the mechanism of the chemical reaction catalyzed. In each of these cases a coenzyme prosthetic group involved in the overall reaction is covalently bound to one of the enzymes in the complex. In the pyruvate dehydrogenase complex, the lipoyl group which serves as an electron carrier as well as an acetyl group carrier is covalently bound to one of the protein components. In the fatty acid synthetase complex, a pantetheine moiety which is covalently linked to the acyl carrier protein bears the catalytic site for acyl group transfer. Thus, in each of these cases a protein component of the system serves as a complex substrate that must interact with two or more enzymes involved in the overall reaction. In such instances, kinetic difficulties might result from the fact that the protein-substrate and the enzymes that act upon it are present in similar concentrations. This is in contrast to the usual situation in which the substrate is present in considerable excess. It is therefore evident that in these unusual cases the overall reaction would be greatly facilitated by increasing the effective concentration of the protein-substrates and their enzymes, which can be accomplished by their association in a multienzyme complex.

Lynen

I think, we should compare these multienzyme complexes with the mitochondria. Compartmentation has a very important effect, I fully agree with you Dr. Stadtman, that by binding a substrate in a complex its effective concentration is much greater than in solution. We have calculated the actual concentration of the intermediates in the fatty acid synthetase complex. It is about 5×10^{-4} M, which is rather high. With respect to the assumption of Dr. Helmreich, that the covalent modification of a component of a multienzyme complex has different regulatory consequences than the modification of a separate enzyme. I do not think so. I believe, that the situation is just the same.

One might look at a multienzyme complex as being like any single enzyme with a quarternary structure.

O. Wieland

An important point that Dr. Helmreich made with respect to the pyruvate dehydrogenase complex is the lack of allosteric control. This points to a difference because these complexes are rather tight and allosteric interactions might be severely restricted.

Henning

A generalization is impossible since the PDH complex in E. coli is allosterically controlled by a remarkable number of effectors, although it is not interconvertible.

Stadtman

It would seem to me that neither the size of a complex nor the number of its subunits will restrict its capacity to be regulated by allosteric mechanisms. The allosteric response elicited by the interaction of a ligand with one subunit could be transmitted to others through protein-protein-interactions. This response might even be concerted and involve simultaneous modification of all subunits in the complex.

Dixon

I would like some clarification on that point. What is the state of the general protein kinase in the glycogen phosphorylase complex? Does the protein kinase convert the complex from state A to B, which then responds to calcium, or is the kinase itself actually a part of the complex. This is compared to the pyruvate dehydrogenase complex where the kinase is itself a part of the complex.

Fischer

The major difference between the glycogen-protein complex and the pyruvate dehydrogenase complex is that the latter contains stoichiometric amounts of enzymes while the former probably exists as a dirty "Gemisch". It contains most of the phosphorylase and phosphorylase phosphatase activities and about 20 percent of the phosphorylase

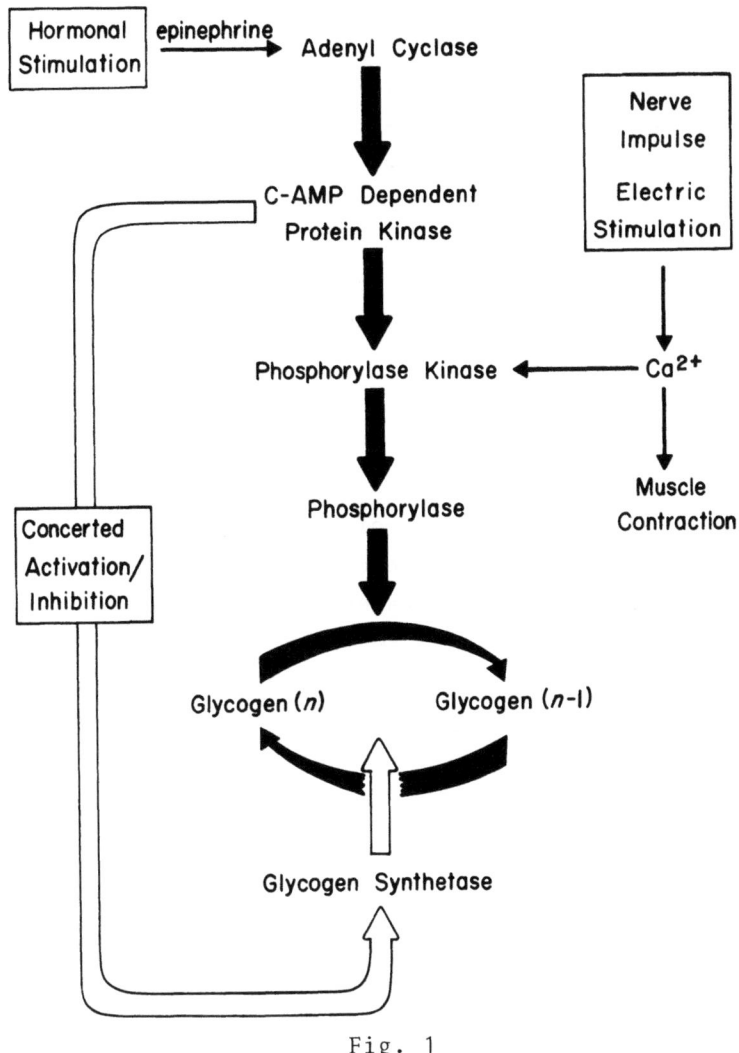

Fig. 1

kinase. As isolated, it is "contaminated" by Ed Krebs' protein kinase since it responds to cyclic AMP, and by elements of the sarcoplasmic reticulum since it responds to epinephrine. Yet it might represent the kind of organization which actually exists in the muscle. It responds rapidly to calcium confirming that phosphorylase kinase is a calcium metallo-enzyme serving as a link between muscle contraction and carbohydrate metabolism.

This brings us back to another question raised by Dr. Cori, namely, why do we have cascades of interconvertible enzymes acting successively on one another? The reason for covalent control has already been summarized: (a) it removes one system from modulation by allosteric effectors; (b) superposition of allosteric to covalent control provides a definite genetic advantage - if one system fails (as is the case of the I-strain mice) the other can take over; (c) covalent interconversion is far more specific than allosteric control since one particular metabolic step can be affected by a specific control enzyme; (d) there is, of course, a huge signal amplification; for instance, considerable amounts of glycogen can be mobilized by minute amounts of hormone or a transmitter such as cyclic AMP. However, perhaps the most important aspect of such a system is that by having three control enzymes acting on one another (adenyl cyclase, the cyclic AMP-dependent protein kinase and phosphorylase kinase), one can link one metabolic pathway to another, e.g., carbohydrate metabolism to muscle contraction and hormonal stimulation (Fig. 1).

Helmreich

With respect to calcium activation of phosphorylase kinase, I wish to make it quite clear that the isolated phosphorylase kinase of E.G. Krebs can be reversibly activated and deactivated by addition or removal of calcium. But was there not a difference between the purified kinase and the enzyme in the glycogen organelle with respect to their respective sensitivity to calcium? The activation of the kinase by calcium in the organelle coming closer to the concentrations of calcium required for initiating muscle contraction.

Krebs

As I recall there is a modest difference in sensitivity to calcium. In either case though, the level of calcium to which the enzyme responds, is well within the range of the calcium concentration that regulates muscle contraction.

Cori

What is the present state with respect to phosphorylase kinase regulation in the liver. Is there a requirement for calcium and activation by phosphorylation?

Krebs

The liver phosphorylase kinase is less well understood than the enzyme from muscle. With respect to a calcium requirement, our own results are confusing. Addition of EGTA inhibits the enzyme only partially.

Fischer

We have likewise not been able to obtain entirely purified liver kinase. The state of affairs with respect to the calcium requirement and phosphorylation of this enzyme is confusing.

Stadtman

I wonder, if we may not leave the phosphorylase system now and turn to another subject and ask the question of why some enzyme systems appear to be covalently regulated and some not. Is there any rationale for this phenomenon?

Krebs

I sometimes think of covalent regulation of enzymes as analogous to power steering in an automobile. I have an old Ford, and when I drive this car down the street, a heavy pull on the steering wheel turns the car abruptly. When I drive a modern car, that is draining off a little of the power from its engine in order to provide power steering, finger tip control of the steering wheel is sufficient to steer the car. I am struck with the analogy between power steering and an enzyme system which drains off a little of the energy to provide signal amplification and smooth control.

Stadtman

One might extend that analogy to the comparative aspects of regulation in microorganisms. Adenylylation of glutamine synthetase has been found in E. coli, Pseudomonas and a variety of other bacteria. It does not occur in other microorganisms such as B. subtilis and clostridia, organisms, which have the same metabolic requirement for glutamine as E. coli. These latter organisms can apparently regulate their glutamine metabolism perfectly well without covalent modification. Perhaps they are like the old Ford and they manage although with some difficulty without powersteering. One cannot say, therefore, that the differences between covalent and noncovalent modification are related to different physiological functions.

Dixon

As long as we are being philosophical, I red on the plane the book "Future Shock" by Toffler in which the super industrial society forces upon us "over choice". In the general field of enzyme regulation it seems to me that enzymes in the hands of biochemists are sometimes in the state of "over choice". There are almost too many regulatory mechanisms available to them. There are almost too many effectors. We should address ourselves to the key question as to what is important and what is not. What is a selective advantage and what is not? We must develop a better test for what is a real selective advantage.

Katunuma

I would like to enlarge our discussion of protein-protein-interactions to include isozymes. I wonder, what are the differences of interactions between the heterologous subunits in isozymes and the interactions discussed so far: between homologous subunits of oligomeric enzymes and between one enzyme and another enzyme.

Frieden

Now, we have been conditioned to think that the enzymatically active tertiary structure is determined by primary sequence. This of course applies to single chain enzymes but perhaps not to subunit enzymes. This becomes very clear if one recalls that the subunits of oligomeric enzymes have little or no activity but gain activity by subunit-subunit interactions in the polymeric form of the enzyme. I believe, that hybridisation experiments the type shown by Dr. Helmreich or hybridisation experiments where one of the subunits is chemically modified are an important tool. From a number of similar types of experiments it appears that the activity is directly correlated with the number of modified chains, whereas allosteric regulation lacks this correlation. In allosteric regulation a small change in anyone of the subunits changes completely the regulatory properties of the whole multisubunitenzyme. I think, that this is important to keep in mind, also with interconvertible enzymes, since some of these do seem to exist in hybrid states.

Fischer

To this general question as to why subunits of oligomeric enzymes are active in some cases and inactive in others, though they all seem to possess the necessary structural features required for activity, I would like to bring to your attention Lazdunski's paper on the "flip-flop" mechanism of alkaline phosphatase. The authors have assumed on the basis of kinetic evidence that this enzyme undergoes an isomerization step, by which one phosphate group which is bound noncovalently to one of the subunits becomes covalently bound and vice versa. This might explain why one subunit is active in one case and inactive in another: One could assume that the subunit which binds substrate covalently is active, while that which binds the substrate noncovalently is inactive. Westheimer and others have shown that some enzymes bind substrates only to one subunit. Many of these may function through a "flip-flop" mechanism.

Frieden

Yes, this is a very important point which should be considered for all enzymes, which show "half site" activity. That is, where only one half of the subunits appear to participate in catalysis at any given time. There are several enzymes, CTP synthetase, glutamic dehydrogenase alkaline phosphatase, acetoacetate decarboxylase and others, which belong to that class. In the case of glutamic dehydrogenase we could show, that one only needs to modify three of six chains in order to abolish excess DPNH inhibition completely (Biochem. 10, 3516, 3527, 1971).

Stadtman

Regulation by covalent modifications involving phosphorylation and subsequent dephosphorylation results in ATPase -activity. With respect to the energetics of adenylylation-deadenylylation, I know, that Dr. Holzer has some thoughts about this that he might share with us.

Holzer

I would like to keep my comments short, because I am actually more interested, if time allows, to discuss some new findings on enzyme interconversion by proteolysis. About two years ago we have shown (Mantel and Holzer, Proc.Nat.Acad.Sci. 65, 660, 1970) that the adenylyl-O-tyrosine bond in adenylylated glutamine synthetase is an energy rich bond, liberating about 10 kcal/mole at hydrolysis. In agreement with this is the recent finding of Stadtman and coworkers that the deadenylylation reaction is not a hydrolytic one, as hitherto thought, but a phosphorolytic one. The energy of the adenylyl-O-tyrosine bond is preserved in the ADP formed by phosphorolysis. These new findings indicate that the energy requirement for a single inactivation/reactivation cycle with glutamine synthetase is only one energy rich phosphate bond (assuming the inorganic pyrophosphate formed during adenylylation is immediately hydrolyzed), in contrast to the two energy rich phosphate bonds required if the back reaction were hydrolytic. (Holzer and Wohlhueter, Advan. Enzyme Regulation 10 (1972) Pergamon Press/ in press). There remains a problem however. Assuming, that the energy of the α-β-phosphate bond in ADP is about the same as that of the adenylyl-O-tyrosine bond, I wonder, how the deadenylylation reaction can actually proceed practically irreversibly towards the formation of ADP.

Stadtman

Hydrolytic cleavage of the α-β-pyrophosphate bond in ATP to yield AMP and PP_i in the presence of Mg^{2+} is accompanied by a standard free energy change of approximately -10 Kcal (Wood, Davis and Lochmuller, J.Biol.Chem. 241, 5692, 1966). From the studies of Mantel and Holzer (Proc.Nat.Acad.Sci., USA 65, 660, 1970) it is clear that the adenylylation of glutamine synthetase proceeds with a $\Delta F'$ of only -0.5 to -1.0 Kcal. It follows therefore that the free energy hydrolysis of the AMP-O-tyrosyl bond of the adenylylated enzyme is -9.0 to -9.5 Kcal. This is -2.0 to -2.5 Kcal greater than the free energy of hydrolysis of the pyrophosphate bond in ADP. Accordingly the phosphorolysis of the adenylylated enzyme to give ADP is quite exergonic (ca. 2.0 to -2.5 Kcal). Therefore the reaction should offer no thermodynamic difficulty.

Lynen

We also must consider the fact, that while thermodynamics play an essential part in the reversibility of enzyme reactions, equally as important are the kinetic properties of the enzyme. For example, it could be that pyrophosphate may just have too low an affinity for the enzyme, and that this is the reason why you cannot measure the reversal with pyrophosphate. I would like to remind you of the two forms of isocitrate dehydrogenase one which works with TPN and is easily reversible and the one which works with DPN and which is not reversible. But the thermodynamics of both reactions are of course the same. Therefore, I think, that association - dissociation of the enzyme substrate complex decides, wether you can reverse a reaction or not.

Stadtman

I agree, Fitzi, except on one point. I think in the case of the isocitrate dehydrogenases it is allosterism which accounts for the difference in reversibility. Consider the reaction, which appears to be irreversible. If you assume that irreversibility is because the K_m for products is too low, then with low amounts of substrate you ought to exceed the equilibrium constant and the reaction should go to completion. This however, does not happen. Why then is the reaction not reversible? The reason may be that in addition to the catalytic site at which the product can react there is also a higher affinity allosteric site on the enzyme for the product, and when the product reacts with this allosteric site it leads to inactivation of the enzyme. In essence it poisons the enzyme so that it cannot catalyze either the forward or the reverse reactions.

Wohlhueter

In addition to energetic advantages, I think we should consider the possible evolutionary advantages of phosphorolytic deadenylylation of glutamine synthetase as opposed to hydrolytic. It was perhaps evolutionarily "easier" to develop a protein interaction system which metamorphosizes a pyrophosphorolytic catalytic site to a phosphorolytic one than to a hydrolytic one.

Segal

I would like to return to the question of the significance of regulation of enzyme rates by covalent modification as compared to modulator regulation. In some cases at least it seems that the interconversion mechanism is a response to a signal external to the cell which transforms the enzyme into a species that is independent of the internal regulatory mechanisms. That is, it is a mechanism for overriding internal controls to meet needs external to the cell in question. Examples are the response of liver phosphorylase to glucagon when the need arises to increase the blood sugar level (Rall, T.W.,Sutherland, E.W., and Wosilait,W.D., J.Biol.Chem., 218, 483, 1956). Here glycogen phosphorolysis is accelerated beyond the needs of the hepatic cells themselves for energy supplies. The turning on of glycogen synthetase in the liver in response to glucose is another example (DeWulf, H., and Hers, H.G., Europ. J. Biochem., 2, 50, 1967). In this case glucose conveys the message that glucose is available in the blood for deposition by the liver as glycogen. Finally with the hormone sensitive lipase of adipose tissue, when an appropriate hormonal stimulus arrives, lipolysis is turned on and provides fatty acids for energy production in other organs of the body (Huttunen, J.K., Steinberg, D., and Mayer, S.E., Proc.Nat.Acad.Sci., $\underline{67}$, 290, 1970).

Now, if I finally may add a philosophical comment. Coming back to Dr. Krebs' story of his two cars. If the old model T Ford breaks down, it will probably be easier to repair it with a few simple tools than would be the case with his air conditioned Cadillac with power steering, power brakes and other gadgets. Maybe we should not take such satisfaction that the mammals in evolution have reached such a high degree of sophistication, because the ultimate goal is fitness for survival. I think this applies not only to individual species, but also to societies as well. The more overly complex a society becomes the greater the danger of a disastrous breakdown.

Stadtman

May be one should look at some of those sophisticated control devices as being sparetires.

Hasilik

It is noteworthy that first covalent interconversion allows a separation, catalytically as well as regulatorily, of the forward and backward conversion.

Gancedo

I think, it is very important, that in interconversion reactions covalent bonds are used. Here, the covalent bonds preserve what has been achieved by noncovalent modification.

Buc

I think, we miss something in the discussion which we should consider with respect to what one should do in the future. We should study the genetics of interconvertible enzyme systems.

Stadtman

I would like to do genetics with the glutamine synthetase system, but it is very difficult to get regulatory mutants while it is relatively easy to obtain mutants with altered metabolic functions.

Madsen

I would just like to say to Dr. Segal that it is not necessarily important how long you live but how well you live. To turn to a specific point, it has been shown that cyclic AMP has a high energy bond. This raises the question as to whether this effector, when it binds to a regulatory protein, might form a covalent bond which may subsequently be split by that protein itself, or by another enzyme.

Krebs

We have found no evidence for a covalent attachment of cyclic AMP to protein kinase.

Stadtman

May be, we should now bring up the topic of enzyme modification by proteolysis.

Holzer

Dr. Katunuma has reported in this symposium, that the specific cleavage of only one peptide bond converts an enzyme to an inactive form, which is then readily digested in the cell. You may look at this kind of specific proteolysis as a covalent modification. The only difference to phosphorylation/dephosphorylation, adenylylation/ deadenylylation etc. is that covalent modification by proteolysis is not reversible. In this context, I would like to mention one experiment done by Dr. T. Katunuma and Dr. Schött at Freiburg. They purified the tryptophan synthase inactivating enzyme about 4000-fold from yeast as described in 1968 by Manney. The inactivating system from yeast is similar to the mammalian system described by Professor Katunuma in so far as pyridoxal-5-phosphate prevents inactivation of the yeast tryptophan synthase by the purified inactivase. Furthermore, a heat-stable and trichloracetic acid precipitable factor has been purified about 1000-fold which inhibits the tryptophan synthase inactivating enzyme. Finally, evidence was obtained for a heat labile factor which inhibits tryptophan synthase. It appears then, in summary, that we are dealing here with a complex regulatory system, which is in some ways similar to the systems described by Professor Katunuma in this symposium. That the tryptophan synthase inactivating system plays an important role in regulatory mechanisms of yeast is indicated by the findings, that during the early exponential growth phase of a batch culture of yeast no tryptophan synthase inactivating activity is observed, whereas in the late growth phase and in the stationary phase a high inactivating activity is present. In our opinion this makes sense, because, in the stationary phase, i.e. in resting yeast cells, tryptophan synthase is no longer necessary and therefore inactivation and degradation of the enzyme is useful for cell economy. We believe that in the near future more of these complex enzyme inactivating systems may be found.

Stadtman

Yes, I think that this may become a fascinating area of research.

Lynen

Is the inactivating enzyme not formed in the log phase or is the inactivating enzyme inhibited, because the inhibitor of the inactivating enzyme is present in the log phase yeast?

Holzer

I do not know yet.

Lynen

If you mix the two extracts, what happens?

Holzer

This is one of the experiments we have not done yet.

Stadtman

I like to comment briefly as to the possible significance of this type of enzymes. You may recall that in Dr. Katunuma's studies he showed that specific enzymes seem to develop during vitamin deficiencies. It seems to me this makes very good sense, because in the case of the vitamin deficiencies, there is really no virtue to have an apo-enzyme around unless you also have the coenzyme to go with it. Therefore it may be, that through scission of the catalytic site, which seems to be the mechanism initiating this proteolysis in the absence of the coenzymes, you start to get rid of the apo-enzymes, which are functionally unimportant, because of their lack of coenzymes. This then enables the cells, to reuse the protein building blocks for other biosyntheses.

Katunuma

I like to assume that the activity of the proteolytic enzymes in rats remains constant and that what varies is the inhibitor concentration. The correlation of tryptophan synthase inactivating activity with the growth curve of yeast in the experiments of Drs. Holzer and Katunuma might be due to changes in the inhibitor concentrations. But there is no direct evidence for the yeast system yet. In my studies with mammalian enzymes, the concentration of inactivating enzyme is about the same in normal and vitamin deficient animals. But the amount of inhibitor is different in the two conditions.

Sols

Instead of saving the building blocks of enzymes that are no longer useful, would it not be a much greater bargain not to synthesize the protein at all. Why evolve a mechanism to destroy the apo-protein rather than preventing its synthesis in the fist place. It has been shown that in biotin deficient rats the level of apo-carboxylase is the same as that of the holoenzyme in normal rats. In biotin deficient rats all the apo-enzymes can be activated on application of biotin within one hour in the absence of de novo protein synthesis. Therefore, I suggest that in the cases where these new proteolytic enzymes are postulated, the level of the apo-enzymes in normal and vitamin deficient animals is measured.

Stadtman

I agree from a teleological point of view but I think you must remember that biochemists have the ability in hind sight to explain a phenomenon but rarely can they predict how a cell would go about doing something or why a cell does something in one way and not in an other way.

Holzer

I would like to offer a compromise solution to Dr. Katunuma's suggestion. I expect that both mechanisms, changes of the inhibitor of the proteolytic enzyme or changes of the proteolytic enzymes themselves are functioning: My speculation is: when yeast does not need an enzyme for a short period of time it changes the inhibitor concentration. When the enzyme is not needed for a long time, then it is the activity of the proteolytic enzymes which determine rate and extent of degradation.

Krebs

If we allow ourselves to speculate as to the possibility that proteolysis is directed by something other than the primary amino acid sequence in a protein, then should we not also consider the possibility of "repair reactions" which could reverse limited proteolysis of this type?

Rabussay

Another example for modifying an enzyme in this way, is the modification of RNA polymerase in B. subtilis, when the organism begins to sporulate. (Losick, R., Shorenstein, R.G., and Sonenshein, A.L., Nature, 227, 910, 1970). During sporulation a peptide from the β-subunit is split off and as a consequence of this modification, the sigma factor (which is needed for activity) seems no longer to bind to the modified core enzyme and transcription of the vegetative genes is stopped.

List of Participants

E.-G. Afting
Biochemisches Institut
der Universität Freiburg

7800   Freiburg   i.Br.
Hermann-Herder-Strasse 7
(GFR)

C. Beaucamp
Boehringer Mannheim GmbH
Werk Tutzing

8132   Tutzing
Bahnhofstrasse 5
(GFR)

H.U. Bergmeyer
Boehringer Mannheim GmbH
Werk Tutzing

8132   Tutzing
Bahnhofstrasse 5
(GFR)

H. Buc
Institute Pasteur
Rue du Docteur Roux

Paris
(France)

M.H. Buc
Institute Pasteur
Rue du Docteur Roux

Paris
(France)

Th. Bücher
Institut für Physiologische Chemie
der Universität München

8000   München   15
Goethestrasse 33
(GFR)

F. Chapeville
Institute of Molecular Biology
Faculty of Sciences
9 Quai St. Bernard

Paris
(France)

C.F. Cori	Harvard University Medical School Boston, Massachusetts 02114 (USA)
K. Decker	Biochemisches Institut der Medizinischen Fakultät 7800 Freiburg i.Br. (GFR)
G.H. Dixon	Department of Biochemistry University of British Columbia Vancouver 8, British Columbia (Canada)
J. Ehrlich	Physiologisch-chemisches Institut der Universität Würzburg 8700 Würzburg Koellikerstrasse 2 (GFR)
M. Eigen	Max-Planck-Institut für Biophysikalische Chemie 3400 Göttingen-Nikolausberg Am Fassberg (GFR)
H. Fasold	Biochemisches Institut der Universität Frankfurt 6000 Frankfurt/Main (GFR)
M. de Fekete	Botanisches Institut Technische Hochschule Darmstadt 6100 Darmstadt (GFR)
K. Feldmann	Physiologisch-chemisches Institut der Universität Würzburg 8700 Würzburg Koellikerstrasse 2 (GFR)
A.R. Ferguson	School of Biological Sciences University of East Anglia Norwich, NOR 88 C (England)

E.H. Fischer   Department of Biochemistry
                                University of Washington
                                Seattle, Washington 98105
                                (USA)

C. Frieden   Department of Biological Chemistry
                                Washington University
                                660 South Kings Highway
                                St. Louis, Missouri
                                (USA)

HJ. v. Funcke   Institut für Diabetesforschung
                                8000   München   23
                                Kölner Platz 1
                                (GFR)

C. Gancedo   Centro de Investigaciones Biologicas
                                Instituto de Enzimologia
                                Velazquez, 144
                                Madrid   6
                                (Spain)

D.J. Graves   Department of Biochemistry
                                and Biophysics
                                Iowa State University
                                Ames, Iowa
                                (USA)

W. Guder   Klinisch-chemisches Institut
                                Städt. Krankenhaus Schwabing
                                8000   München   23
                                Kölner Platz 1
                                (GFR)

G. Hartmann   Biochemisches Institut
                                der Universität Würzburg
                                8700   Würzburg
                                (GFR)

R. Haschke   Physiologisch-chemisches Institut
                                der Universität Würzburg
                                8700   Würzburg
                                Koellikerstrasse 2
                                (GFR)

A. Hasilik   Biochemisches Institut
                                der Universität Freiburg
                                7800   Freiburg   i.Br.
                                Hermann-Herder-Strasse 7
                                (GFR)

O. Hayaishi   Department of Medical Chemistry
                            University Faculty of Medicine
                            Yoshida, Sakyo-ku-Kyoto
                            (Japan)

L. Heilmeyer   Physiologisch-chemisches Institut
                            der Universität Würzburg
                            8700   Würzburg
                            Koellikerstrasse 2
                            (GFR)

P.C. Heinrich   Biochemisches Institut
                            der Universität Freiburg
                            7800   Freiburg i.Br.
                            Hermann-Herder-Strasse 7
                            (GFR)

E. Helmreich   Physiologisch-chemisches Institut
                            der Universität Würzburg
                            8700   Würzburg
                            Koellikerstrasse 2
                            (GFR)

U. Henning   Max-Planck-Institut für Biologie
                            7400   Tübingen
                            Corrensstrasse 38
                            (GFR)

B. Hess   Max-Planck-Institut für
                            Ernährungsphysiologie
                            4600   Dortmund
                            Rheinlanddamm 201
                            (GFR)

R.L. Hill   Department of Biochemistry
                            Duke University Medical Center
                            Durham, North Carolina 27706
                            (USA)

H. Holzer   Biochemisches Institut
                            der Universität Freiburg
                            7800   Freiburg i.Br.
                            Hermann-Herder-Strasse 7
                            (GFR)

F. Hucho   Fachbereich Biologie
                            der Universität Konstanz
                            7750   Konstanz
                            Jacob-Burckhardt-Strasse
                            (GFR)

F.M. Huennekens	Department of Biochemistry Scripps Clinic and Research Foundation 476 Prospect Street La Jolla, California 92037 (USA)
N. Katunuma	Department of Enzyme Chemistry Institute for Enzyme Research School of Medicine Tokushima University Tokushima (Japan)
T. Katunuma	Biochemisches Institut der Universität Freiburg 7800 Freiburg i.Br. Hermann-Herder-Strasse 7 (GFR)
N.A. Kiselev	Institute of Crystallography Academy of Sciences of the USSR Leninsky prospekt, 59 Moscow B-333 (USSR)
M. Klockow	Chemische Forschung Biochemische Abteilung E. Merck 6100 Darmstadt 2 Postfach 4119
J. Knappe	Organisch-chemisches Institut der Universität Heidelberg 6900 Heidelberg Tiergartenstrasse (GFR)
H. Kolb	Institut für Diabetesforschung 8000 München 23 Kölner Platz 1 (GFR)
E.G. Krebs	Department of Biological Chemistry University of California Davis, California 95616 (USA)
N.B. Livanova	Bach Institute of Biochemistry USSR Academy of Sciences Leninsky Prospect 33 Moscow (USSR)

G. Löffler   Institut für Diabetesforschung

                               8000   München   23
                               Kölner Platz 1
                               (GFR)

A. Lynen   Biochemisches Institut
                               der Universität Freiburg

                               7800   Freiburg   i.Br.
                               Hermann-Herder-Strasse 7
                               (GFR)

F. Lynen   Max-Planck-Institut für Zellchemie

                               8000   München   2
                               Karlstrasse 23 - 25
                               (GFR)

N. Madsen   University of Alberta
                               Department of Biochemistry

                               Edmonton 7
                               (Canada)

D. Mecke   Biochemisches Institut
                               der Universität Freiburg

                               7800   Freiburg   i.Br.
                               Hermann-Herder-Strasse 7
                               (GFR)

G. Michal   Boehringer Mannheim GmbH
                               Werk Tutzing

                               8132   Tutzing
                               Bahnhofstrasse 5
                               (GFR)

J. Neeff   Biochemisches Institut
                               der Universität Freiburg

                               7800   Freiburg   i.Br.
                               Hermann-Herder-Strasse 7
                               (GFR)

D. Palm   Physiologisch-chemisches Institut
                               der Universität Würzburg

                               8700   Würzburg
                               Koellikerstrasse 2
                               (GFR)

N. Palmer   Max-Planck-Institut für
                               Ernährungsphysiologie

                               4600   Dortmund
                               Rheinlanddamm 201
                               (GFR)

Ch. Patzelt　　　　　　　　　　Institut für Diabetesforschung

　　　　　　　　　　　　　　　8000　　München　　23
　　　　　　　　　　　　　　　Kölner Platz 1
　　　　　　　　　　　　　　　(GFR)

D. Pette　　　　　　　　　　　Fachbereich Biologie
　　　　　　　　　　　　　　　Universität Konstanz

　　　　　　　　　　　　　　　7750　　Konstanz
　　　　　　　　　　　　　　　Postfach 733
　　　　　　　　　　　　　　　(GFR)

T. Pfeuffer　　　　　　　　　Physiologisch-chemisches Institut
　　　　　　　　　　　　　　　der Universität Würzburg

　　　　　　　　　　　　　　　8700　　Würzburg
　　　　　　　　　　　　　　　Koellikerstrasse 2
　　　　　　　　　　　　　　　(GFR)

R. Portenhauser　　　　　　　Institut für Diabetesforschung

　　　　　　　　　　　　　　　8000　　München　　23
　　　　　　　　　　　　　　　Kölner Platz 1
　　　　　　　　　　　　　　　(GFR)

D. Rabussay　　　　　　　　　Max-Planck-Institut für Biochemie

　　　　　　　　　　　　　　　8000　　München　　15
　　　　　　　　　　　　　　　Goethestrasse 31
　　　　　　　　　　　　　　　(GFR)

L.J. Reed　　　　　　　　　　Clayton Foundation
　　　　　　　　　　　　　　　Biochemical Institute
　　　　　　　　　　　　　　　University of Texas at Austin

　　　　　　　　　　　　　　　Austin, Texas 78712
　　　　　　　　　　　　　　　(USA)

H. Reinauer　　　　　　　　　Institut für Physiologische Chemie
　　　　　　　　　　　　　　　der Universität Düsseldorf

　　　　　　　　　　　　　　　4000　　Düsseldorf
　　　　　　　　　　　　　　　(GFR)

J. Saari　　　　　　　　　　　Institute Pasteur
　　　　　　　　　　　　　　　Rue du Docteur Roux

　　　　　　　　　　　　　　　Paris
　　　　　　　　　　　　　　　(France)

K. Schnackerz　　　　　　　　Physiologisch-chemisches Institut
　　　　　　　　　　　　　　　der Universität Würzburg

　　　　　　　　　　　　　　　8700　　Würzburg
　　　　　　　　　　　　　　　Koellikerstrasse 2
　　　　　　　　　　　　　　　(GFR)

H.L. Segal	State University of New York at Buffalo Department of Biology Health Sciences Building Buffalo, New York 14214 (USA)
W. Seubert	Physiologisch-chemisches Institut der Universität Göttingen 3400 Göttingen Humboldtallee 7 (GFR)
Sh. Shaltiel	Weizmann Institute Department of Chemical Immunology Rehovot (Israel)
E. Siess	Institut für Diabetesforschung 8000 München 23 Kölner Platz 1 (GFR)
A.P. Sims	School of Biological Sciences University of East Anglia Norwich, NOR 88 C (England)
A. Sols	Centro de Investigaciones Biologicas Instituto de Enzimologia Velazquez, 144 Madrid 6 (Spain)
E.R. Stadtman	Laboratory of Biochemistry National Heart Institute NIH Bethesda, Maryland 20014 (USA)
W. Stalmans	Université Catholique de Louvain Laboratoire de Chimie Physiologique 6, Dekenstraat 3000 Louvain (Belgique)
S. Velick	Department of Biochemistry University of Utah Medical School Salt Lake City, Utah (USA)

K. Wallenfels   Lehrstuhl Biochemie
                      Chemisches Laboratorium der Universität

                      7800   Freiburg   i.Br.
                      Albertstrasse 21
                      (GFR)

L. Weiss   Klinisch-chemisches Institut
                      Städt. Krankenhaus Schwabing

                      8000   München   23
                      Kölner Platz 1
                      (GFR)

J.M. Wiame   Université Libre
                      Department of Microbiology

                      Bruxelles
                      (Belgique)

O. Wieland   Institut für Diabetesforschung

                      8000   München   23
                      Kölner Platz 1
                      (GFR)

Th. Wieland   Max-Planck-Institut für
                      Medizinische Forschung
                      Abteilung Chemie

                      6900   Heidelberg
                      Jahnstrasse 29
                      (GFR)

H. Winkler   Max-Planck-Institut für
                      Biophysikalische Chemie

                      3400 Göttingen-Nikolausberg
                      Am Fassberg
                      (GFR)

R. Wohlhueter   Biochemisches Institut
                      der Universität Freiburg

                      7800   Freiburg i.Br.
                      Hermann-Herder-Strasse 7
                      (GFR)

D. Wolf   Biochemisches Institut
                      der Universität Freiburg

                      7800   Freiburg   i.Br.
                      Hermann-Herder-Strasse 7
                      (GFR)

H. Ziegler   Technische Hochschule München
                                Institut für Botanik

                                8000   München   2
                                Arcisstrasse 21
                                (GFR)

I. Ziegler   Technische Hochschule München
                                Institut für Botanik

                                8000   München   2
                                Arcisstrasse 21
                                (GFR)

G. Zubay   Department of Biological Sciences
                                Columbia University
                                Schermerhorn Hall

                                New York, N.Y. 10027
                                (USA)

Funktionelle und morphologische Organisation der Zelle

Herausgeber: P. Karlson
91 Abbildungen. IV, 253 Seiten (9 Beiträge in Deutsch und 6 Beiträge in Englisch). 1963. (1. wissenschaftliche Konferenz der Gesellschaft Deutscher Naturforscher und Ärzte in Rottach-Egern 1962)

Sekretion und Exkretion der Zelle

Funktionelle und morphologische Organisation der Zelle
180 Abbildungen. XII, 404 Seiten (100 Seiten in Englisch). 1965 (2. wissenschaftliche Konferenz der Gesellschaft Deutscher Naturforscher und Ärzte Schloß Reinhardsbrunn bei Friedrichroda 1964)

Probleme der biologischen Reduplikation
Problems of Reduplication in Biology

Funktionelle und morphologische Organisation der Zelle
142 Abbildungen. VIII, 412 Seiten (96 Seiten in Englisch). 1966 (3. wissenschaftliche Konferenz der Gesellschaft Deutscher Naturforscher und Ärzte in Semmering bei Wien 1965)

Molecular Genetics

Editors: H. G. Wittman and H. Schuster
141 figures. VIII, 341 pages. 1968. (4. wissenschaftliche Konferenz der Gesellschaft Deutscher Naturforscher und Ärzte Berlin 1967)

Membranaspekte der Immunologie

Von H. Fischer, E. Rüde und D. Sellin
(5. wissenschaftliche Konferenz Deutscher Naturforscher und Ärzte in Titisee bei Freiburg 1969)
Erschienen in „Die Naturwissenschaften" 57. Jahrgang 1970, Heft 11

Springer-Verlag
Berlin · Heidelberg · New York

London · München · Paris · Sydney · Tokyo · Wien

SPRINGER-VERLAG
BERLIN·HEIDELBERG·NEW YORK

Thomas E. Barman

ENZYME HANDBOOK

By Thomas E. Barman, Ph. D., University of Reading,
The National Institute for Research in Dairying,
Shinfield, Reading/England

In two volumes,
not sold separately
Volume I:
XI, 499 pages. 1969
Volume II:
III, 428 pages. 1969

The Enzyme Handbook provides in a concise form molecular data on about 800 enzymes. The enzymes are arranged according to the Recommendations (1964) of the International Union of Biochemistry on the Nomenclature and Classification of Enzymes and the most important of their molecular properties are considered, namely molecular weight (including subunit and sequence data), specific activity, specificity and kinetic properties. Also included are data on the reversibility of enzyme-catalysed reactions and, where appropriate, light absorption characteristics of substrates and products. Some 120 enzymes discovered after publication of the Enzyme Commission List of Enzymes in 1964 have been classified and given numbers according to the Enzyme Commission recommendations.

■ **Prospectus on request**

The author, who has experience in both the molecular and metabolic aspects of enzymology, has made a thorough search of the literature and the book should constitute a ready reference catalogue of molecular enzymology.

MIX
Papier aus verantwortungsvollen Quellen
Paper from responsible sources
FSC® C105338

If you have any concerns about our products,
you can contact us on
ProductSafety@springernature.com

In case Publisher is established outside the EU,
the EU authorized representative is:
**Springer Nature Customer Service Center GmbH
Europaplatz 3, 69115 Heidelberg, Germany**

Printed by Libri Plureos GmbH
in Hamburg, Germany